Continuous Lattices and Their Applications

LECTURE NOTES

IN PURE AND APPLIED MATHEMATICS

1. *N. Jacobson,* Exceptional Lie Algebras
2. *L. -Å. Lindahl and F. Poulsen,* Thin Sets in Harmonic Analysis
3. *I. Satake,* Classification Theory of Semi-Simple Algebraic Groups
4. *F. Hirzebruch, W. D. Newmann, and S. S. Koh,* Differentiable Manifolds and Quadratic Forms (out of print)
5. *I. Chavel,* Riemannian Symmetric Spaces of Rank One (out of print)
6. *R. B. Burckel,* Characterization of C(X) Among Its Subalgebras
7. *B. R. McDonald, A. R. Magid, and K. C. Smith,* Ring Theory: Proceedings of the Oklahoma Conference
8. *Y.-T. Siu,* Techniques of Extension on Analytic Objects
9. *S. R. Caradus, W. E. Pfaffenberger, and B. Yood,* Calkin Algebras and Algebras of Operators on Banach Spaces
10. *E. O. Roxin, P.-T. Liu, and R. L. Sternberg,* Differential Games and Control Theory
11. *M. Orzech and C. Small,* The Brauer Group of Commutative Rings
12. *S. Thomeier,* Topology and Its Applications
13. *J. M. Lopez and K. A. Ross,* Sidon Sets
14. *W. W. Comfort and S. Negrepontis,* Continuous Pseudometrics
15. *K. McKennon and J. M. Robertson,* Locally Convex Spaces
16. *M. Carmeli and S. Malin,* Representations of the Rotation and Lorentz Groups: An Introduction
17. *G. B. Seligman,* Rational Methods in Lie Algebras
18. *D. G. de Figueiredo,* Functional Analysis: Proceedings of the Brazilian Mathematical Society Symposium
19. *L. Cesari, R. Kannan, and J. D. Schuur,* Nonlinear Functional Analysis and Differential Equations: Proceedings of the Michigan State University Conference
20. *J. J. Schäffer,* Geometry of Spheres in Normed Spaces
21. *K. Yano and M. Kon,* Anti-Invariant Submanifolds
22. *W. V. Vasconcelos,* The Rings of Dimension Two
23. *R. E. Chandler,* Hausdorff Compactifications
24. *S. P. Franklin and B. V. S. Thomas,* Topology: Proceedings of the Memphis State University Conference
25. *S. K. Jain,* Ring Theory: Proceedings of the Ohio University Conference
26. *B. R. McDonald and R. A. Morris,* Ring Theory II: Proceedings of the Second Oklahoma Conference
27. *R. B. Mura and A. Rhemtulla,* Orderable Groups
28. *J. R. Graef,* Stability of Dynamical Systems: Theory and Applications
29. *H.-C. Wang,* Homogeneous Branch Algebras
30. *E. O. Roxin, P.-T. Liu, and R. L. Sternberg,* Differential Games and Control Theory II
31. *R. D. Porter,* Introduction to Fibre Bundles
32. *M. Altman,* Contractors and Contractor Directions Theory and Applications
33. *J. S. Golan,* Decomposition and Dimension in Module Categories
34. *G. Fairweather,* Finite Element Galerkin Methods for Differential Equations
35. *J. D. Sally,* Numbers of Generators of Ideals in Local Rings
36. *S S. Miller,* Complex Analysis: Proceedings of the S.U.N.Y. Brockport Conference
37. *R. Gordon,* Representation Theory of Algebras: Proceedings of the Philadelphia Conference
38. *M. Goto and F. D. Grosshans,* Semisimple Lie Algebras
39. *A. I. Arruda, N. C. A. da Costa, and R. Chuaqui,* Mathematical Logic: Proceedings of the First Brazilian Conference

40. *F. Van Oystaeyen,* Ring Theory: Proceedings of the 1977 Antwerp Conference
41. *F. Van Oystaeyen and A. Verschoren,* Reflectors and Localization: Application to Sheaf Theory
42. *M. Satyanarayana,* Positively Ordered Semigroups
43. *D. L. Russell,* Mathematics of Finite-Dimensional Control Systems
44. *P.-T. Liu and E. Roxin,* Differential Games and Control Theory III: Proceedings of the Third Kingston Conference, Part A
45. *A. Geramita and J. Seberry,* Orthogonal Designs: Quadratic Forms and Hadamard Matrices
46. *J. Cigler, V. Losert, and P. Michor,* Banach Modules and Functors on Categories of Banach Spaces
47. *P.-T. Liu and J. G. Sutinen,* Control Theory in Mathematical Economics: Proceedings of the Third Kingston Conference, Part B
48. *C. Byrnes,* Partial Differential Equations and Geometry
49. *G. Klambauer,* Problems and Propositions in Analysis
50. *J. Knopfmacher,* Analytic Arithmetic of Algebraic Function Fields
51. *F. Van Oystaeyen,* Ring Theory: Proceedings of the 1978 Antwerp Conference
52. *B. Kedem,* Binary Time Series
53. *J. Barros-Neto and R. A. Artino,* Hypoelliptic Boundary-Value Problems
54. *R. L. Sternberg, A. J. Kalinowski, and J. S. Papadakis,* Nonlinear Partial Differential Equations in Engineering and Applied Science
55. *B. R. McDonald,* Ring Theory and Algebra III: Proceedings of the Third Oklahoma Conference
56. *J. S. Golan,* Structure Sheaves over a Noncommutative Ring
57. *T. V. Narayana, J. G. Williams, and R. M. Mathsen,* Combinatorics, Representation Theory and Statistical Methods in Groups: YOUNG DAY Proceedings
58. *T. A. Burton,* Modeling and Differential Equations in Biology
59. *K. H. Kim and F. W. Roush,* Introduction to Mathematical Consensus Theory
60. *J. Banas and K. Goebel,* Measures of Noncompactness in Banach Spaces
61. *O. A. Nielson,* Direct Integral Theory
62. *J. E. Smith, G. O. Kenny, and R. N. Ball,* Ordered Groups: Proceedings of the Boise State Conference
63. *J. Cronin,* Mathematics of Cell Electrophysiology
64. *J. W. Brewer,* Power Series Over Commutative Rings
65. *P. K. Kamthan and M. Gupta,* Sequence Spaces and Series
66. *T. G. McLaughlin,* Regressive Sets and the Theory of Isols
67. *T. L. Herdman, S. M. Rankin, III, and H. W. Stech,* Integral and Functional Differential Equations
68. *R. Draper,* Commutative Algebra: Analytic Methods
69. *W. G. McKay and J. Patera,* Tables of Dimensions, Indices, and Branching Rules for Representations of Simple Lie Algebras
70. *R. L. Devaney and Z. H. Nitecki,* Classical Mechanics and Dynamical Systems
71. *J. Van Geel,* Places and Valuations in Noncommutative Ring Theory
72. *C. Faith,* Injective Modules and Injective Quotient Rings
73. *A. Fiacco,* Mathematical Programming with Data Perturbations I
74. *P. Schultz, C. Praeger, and R. Sullivan,* Algebraic Structures and Applications Proceedings of the First Western Australian Conference on Algebra
75. *L. Bican, T. Kepka, and P. Nemec,* Rings, Modules, and Preradicals
76. *D. C. Kay and M. Breen,* Convexity and Related Combinatorial Geometry: Proceedings of the Second University of Oklahoma Conference
77. *P. Fletcher and W. F. Lindgren,* Quasi-Uniform Spaces
78. *C.-C. Yang,* Factorization Theory of Meromorphic Functions
79. *O. Taussky,* Ternary Quadratic Forms and Norms
80. *S. P. Singh and J. H. Burry,* Nonlinear Analysis and Applications
81. *K. B. Hannsgen, T. L. Herdman, H. W. Stech, and R. L. Wheeler,* Volterra and Functional Differential Equations

82. *N. L. Johnson, M. J. Kallaher, and C. T. Long,* Finite Geometries: Proceedings of a Conference in Honor of T. G. Ostrom
83. *G. I. Zapata,* Functional Analysis, Holomorphy, and Approximation Theory
84. *S. Greco and G. Valla,* Commutative Algebra: Proceedings of the Trento Conference
85. *A. V. Fiacco,* Mathematical Programming with Data Perturbations II
86. *J.-B. Hiriart-Urruty, W. Oettli, and J. Stoer,* Optimization: Theory and Algorithms
87. *A. Figa Talamanca and M. A. Picardello,* Harmonic Analysis on Free Groups
88. *M. Harada,* Factor Categories with Applications to Direct Decomposition of Modules
89. *V. I. Istrăţescu,* Strict Convexity and Complex Strict Convexity: Theory and Applications
90. *V. Lakshmikantham,* Trends in Theory and Practice of Nonlinear Differential Equations
91. *H. L. Manocha and J. B. Srivastava,* Algebra and Its Applications
92. *D. V. Chudnovsky and G. V. Chudnovsky,* Classical and Quantum Models and Arithmetic Problems
93. *J. W. Longley,* Least Squares Computations Using Orthogonalization Methods
94. *L. P. de Alcantara,* Mathematical Logic and Formal Systems
95. *C. E. Aull,* Rings of Continuous Functions
96. *R. Chuaqui,* Analysis, Geometry, and Probability
97. *L. Fuchs and L. Salce,* Modules Over Valuation Domains
98. *P. Fischer and W. R. Smith,* Chaos, Fractals, and Dynamics
99. *W. B. Powell and C. Tsinakis,* Ordered Algebraic Structures
100. *G. M. Rassias and T. M. Rassias,* Differential Geometry, Calculus of Variations, and Their Applications
101. *R.-E. Hoffmann and K. H. Hofmann,* Continuous Lattices and Their Applications
102. *J. H. Lightbourne, III, and S. M. Rankin, III,* Physical Mathematics and Nonlinear Partial Differential Equations

Other Volumes in Preparation

Continuous Lattices and Their Applications

Edited by

Rudolf-E. Hoffmann
Universität Bremen
Bremen, Federal Republic of Germany

Karl H Hofmann
Technische Hochschule Darmstadt
Darmstadt, Federal Republic of Germany

CRC Press
Taylor & Francis Group
Boca Raton London New York

CRC Press is an imprint of the
Taylor & Francis Group, an **informa** business

CRC Press
Taylor & Francis Group
6000 Broken Sound Parkway NW, Suite 300
Boca Raton, FL 33487-2742

First issued in hardback 2017

ISBN-13: 978-0-8247-7331-1 (pbk)
ISBN-13: 978-1-138-44185-9 (hbk)

Library of Congress Cataloging in Publication Data
Main entry under title:

Continuous lattices and their applications.

(Lecture notes in pure and applied mathematics ; 101)
Proceedings of a conference held at the University of Bremen, July 2-3, 1982.
Bibliography: p.
Includes index.
1. Lattices, Continuous--Congresses. I. Hoffmann, R.-E. (Rudolf-Eberhard), [date]. II. Hofmann, Karl Heinrich. III. Series: Lecture notes in pure and applied mathematics ; v. 101.
QA171.5.C68 1985 511.3'3 85-10382
ISBN 0-8247-7331-4

Visit the Taylor & Francis Web site at
http://www.taylorandfrancis.com

and the CRC Press Web site at
http://www.crcpress.com

Preface

This volume contains the proceedings of a (third) conference on categorical and topological aspects of continuous lattices held at the Mathematics Department of the University of Bremen, July 2-3, 1982. In addition to the papers read at the conference three further contributions by authors who had been unable to participate have been included.

The focus of the articles in this volume is on the notion of a continuous lattice, which has its roots in Dana Scott's work on a mathematical theory of computation some 16 years ago (1969). Scott recognized its significance in various other branches of mathematics and theoretical computer science. Most notably, continuous lattices provide models for the λ-calculus (Scott 1972). On the other hand, they also figure prominently in what has been called "categorical topology." In his effort to characterize T_0-spaces that are injective with regard to all embeddings, Scott (1972) found that they are precisely the continuous lattices endowed with a (non-Hausdorff) topology now known as the Scott topology. Earlier, B. Day and G. M. Kelly (1970) showed that the "cartesian," or "exponentiable," topological spaces X (i.e., those X for which $X \times -$ has a right adjoint — which necessarily must be an internal "hom" [X,-] — or, equivalently, by the special adjoint functor theorem, which enjoy the property that $X \times f$ is an identification mapping whenever f is one) are precisely those for which $\underline{O}(X)$, the lattice of open subsets of X, is a continuous lattice.

It had been known previously that all locally quasi-compact spaces (in which every point has a neighborhood basis consisting of quasi-compact sets) have the above-mentioned property with respect to identification mappings, and that among Hausdorff spaces these are the only exponentiable spaces (E. A. Michael, K. Baker 1968). The (later) discovery (K. H. Hofmann and J. D. Lawson 1978) that every distributive continuous lattice arises in the form of such an $\underline{O}(X)$ for some space X and that X can be chosen as a locally quasi-compact (sober) space has shown that there are strong ties between (distributive) continuous lattice theory and the study of locally quasi-compact spaces [which previously arose in the investigation of the spectrum of a C^*-algebra (J. M. G. Fell, J. Dixmier)]. The above-mentioned equivalence also indicates that continuous lattice theory should have some connections with the study of function spaces. Indeed, such connections exist in several directions.

In a different context, continuous lattices were discovered by K. H. Hofmann and A. R. Stralka (1975): (complete) lattices carrying a compact Hausdorff topology such that the binary meet $\wedge$ is (jointly) continuous with regard to the product topology and such that the continuous $\wedge$-semilattice homomorphisms into the unit interval (with its min-semilattice structure and its euclidean topology) separate the points turn out to be precisely the continuous lattices. Their topology is uniquely determined by the lattice structure.

In these contexts, several functors arise making a continuous lattice into a compact Hausdorff space and a (non-Hausdorff) T_0-space, respectively. The latter functor as well as its composite with the "underlying set" functor had been investigated earlier by A. Day (1975), who showed that both are "monadic" or "tripleable," i.e., the category of algebras (in the sense of S. Eilenberg and J. C. Moore, 1966) for a monad or triple; he also identified the monads. The functor parts of these – the space of open filters of a space and the set of all filters of a set, respectively – had been previously studied in general topology, but no one was then aware of their later integration into a monad. Moreover, Day also established an equation characterizing continuous lattices among all complete lattices, which is analogous to (but different from) complete distributivity. Day's monads were soon joined by other monads arising in continuous lattice theory: a "Vietoris monad" (O. Wyler), the monad of open prime filters (H. Simmons), the monad of closed subsets in the weak topology (R.-E. Hoffmann), and still further such monads have been discovered in recent years.

Some of the notions Scott had in mind when he invented continuous lattices later suggested the study of the more general concept of a "continuous poset." Primarily initiated by theoretical computer scientists, the significance of this structure was recognized both from a topological and a lattice-theoretical point of view: These posets yield the (reasonably defined) projective objects in the category of sober spaces (R.-E. Hoffmann 1981), and they are – via the "prime spectrum" functor – in bijective correspondence with the completely distributive lattices (R.-E. Hoffmann 1981, J. D. Lawson 1979). Interest in this class of lattices arose from the early work of A. Tarski on the foundations of mathematics. The known fact that complete distributivity is a self-dual condition (G. N. Raney 1952) leads in a natural way to the notion of a (continuous) dual of a continuous poset.

The final piece of work to be mentioned in this introduction remained for a long time unrecognized in its intimate relationship to other branches of the theory of continuous lattices. In categories of modules over a ring the search for injective objects is connected with the construction of an injective hull. This is, by definition, an "essential" embedding of an object into an injective object and may be characterized as a greatest essential extension. Although in the category of T_0-spaces and continuous maps every space can be embedded into an injective T_0-space (indeed, into a power of the Sierpinski two-point space), B. Banaschewski (1977) showed – in an ingenious proof relying on his earlier theory of "strict" extensions (1964) – that not every object has an injective hull. Specifically, every object has a unique greatest essential extension (which fails to be an injective hull for all nondiscrete T_1-spaces). On the other hand, some spaces do have an injective hull: continuous posets endowed with their Scott topology (R.-E. Hoffmann 1981). (Contrary to a previous erroneous claim, some other sober spaces share this property.) The essentially complete T_0-spaces (i.e., those for which the identity is the greatest essential extension) have turned out to be those T_0-spaces X carrying a complete lattice structure (which is uniquely determined) such that for every index set J the J-indexed supremum operation $X^J \to X$ is a continuous map for the product topology (R.-E. Hoffmann 1979). The articles of this volume augment and refine the interrelations here outlined.

The editors of this volume would like to express their thanks to the participants for a lively and successful conference, to the authors for contributing to this volume, to the referees of the articles for their invaluable assistance, to Dana Scott and his collaborators for their willingness to type the bibliography on the word processing system of the Department of Computer Science of Carnegie-Mellon University, to Edwin Hewitt for encouraging the publication of this volume, and, finally, to Marcel Dekker, Inc., for both patience and care in the preparation of this book.

Rudolf-E. Hoffmann

Karl H. Hofmann

Contents

Contributors

BERNHARD BANASCHEWSKI McMaster University,
Hamilton, Ontario, Canada

MARCEL ERNÉ Mathematical Institute, Universität Hannover,
Hannover, Federal Republic of Germany

HARTMUT GATZKE* Mathematical Institute, Universität Hannover,
Hannover, Federal Republic of Germany

GERHARD GIERZ University of California,
Riverside, California

RUDOLF-E. HOFFMANN Universität Bremen,
Bremen, Federal Republic of Germany

KARL H. HOFMANN† Tulane University,
New Orleans, Louisiana

JOHN ISBELL State University of New York at Buffalo,
Buffalo, New York

PETER T. JOHNSTONE University of Cambridge,
Cambridge, England

PANOS TH. LAMBRINOS School of Engineering, Democritus University of
Thrace, Xanthi, Greece

JIMMIE D. LAWSON Louisiana State University,
Baton Rouge, Louisiana

MICHAEL MISLOVE Tulane University,
New Orleans, Louisiana

ARESKI NAIT ABDALLAH‡ University of Waterloo,
Waterloo, Ontario, Canada

---

*Current affiliation: Siemens AG, München-Perlach, Federal Republic of Germany

†Current affiliation: Technische Hochschule Darmstadt, Darmstadt, Federal Republic of Germany

‡Current affiliation: Unit for Computer Science, McMaster University, Hamilton, Ontario, Canada

BASIL PAPADOPOULOS School of Engineering, Democritus University of Thrace, Xanthi, Greece

HILARY A. PRIESTLEY Mathematical Institute, University of Oxford, Oxford, England

FRIEDHELM SCHWARZ* Mathematical Institute, Universität Hannover, Hannover, Federal Republic of Germany

DANA S. SCOTT Carnegie-Mellon University, Pittsburgh, Pennsylvania

MANUELA SOBRAL Universidade de Coimbra, Coimbra, Portugal

ALBERT STRALKA University of California, Riverside, California

MARCEL van de VEL Free University of Amsterdam, Amsterdam, The Netherlands

SIBYLLE WECK* Mathematical Institute, Universität Hannover, Hannover, Federal Republic of Germany

*Current affiliation: University of Toledo, Toledo, Ohio

Continuous Lattices and Their Applications

1

On the Topologies of Injective Spaces

BERNHARD BANASCHEWSKI
McMaster University
Hamilton, Ontario
Canada

This paper presents a characterization of the lattices of open sets of injective T_0-spaces, and thus of the Scott topologies of continuous lattices (Scott [10]; see also Compendium [3], chapter 2), in purely lattice theoretic terms, i.e., by conditions involving only the join and meet operations of the lattice. We remark that this differs conceptually from the earlier characterization in Banaschewski [2] which employs completely prime filters.

In any complete lattice L, let $a \lhd b$ (read perhaps as "a is totally below b") mean that, for any subset S of L, $b \leq \bigvee S$ implies that $a \leq t$ for some $t \in S$, and call L supercontinuous iff $x = \bigvee y \ (y \lhd x)$ for all $x \in L$. Further, call L stably supercontinuous iff, in addition, $a \lhd b$ and $a \lhd c$ implies $a \lhd b \wedge c$, and $e \lhd e$ for the top (= unit) e of L.

By way of apology, we observe that the term "supercontinuous" is motivated by the standard terms "supercompact" (for topological spaces) and "continuous" (for lattices).

REMARK 1 The above relation was originally introduced by Raney [10] although the terminology used here is ours. Raney proved that a complete lattice is completely distributive iff it is supercontinuous. On the other hand, the work of Hoffmann [4,5] and Lawson [7] shows that the supercontinuous lattices are exactly the topologies of projective sober spaces or, alternatively, the Scott topologies of continuous partially ordered sets. The present paper does not depend on any of these results.

We note a few instances of supercontinuous lattices:

EXAMPLE 1 In the power set lattice $\mathcal{P}(E)$ of any set E, $A \lhd B$ iff $A \subseteq \{x\}$ for some $x \in B$. Since $X = \bigcup\{x\}$ $(x \in X)$ for each subset X of E, this shows $\mathcal{P}(E)$ is supercontinuous. However, if E has two distinct elements b and c, then $\emptyset \lhd \{b\}$ and $\emptyset \lhd \{c\}$, but $\emptyset \lhd \{b\} \wedge \{c\} = \emptyset$ does not hold: $\emptyset = \bigcup\{x\}$ $(x \in \emptyset)$, yet there is no $x \in \emptyset$. Hence $\mathcal{P}(E)$ is not stably supercontinuous in this case.

EXAMPLE 2 Any complete chain C with supercompact unit (i.e., $e \lhd e$) is stably supercontinuous. To begin with, note that $a < b$ implies $a \lhd b$ in any complete chain C: if $b \leq \bigvee S$ and no $t \in S$ is above a, then $t \leq a$ for all $t \in S$ and hence $\bigvee S \leq a$, a contradiction. Now, for any $a \in C$, either $a = \bigvee x$ $(x < a)$ or $\bigvee x$ $(x < a) < a$, and thus $a = \bigvee x$ $(x \lhd a)$ or $a \lhd a$, where the latter also implies $a = \bigvee x$ $(x \lhd a)$. Hence C is supercontinuous. Further, let $a \lhd b$ and $a \lhd c$ in C. If $a < b$ and $a < c$, then also $a < b \wedge c$; and hence $a \lhd b \wedge c$; on the other hand, if $a = b$ (say), then $a \lhd a$ and $a = b \wedge c$ so that, again, $a \lhd b \wedge c$. It follows that C is stably supercontinuous iff $e \lhd e$.

EXAMPLE 3 Any finite distributive lattice L is supercontinuous because each $a \in L$ is a join of $\vee$-irreducible elements of L which in turn are characterized as the $x \in L$ such that $x \lhd x$. In particular, $a \lhd b$ holds in L iff $a \leq x \leq b$ for some $\vee$-irreducible x, and hence L is stably supercontinuous iff (i) $x \wedge y$ is $\vee$-irreducible for any $\vee$-irreducibles x and y, and (ii) $e \lhd e$. Considering the description of finite distributive lattices by downsets of partially ordered sets, one sees that these lattices are characterized, up to isomorphism, as the lattices of downsets of finite $\wedge$-semilattices with unit.

EXAMPLE 4 In one direction the foregoing example allows considerable generalization: the ideal lattice of any monoid M is supercontinuous and will be stably supercontinuous iff any intersection of two principal ideals is again principal. This certainly holds if M is commutative and idempotent, which is the preceding case but without finiteness.

EXAMPLE 5 If $h: L \to M$ is any $\bigvee$-homomorphism with left inverse $\bigvee$-homomorphism $g: M \to L$ and M is supercontinuous, then L is supercontinuous. This is a direct consequence of the obvious fact that $y \lhd h(x)$ in M implies $g(y) \lhd x$ in L. Moreover, if M is stably supercontinuous, and both g and h also preserve finite meets, then L is also stably supercontinuous, by the following argument: Since $b = \bigvee g(y)$ $[y \lhd h(b)]$ for any $b \in L$, $a \lhd b$ in L implies that $a \leq g(y)$ for some $y \lhd h(b)$ in M. Analogously, if also $a \lhd c$ in L, then $a \leq g(z)$ for some $z < h(c)$ in M, and therefore

$$a \leq g(y) \wedge g(z) = g(y \wedge z)$$

where

$$y \wedge z \lhd h(b) \wedge h(c) = h(b \wedge c)$$

by the hypotheses on g, M, and h. The latter implies $g(y \wedge z) \lhd b \wedge c$ in L, and hence we have $a \lhd b \wedge c$. The additional condition, $e \lhd e$ in L, results from the fact that $e \lhd e$ in M and $e = h(e)$ implies $g(e) \lhd e$ in L which says that $e \lhd e$ in L.

REMARK 2 Any supercontinuous lattice L is continuous because $a \lhd b$ evidently implies $a \ll b$, the usual "way below" relation, since the latter is defined exactly like the former but with the sets S restricted to be up-directed. Also, any stably supercontinuous lattice is stably continuous, meaning L has the property that $a \ll b$ and $a \ll c$ implies $a \ll b \wedge c$, and $e \ll e$. This results because, in such L, $a \ll b$ iff $a \leq x_1 \vee \cdots \vee x_n$ for some $x_i \lhd b$. On the other hand, though, a supercontinuous lattice may well be stably continuous without being stably supercontinuous: any finite distributive lattice is trivially stably continuous (because all its elements are compact), and supercontinuous by Example 3, but need not be stably supercontinuous.

Most of the discussion below deals with completely prime filters, i.e., the filters F in L such that $\bigvee S \in F$ implies $S \cap F \neq \emptyset$ for all subsets S of L.

For any supercontinuous lattice L, the following hold:

(1) $\lhd$ interpolates, i.e., if $a \lhd c$ in L, then also $a \lhd b$ and $b \lhd c$ for some $b \in L$.

(2) A filter F in L is completely prime iff, for any $c \in F$ there exists an $a \lhd c$ in F.

(3) $a \lhd b$ iff there exists a completely prime filter P in L such that $b \in P$ and $P \subseteq \uparrow a$ (where, as usual, $\uparrow a = \{x | a \leq x \in L\}$).

(4) L is a spatial local lattice, i.e., the completely prime filters in L separate the elements of L, i.e., if $a \not\leq b$, then there exists a completely prime filter P in L for which $a \in P$ and $b \notin P$.

Finally, if L is stably supercontinuous, then:

(5) Any finite join of completely prime filters, in the lattice of all filters in L, is again completely prime.

Of these assertions, (1) holds because, for each $x \in L$,

$$x = \bigvee y \ (y \prec x) = \bigvee z \ (z \lhd y \text{ for some } y \lhd x)$$

by supercontinuity, and hence $a \lhd c$ implies $a \leq z$ for some z such that $z \lhd b$ for some $b \lhd c$, and $a \leq z \lhd b$ implies $a \lhd b$. (2) is immediately obvious from the definitions. Regarding (3), if $a \lhd b$, then one has a sequence $b \rhd b_1 \rhd b_2 \rhd \cdots \geq a$ by successive application of (1), and the filter P generated by $b_1, b_2, \ldots$ is then a completely prime filter of the desired kind. This proves the forward implication, and the converse is immediate in <u>any</u> complete lattice. (4) results from (3) and the fact that $a \nleq b$ implies the existence of a $c \lhd a$ for which $c \nleq b$: if $a \in P \subseteq \uparrow c$ for some completely prime filter P as stated in (3), then $b \notin P$ and thus P separates a and b. Finally, concerning (5), one notes that $P \vee Q$ is generated by all $a \wedge b$, $a \in P$ and $b \in Q$, and if $c \lhd a$ in P and $d \lhd b$ in Q, then $c \wedge d \in P \vee Q$ and $c \wedge d \lhd a \wedge b$ by the present hypothesis. By (2) this shows that $P \vee Q$ is completely prime whenever P and Q are. On the other hand, the empty join in this context is $\{e\}$ and since also $e \lhd e$ in the present case this is a completely prime filter.

PROPOSITION The topologies of injective T_0-spaces are, up to isomorphism, exactly the stably supercontinuous lattices.

Proof: ($\Rightarrow$) We use the characterization of injective T_0-spaces given in Banaschewski [1]: A T_0-space X is injective iff (i) the join of any neighborhood filters $\mathfrak{D}(x) = \{U \mid x \in U \in \mathfrak{D}X\}$ in the filter lattice of the lattice $\mathfrak{D}X$ of all open sets of X is again a neighborhood filter, and (ii) for any $U \in \mathfrak{D}X$ and any $x \in U$, there exists a $W \in \mathfrak{D}(x)$ and some $y \in X$ such that $U \in \mathfrak{D}(y) \subseteq \uparrow W$. The latter implies $W \lhd U$ as observed above, and since $x \in U$ is arbitrary, it follows that $U = \bigcup W \ (W \lhd U)$, i.e., $\mathfrak{D}X$ is supercontinuous. Next, if $U \lhd V$ and $U \lhd W$ in $\mathfrak{D}X$, then it follows† from (ii) that $V \in \mathfrak{D}(x) \subseteq \uparrow U$ and $W \in \mathfrak{D}(y) \subseteq \uparrow U$ for some $x,y \in X$; further, by the condition (i) just quoted there is a $z \in X$ such that $\mathfrak{D}(z) = \mathfrak{D}(x) \vee \mathfrak{D}(y)$ and consequently $V \cap W \in \mathfrak{D}(z) \subseteq \uparrow U$, showing that $U \lhd V \cap W$. Finally, the empty

---

†(ii) means, as noted, that any V is the union of the $\tilde{W}$ such that $V \in \mathfrak{D}(x) \subseteq \uparrow \tilde{W}$ for some $x \in X$. Hence, if $U \lhd V$, then (just by the definition of $\lhd$) $U \subseteq \tilde{W}$ for some $\tilde{W}$ such that $V \in \mathfrak{D}(x) \subseteq \uparrow \tilde{W}$ for some x, and then also $V \in \mathfrak{D}(x) \subseteq \uparrow U$.

join of neighborhood filters is $\{X\}$ and hence, again by (i), there is a point $x \in X$ such that $\mathfrak{D}(x) = \{X\}$, which shows that $X \lhd X$. In all, this proves that $\mathfrak{D}X$ is stably supercontinuous.

($\Leftarrow$) If L is stably supercontinuous, then by (4) it is isomorphic to the lattice of open sets of its spectrum ΨL with the completely prime filters P in L as its points and the sets $\Psi_a = \{P | a \in P \in \Psi L\}$, $a \in L$, as its open sets. Now, since $a \in P$ iff $P \in \Psi_a$ for any $a \in L$ and $P \in \Psi L$, the isomorphism $L \to \mathfrak{D}(\Psi L)$ given by $a \rightsquigarrow \Psi_a$ makes the completely prime filters of L correspond to the analogous filters of $\mathfrak{D}(\Psi L)$, and hence the latter are exactly the open neighborhood filters of ΨL. This shows that these filters are closed under up-directed unions. Further, by (5), any finite join of completely prime filters is again a completely prime filter, and therefore the same now holds for arbitrary joins. Hence ΨL satisfies the condition (i) quoted above. Regarding (ii), if $P \in \Psi_a$, then $a \in P$; hence $c \lhd a$ for some $c \in P$ by (2) and therefore there exists a completely prime filter Q such that $a \in Q \subseteq \uparrow c$ by (3), which is exactly what (ii) amounts to for ΨL. In all, this shows ΨL is injective.

REMARK 3 The condition (ii) employed in the preceding proof appears in Banaschewski [1] in a slightly different form. Using $\Gamma_0 S = \cap \Gamma\{x\}$ $(x \in S)$ (Γ the closure operator of the space), the original version in [1] reads: for any $U \in \mathfrak{D}(x)$ $(x \in X)$ there exists a $W \in \mathfrak{D}(x)$ such that $U \cap \Gamma_0 W \neq \emptyset$. Now, $y \in \Gamma_0 W$ obviously holds iff $\mathfrak{D}(y) \subseteq \uparrow W$, and hence the present formulation is equivalent to that in [1]. Moreover, the above proof actually shows two separate things: First, $\mathfrak{D}X$ is supercontinuous iff X satisfies (ii), and secondly, given $\mathfrak{D}X$ is supercontinuous, it is stably so iff (i) holds for X.

REMARK 4 An alternative proof of the above proposition could be based on the result in Banaschewski [2] that the topologies of injective T_0-spaces are characterized by (5) together with the condition: for each $x \in L$, $x = \bigvee \bigwedge P$ $(x \in P \in \Psi L)$. Such a proof might be marginally shorter than that given here, but since we find the present result rather more interesting than the earlier one in [2] it seemed preferable to derive it directly from our original topological characterization of injective T_0-spaces.

REMARK 5 In the category Loc $\mathfrak{L}$ of local lattices (= frames) and $\bigvee\wedge$-homomorphisms, the regular epimorphisms (= coequalizers) are evidently the onto maps, and hence the <u>regular projectives</u> in Loc $\mathfrak{L}$ are exactly the retracts of free local lattices. Now, the latter are precisely (Isbell [6]) the topologies of Sierpinski cubes which, in turn, are injective T_0-spaces. As a result, by our Example 5 and the Proposition, the regular projectives in Loc $\mathfrak{L}$ are stably supercontinuous. Conversely, for any such lattice L, the spectrum ΨL is an injective T_0-space by the Proposition, hence a retract of some Sierpinski cube, and therefore $L \cong \mathcal{D}(\Psi L)$ is a retract of some free local lattice, i.e., regular projective. Therefore we have the following result, entirely concerned with the category Loc $\mathfrak{L}$: <u>The stably supercontinuous lattices are exactly the regular projectives in</u> Loc $\mathfrak{L}$. Note this has the interesting aspect that, in the equational class Loc $\mathfrak{L}$ of (infinitary) algebras, regular projectivity is internally characterized by a condition expressed in terms of the fundamental operations $\bigvee$ and $\wedge$. It might be added to this that Loc $\mathfrak{L}$ does indeed have epimorphisms which are not onto maps, e.g., the embedding of the topology of a space X with exactly one nonisolated point into the lattice of all subsets of X, and hence some regular projectives in Loc $\mathfrak{L}$, such as the free ones, fail to be (ordinary) projectives. We do not know what, if any, the latter are.

REMARK 6 The following argument shows directly that any stably supercontinuous L is a regular projective local lattice, once one knows it is indeed such a lattice. Let F(E) be the free local lattice on the underlying set E of L, i.e., the lattice of all upsets in the lattice Fin $\mathcal{P}(E)$ of finite subsets of E, freely generated by the principal upsets $\uparrow\{x\} \subseteq$ Fin $\mathcal{P}(E)$, $x \in E$, with canonical homomorphism h: $F(E) \to L$ where $h(\mathcal{A}) = \bigvee\bigwedge S (S \in \mathcal{A})$. Now define f: $L \to F(E)$ by

$$f(x) = \{A \mid A \in \text{Fin}\,\mathcal{P}(E),\ \bigwedge A \lhd x\}$$

Clearly, this is an upset in Fin $\mathcal{P}(E)$, i.e., belongs to F(E). Moreover, for any $S \subseteq L$, $f(\bigvee S) = \bigcup f(t)\ (t \in S)$, the nontrivial inclusion because $\bigwedge A \lhd \bigvee S$ implies $\bigwedge A \lhd c \lhd \bigvee S$ for some $c \in L$; hence $c \leq t$ for some $t \in S$ and then $\bigwedge A \lhd t$, i.e., $A \in f(t)$, for this t. Further, if $A \in f(x) \cap f(y)$, then $\bigwedge A \lhd x$ and $\bigwedge A \lhd y$; hence $\bigwedge A \lhd x \wedge y$ and thus $A \in f(x \wedge y)$, giving the nontrivial part of the identity $f(x \wedge y) = f(x) \cap f(y)$. Finally, f(e) consists of all $A \in$ Fin $\mathcal{P}(E)$ since $e \lhd e$. This shows that f is a local lattice homomorphism, and since $hf(x) = \bigvee\bigwedge A\ (\bigwedge A \lhd x) = x$, it is right inverse to h, i.e., L is a retract of F(E) and hence regular projective in Loc $\mathfrak{L}$. With this fact one can arrange an alternative proof of the Proposition by first establishing the result of the preceding Remark and then using the argument that the regular projectives in Loc $\mathfrak{L}$ are, up to isomorphism, the topologies of injective T_0-spaces. This approach involves showing first that stably supercompact lattices are in fact local [proving (4) will do, but there may be a more direct way] and then that free local lattices are stably supercontinuous, the remainder of the proof given by the argument just presented and the second part of Example 5. Evidently, this line of reasoning is independent of the topological characterization of injective T_0-spaces; indeed, if one derives the Proposition in this manner one could then, in turn, use it to obtain that characterization. In conclusion, we note that the (fairly obvious) characterization of the topologies of injective T_0-spaces as the regular projective local lattices is also given in Sobral [11], and the (less obvious) equivalence of regular projectivity in Loc $\mathcal{L}$ and stable supercontinuity is mentioned, without proof, in Niefield [8].

REMARK 7 Given the Hoffmann-Lawson characterization [5,7] of the supercontinuous lattices as the Scott topologies of continuous partially ordered sets, the difference between mere supercontinuity and stable supercontinuity corresponds exactly to the specialization from continuous partially ordered sets to continuous lattices. In a similar vein, since Niefield [8] proves that the supercontinuous lattices are exactly the regular projectives in the category Sup $\mathcal{L}$ of complete lattices and $\bigvee$-homomorphisms, that difference corresponds precisely to the difference between regular projectivity in Sup $\mathcal{L}$ and in its subcategory Loc $\mathcal{L}$. This may sound somewhat curious at first since there seems to be no a priori reason that the regular projectives in the smaller category should retain their property in the larger one, but in actual fact there is such a reason: each free object in Loc $\mathcal{L}$ happens to be a retract in Sup $\mathcal{L}$ of a free object in Sup $\mathcal{L}$, the retraction being the map from the power set of the set Fin $\mathcal{P}(E)$ of all finite subsets of a set E to the lattice of all upsets of Fin $\mathcal{P}(E)$ which sends each subset of Fin $\mathcal{P}(E)$ to the upset it generates.

ACKNOWLEDGMENTS

This work was done during my visit to the University of Cape Town in June 1982. Financial assistance from that institution and the Council of Scientific and Industrial Research as well as the kind hospitality of the Topology Research Group at the University of Cape Town are gratefully acknowledged. Additional thanks go, as usual, to the Natural Sciences and Engineering Research Council of Canada for its ongoing support. Finally, I am much indebted to the editors of this volume for a number of most helpful comments.

REFERENCES

1. B. Banaschewski, Essential extensions of T_0-spaces. Gen. Top. Appl. 7 (1977), 233-246.
2. B. Banaschewski, The duality of distributive continuous lattices. Can. J. Math. 22 (1980), 385-394.
3. G. Gierz, K. H. Hofmann, K. Keimel, J. D. Lawson, M. Mislove, D. S. Scott, A Compendium of Continuous Lattices. Springer-Verlag, Berlin-Heidelberg-New York, 1980.
4. R.-E. Hoffmann, Projective sober spaces. Lecture Notes in Math. 871, Springer-Verlag, Berlin-Heidelberg-New York, 1981, 125-158.
5. R.-E. Hoffmann, Continuous posets, prime spectra of completely distributive complete lattices, and Hausdorff compactifications. Ibid., 159-208.
6. J. R. Isbell, Atomless parts of spaces. Math. Scand. 31 (1972), 5-32.
7. J. D. Lawson, The duality of continuous posets. Houston J. Math. 5 (1979), 357-394.
8. S. B. Niefield, Exactness and projectivity. Lecture Notes in Math. 962, Springer-Verlag, Berlin-Heidelberg-New York, 1982, 221-227.
9. G. N. Raney, A subdirect-union representation for completely distributive complete lattices. Proc. Amer. Math. Soc. 4 (1953), 518-522.
10. D. S. Scott, Continuous lattices. Lecture Notes in Math. 274, Springer-Verlag, Berlin-Heidelberg-New York, 1972, 97-136.
11. M. Sobral, Restricting the comparison functor of an adjunction to projective objects. Quaest. Math. (to appear).

2

Convergence and Continuity in Partially Ordered Sets and Semilattices

MARCEL ERNÉ and HARTMUT GATZKE*

Mathematical Institute
Universität Hannover
Hannover, Federal Republic of Germany

0. INTRODUCTION AND SURVEY

Given a function τ associating to each (semi)lattice S a certain topology $\tau(S)$, it is natural to ask whether these topologies are compatible with the meet operations. More precisely, one may raise the following problems:

(1) Are the unary operations $\wedge_x : S \to S$, $y \mapsto x \wedge y$ continuous with respect to $\tau(S)$?

(2) Is the binary meet operation $\wedge : S \times S \to S$ continuous with respect to $\tau(S \times S)$ and $\tau(S)$ ("τ-continuous")?

(3) Under what assumptions is $(S,\tau(S))$ a topological semilattice [i.e., is meet continuous with respect to $\tau(S) \times \tau(S)$ and $\tau(S)$]?

In this paper we are mainly concerned with some special choices of τ which are of particular interest for the theory of topological lattices and semilattices, namely, the <u>upper topology</u> $\upsilon(S)$, the <u>Scott topology</u> $\sigma(S)$, the <u>Lawson topology</u> $\lambda(S)$, the <u>lim inf topology</u> $\xi(S)$ (and their duals, denoted by an additional $\tilde{}$); further, "self-dual" topologies such as the <u>interval topology</u> $\Omega_\iota(S)$, the <u>Bi-Scott topology</u> $\Omega_\sigma(S)$, the <u>order topology</u> $\Omega(S)$, and the <u>convex order topology</u> $\Omega_\kappa(S)$.

For $\tau \in \{\sigma,\lambda,\xi,\Omega_\sigma,\Omega_\kappa,\Omega\}$, questions 1 and 2 can be answered affirmatively if and only if the semilattice is meet continuous. But as there are meet-continuous semilattices (moreover, complete lattices) for which meet is not jointly continuous with respect to the above choices of τ (see

*Current affiliation: Siemens AG, München-Perlach, Federal Republic of Germany

Section 4), we have to look for stronger conditions than meet continuity in order to solve problem (3). Apparently, this problem is closely related to the following one:

(4) When is $\tau(S \times S)$ identical with the product topology $\tau(S) \times \tau(S)$?

It turns out that for most of the relevant choices of τ the inclusion $\tau(S) \times \tau(S) \subseteq \tau(S \times S)$ is fulfilled, so we are more interested in the inverse inclusion. The analysis of this problem for the more general concept of groupoids leads to a very useful result (Theorem 3.2), which for bounded semilattices reads as follows: For a bounded meet-semilattice S and fixed $\tau \in \{\sigma,\lambda,\xi,\Omega_\sigma,\Omega_\kappa,\Omega\}$, the "square" $(S \times S,\tau(S \times S))$ is a topological semilattice iff $\tau(S \times S) \subseteq \tau(S) \times \tau(S)$ and S is meet continuous. (For the case $\tau = \sigma$, see also Isbell [12].) A related unsolved problem is this: Are there semilattices S for which $(S,\tau(S))$ is a topological semilattice although $\tau(S \times S)$ differs from the product topology $\tau(S) \times \tau(S)$?

In Section 4 we investigate the significance of Hausdorff's separation axiom in connection with the problems mentioned above. We obtain a new necessary and sufficient condition for a complete lattice to be continuous, namely, that its square is a topological meet semilattice with respect to the lim inf topology (Corollary 4.6). Best possible results will be obtained in the case of Boolean lattices because they are not only join and meet continuous but may also be considered as Boolean rings, so that the machinery of topological groups (resp., rings) applies to this situation. We demonstrate that on a Boolean lattice L with non-Hausdorff order topology $\Omega(L)$ there exists no topology coarser than $\Omega(L)$ making the binary meet continuous and $\{0\}$ closed (Theorem 4.10). Our studies on Boolean lattices will be continued in a forthcoming paper.

Several of our general results on continuity and convergence in partially ordered sets have been established earlier for complete lattices (see, for example, [11] and [18]). Our filter-theoretical approach to intrinsic topologies on posets, however, seems to be rather fruitful and enables us to generalize most of the known results from complete lattices to arbitrary (or at least to bounded) posets and to disclose the interplay and similarity between various topological and convergence-theoretical notions discussed in the <u>Compendium of Continuous Lattices</u> [11]. In particular, we shall obtain several characterizations of continuous isotone maps between partially ordered sets equipped with the above-mentioned topologies. These general results will then apply to the more special setting of unary and binary meet operations, considered as isotone maps.

1. CONVERGENCE STRUCTURES AND TOPOLOGIES ON PARTIALLY ORDERED SETS

Let P be a partially ordered set ("poset"). The dual of P is denoted by $P^{\sim}$. For $y \in P$, the set $\downarrow y = \{x \in P : x \leq y\}$ is called the principal ideal generated by y, and for $Y \subseteq P$, we denote by $\downarrow Y$ the lower set generated by Y, i.e., $\downarrow Y = \bigcup \{\downarrow y : y \in Y\}$. Principal dual ideals $\uparrow y$ and upper sets $\uparrow Y$ are defined dually. Furthermore, $Y_{\downarrow} := \bigcap \{\downarrow y : y \in Y\}$ (resp., $Y^{\uparrow} := \bigcap \{\uparrow y : y \in Y\}$) is the set of all lower (resp., upper) bounds for Y. We write $x = \bigwedge Y$ (resp., $x = \bigvee Y$) to indicate that x is the greatest lower (resp., least upper) bound for Y (meet, resp., join). Principal ideals, principal dual ideals, and sets of the form $[y,z] = \uparrow y \cap \downarrow z$ $(y,z \in P)$ are called (closed) intervals. More generally, every intersection of an upper and a lower set is called convex. Thus $Y \subseteq P$ is convex iff $[y,z] \subseteq Y$ for all $y,z \in Y$. For an arbitrary subset Y of P, the set $\kappa Y = \uparrow Y \cap \downarrow Y$ is the least convex set containing Y, called the convex hull of Y.

A set $Y \subseteq P$ is (up-)directed if every finite subset of Y has an upper bound in Y (in particular, Y must be nonempty). Down-directed sets are defined dually. In the Compendium of Continuous Lattices [11] a poset is called continuous if it is up complete (i.e., each directed subset has a join) and for each $x \in P$ the way-below set

$$\Downarrow x = \bigcap \{Y \subseteq P : Y \text{ is a directed lower set with } x \leq \bigvee Y\}$$

is directed and has join x. In the absence of the required up-directed joins, there are several different possibilities to extend this concept, as was demonstrated in [5]. For the present context, we adopt the following definition: A poset P is called continuous if for every $x \in P$ there exists a least directed lower set Y_x possessing a join such that $x \leq \bigvee Y_x$ (s_3-continuity in the sense of [5]). Obviously, this definition agrees with the usual one if P is up complete, but it has the advantage of greater generality. For example, every chain (i.e., linearly ordered set) is continuous in the extended sense, while an up-complete chain with least element must already be complete. A slight modification of the previous definition leads to a property which is equivalent to meet continuity (upper continuity in the sense of Crawley and Dilworth [1]) when restricted to complete lattices. Call a poset P weakly upper continuous if for every directed subset Y of P possessing a join and for every $x \in P$ with $x \leq \bigvee Y$, there exists a directed (lower) set $Y_x \subseteq Y$ such that $x = \bigvee Y_x$. Of course, every continuous poset has this property. A straightforward computation shows

1.1 LEMMA A $\wedge$-semilattice S is weakly upper continuous iff for every directed set $Y \subseteq S$ possessing a join and for every $x \in S$,

$$x \wedge \bigvee Y = \bigvee\{x \wedge y : y \in Y\}$$

In view of this characterization, we shall speak of meet-continuous (written $\wedge$-continuous) semilattices. Join-continuous (written $\vee$-continuous) semilattices are defined dually.

The notions of limit inferior and limit superior, familiar from the reals, generalize to arbitrary posets as follows (cf. Kent [14,15,16], Erné [3,4(III),5], Erné and Weck [6]). Given a (set-theoretical) filter $\mathfrak{F}$ on a poset P, define

$$\mathfrak{F}_{\downarrow} := \bigcup\{F_{\downarrow} : F \in \mathfrak{F}\} = \{x \in P : \uparrow x \in \mathfrak{F}\}$$

$$\mathfrak{F}^{\uparrow} := \bigcup\{F^{\uparrow} : F \in \mathfrak{F}\} = \{x \in P : \downarrow x \in \mathfrak{F}\}$$

We say x is the limit inferior of $\mathfrak{F}$, in symbols $x = \underline{\lim}\, \mathfrak{F}$, if $x = \bigvee \mathfrak{F}_{\downarrow}$. Dually, x is the limit superior of $\mathfrak{F}$, written $x = \overline{\lim}\, \mathfrak{F}$, if $x = \bigwedge \mathfrak{F}^{\uparrow}$. Notice that these statements are meaningful in arbitrary posets although the terms $\underline{\lim}\, \mathfrak{F}$ and $\overline{\lim}\, \mathfrak{F}$ need not be defined for all filters $\mathfrak{F}$. However, in a complete lattice, we have

$$\underline{\lim}\, \mathfrak{F} = \bigvee\{\bigwedge F : F \in \mathfrak{F}\} \qquad \overline{\lim}\, \mathfrak{F} = \bigwedge\{\bigvee F : F \in \mathfrak{F}\}$$

Now we may define the following "one-sided" types of convergence:

Upper convergence:

$\mathfrak{F} \xrightarrow[\upsilon]{} x$ iff $x \in \mathfrak{G}^{\uparrow}{}_{\downarrow}$ for every (ultra)filter $\mathfrak{G}$ finer than $\mathfrak{F}$

Weak Scott convergence:

$\mathfrak{F} \xrightarrow[\underline{s}]{} x$ iff $x \in \mathfrak{F}_{\downarrow}{}^{\uparrow}{}_{\downarrow}$ (s_1-convergence in the sense of [5]; see also [22])

Strong Scott convergence:

$\mathfrak{F} \xrightarrow[\underline{S}]{} x$ iff there exists a directed subset Y of $\mathfrak{F}_{\downarrow}$ possessing a join such that $x \leq \bigvee Y$ (s_3-convergence in the sense of [5])

It is easy to see that

$$\mathfrak{F} \xrightarrow[\underline{S}]{} x \Rightarrow \mathfrak{F} \xrightarrow[\underline{s}]{} x \Rightarrow \mathfrak{F} \xrightarrow[\upsilon]{} x$$

and, in complete lattices,

$$\mathfrak{F} \xrightarrow[\underline{S}]{} x \Leftrightarrow \mathfrak{F} \xrightarrow[\underline{s}]{} x \Leftrightarrow x \leq \underline{\lim}\, \mathfrak{F}$$

We write

$$\mathfrak{F} \xrightarrow[\upsilon^{\sim}]{} x$$

$$\mathfrak{F} \xrightarrow[\underline{s}^{\sim}]{} x$$

$$\mathfrak{F} \xrightarrow[\underline{S}^{\sim}]{} x$$

to indicate that, respectively,

$$\mathfrak{F} \xrightarrow[\upsilon]{} x$$

$$\mathfrak{F} \xrightarrow[\underline{s}]{} x$$

$$\mathfrak{F} \xrightarrow[\underline{S}]{} x$$

in the dual poset $P^{\sim}$. Combining these convergence structures pairwise, we arrive at the following "two-sided" types of convergence:

Interval convergence:

$$\mathfrak{F} \xrightarrow[\Omega_{\iota}]{} x \quad \text{iff } \mathfrak{F} \xrightarrow[\upsilon]{} x \text{ and } \mathfrak{F} \xrightarrow[\upsilon^{\sim}]{} x \qquad \text{(cf. Kent [16])}$$

Weak lim inf convergence:

$$\mathfrak{F} \xrightarrow[\underline{\ell}]{} x \quad \text{iff } \mathfrak{F} \xrightarrow[\underline{s}]{} x \text{ and } \mathfrak{F} \xrightarrow[\upsilon^{\sim}]{} x$$

Strong lim inf convergence:

$$\mathfrak{F} \xrightarrow[\underline{L}]{} x \quad \text{iff } \mathfrak{F} \xrightarrow[\underline{S}]{} x \text{ and } \mathfrak{F} \xrightarrow[\upsilon^{\sim}]{} x$$

Weak lim sup convergence:

$$\mathfrak{F} \xrightarrow[\underline{\ell}^{\sim}]{} x \quad \text{iff } \mathfrak{F} \xrightarrow[\underline{s}^{\sim}]{} x \text{ and } \mathfrak{F} \xrightarrow[\upsilon]{} x$$

Strong lim sup convergence:

$$\mathfrak{F} \xrightarrow[\underline{L}^{\sim}]{} x \quad \text{iff } \mathfrak{F} \xrightarrow[\underline{S}^{\sim}]{} x \text{ and } \mathfrak{F} \xrightarrow[\upsilon]{} x$$

Weak order convergence:

$$\mathfrak{F} \xrightarrow[\underline{o}]{} x \quad \text{iff } \mathfrak{F} \xrightarrow[\underline{s}]{} x \text{ and } \mathfrak{F} \xrightarrow[\underline{s}^{\sim}]{} x$$

Strong order convergence:

$$\mathfrak{F} \xrightarrow[\underline{O}]{} x \quad \text{iff } \mathfrak{F} \xrightarrow[\underline{S}]{} x \text{ and } \mathfrak{F} \xrightarrow[\underline{S}^{\sim}]{} x$$

The terminology "lim inf convergence" is motivated by the following observation:

$$\mathfrak{F} \xrightarrow[\underline{\ell}]{} x \quad \text{iff } x = \underline{\lim}\, \mathfrak{G} \text{ for every filter } \mathfrak{G} \text{ finer than } \mathfrak{F}$$

Furthermore, we have the following equivalences:

$$\mathfrak{F} \xrightarrow[\underline{L}]{} x \quad \text{iff } \mathfrak{F} \xrightarrow[\underline{\ell}]{} x \text{ and } x = \bigvee Y \text{ for some directed } Y \subseteq \mathfrak{F}_{\downarrow}$$

$$\mathfrak{F} \xrightarrow[\underline{o}]{} x \quad \text{iff } x = \underline{\lim}\, \mathfrak{F} = \overline{\lim}\, \mathfrak{F}$$

$$\mathfrak{F} \xrightarrow[\underline{O}]{} x \quad \text{iff } x = \bigvee Y = \bigwedge Z \text{ for some up-directed } Y \subseteq \mathfrak{F}_{\downarrow} \text{ and some down-directed } Z \subseteq \mathfrak{F}^{\uparrow}$$

Notice that on lattices the weak types of two-sided convergence relations agree with the corresponding strong types. However, weak and strong Scott convergence may differ even on lattices (e.g., on $\mathbb{R}^2$).

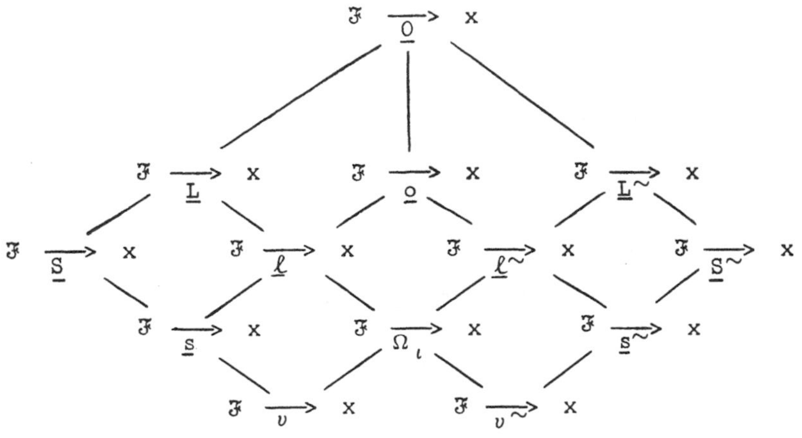

Diagram 1

For an ultrafilter $\mathcal{U}$ on a complete lattice, we have

$$\mathcal{U} \xrightarrow[\upsilon]{} x \Leftrightarrow x \le \overline{\lim}\, \mathcal{U} \qquad \mathcal{U} \xrightarrow[\upsilon^{\sim}]{} x \Leftrightarrow x \ge \underline{\lim}\, \mathcal{U}$$

$$\mathcal{U} \xrightarrow[\underline{L}]{} x \Leftrightarrow x = \underline{\lim}\, \mathcal{U} \qquad \mathcal{U} \xrightarrow[\underline{L}^{\sim}]{} x \Leftrightarrow x = \overline{\lim}\, \mathcal{U} \qquad (\underline{L} = \underline{\ell})$$

$$\mathcal{U} \xrightarrow[\underline{S}]{} x \Leftrightarrow x \le \underline{\lim}\, \mathcal{U} \qquad \mathcal{U} \xrightarrow[\underline{S}^{\sim}]{} x \Leftrightarrow x \ge \overline{\lim}\, \mathcal{U} \qquad (\underline{S} = \underline{s})$$

Each of the previously defined relations between filters and points is a <u>convergence relation</u> in the sense of [6] (see also Kent [14]); that is, every principal ultrafilter $\dot{x} = \{Y \subseteq P : x \in Y\}$ converges to x, and if $\mathfrak{F}$ converges to x, then so does every filter finer than $\mathfrak{F}$. Given an arbitrary convergence relation C over a set X, one defines a topology $\mathfrak{T}_C$ by calling a set U open if every (ultra)filter converging to some $x \in U$ contains the set U. Equivalently, one may call a set A closed if for every (ultra)filter containing A and converging to x, the point x also belongs to A. It is easy to see that $\mathfrak{T}_C$ is the finest topology $\mathfrak{T}$ such that C-convergence implies topological convergence. Obviously, the assignment $C \mapsto \mathfrak{T}_C$ is antitone, i.e., $C \subseteq C' \Rightarrow \mathfrak{T}_{C'} \subseteq \mathfrak{T}_C$, and in particular, $\mathfrak{T}_C \vee \mathfrak{T}_D \subseteq \mathfrak{T}_{C \cap D}$; but, in general, proper inclusion holds.

For the above types of convergence relations, this general concept of "topological modification" leads to the following topologies on a partially ordered set P:

Convergence relation	Notation	Topological modification	Notation
Upper convergence	$\xrightarrow[\upsilon]{}$	Upper topology	$\upsilon(P)$
Lower convergence	$\xrightarrow[\tilde{\upsilon}]{}$	Lower topology	$\tilde{\upsilon}(P)$
Interval convergence	$\xrightarrow[\Omega_\iota]{}$	Interval topology	$\Omega_\iota(P)$
Strong lim inf convergence	$\xrightarrow[\underline{L}]{}$	lim inf topology	$\xi(P)$
Strong lim sup convergence	$\xrightarrow[\underline{\tilde{L}}]{}$	lim sup topology	$\tilde{\xi}(P)$
Strong Scott convergence	$\xrightarrow[\underline{S}]{}$	Scott topology	$\sigma(P)$
Strong dual Scott convergence	$\xrightarrow[\underline{\tilde{S}}]{}$	Dual Scott topology	$\tilde{\sigma}(P)$
Strong order convergence	$\xrightarrow[\underline{O}]{}$	Order topology	$\Omega(P)$

REMARK In the <u>Compendium</u> [11], the lower topology $\tilde{\upsilon}(P)$ is denoted by $\omega(P)$. From the convergence-theoretical point of view, it might be more suggestive to interchange the notations of upper and lower convergence. However, the name "upper topology" refers to the fact that $\upsilon(P)$ is the coarsest topology on P for which every principal ideal is a <u>closed</u> set and, consequently, every open set is an upper set.

A convergence relation C is said to be topological if C-convergence agrees with convergence in the topology $\mathfrak{J}_C$. Examples of topological convergence relations are upper, lower, and interval convergence. For the other convergence relations, this is not the case in general. For example, it has been shown in [5] that

$\underline{S}$-convergence is topological on P iff P is a continuous poset

For complete lattices, this is a central result in the theory of continuous lattices (see Scott [20] and the Compendium [11(II.1)]).

In the present context, we prefer to work with the strong types of convergence because they admit a smoother characterization of the related topologies. For analogous results concerning weak order convergence, resp., Scott convergence, the reader is referred to the forerunners [4], [5], [6], and [22].

Each of the above topologies may be described without employing filters. A set A is closed in the upper (resp., lower) topology iff it is an intersection of finite unions of principal ideals (resp., dual principal ideals); A is closed in the interval topology iff it is an intersection of finite unions of intervals (cf. Frink [9], Kent [16]); A is closed in the Scott topology iff it is a lower set and closed under directed joins; A is closed in the order topology iff for every up-directed set Y and every down-directed set Z with $x = \bigvee Y = \bigwedge Z \notin A$, there exist $y \in Y$ and $z \in Z$ such that $[y,z] \cap A = \emptyset$ (see [4(III)] and [6]).

A characterization of the lim inf topology via directed sets is a bit more subtle (and seems to have been neglected even in the case of complete lattices). The following notation will be convenient: For subsets A and Y of a poset P, set

$$Y_A^{\uparrow} := \bigcap\{\uparrow F : F \text{ finite}, \uparrow y \cap A \subseteq \uparrow F \text{ for some } y \in Y\}$$

Notice that

$$Y_A^{\uparrow} \subseteq Y^{\uparrow} = Y_P^{\uparrow}$$

and that $Y_A^{\uparrow}$ is empty unless $Y \subseteq \downarrow A$.

Now the lim inf topology may be described as follows:

1.2 PROPOSITION A set $A \subseteq P$ is closed in $\xi(P)$ iff for every directed set $Y(\subseteq \downarrow A)$, $x = \bigvee Y \in Y_A^{\uparrow}$ implies $x \in A$.

Proof: First, suppose A is ξ-closed, and let Y be a directed set with $x = \bigvee Y \in Y_A^\uparrow$. Then the system

$$\mathcal{B} = \{\uparrow y \cap A \setminus \uparrow F : y \in Y,\ F \text{ finite},\ x \notin \uparrow F\}$$

is a filterbase (because $x \in Y_A^\uparrow$ and $x \notin \uparrow F$ implies $\uparrow y \cap A \not\subseteq \uparrow F$ for all $y \in Y$). Choosing an ultrafilter $\mathcal{U}$ containing $\mathcal{B}$, we obtain $A \in \mathcal{U}$ and $\mathcal{U} \xrightarrow[L]{} x$. Indeed, $\uparrow y \in \mathcal{U}$ for all $y \in Y$ means $Y \subseteq \mathcal{U}_\downarrow$; moreover, x is an upper bound of $\mathcal{U}_\downarrow$: $\uparrow z \in \mathcal{U}$ implies $x \in \uparrow z$ (otherwise $\uparrow y \cap A \setminus \uparrow z \in \mathcal{B}$ for $y \in Y$, and, in particular, $P \setminus \uparrow z \in \mathcal{U}$). Hence, $x = \bigvee Y = \bigvee \mathcal{U}_\downarrow$. Now, using the fact that A is ξ-closed, we conclude $x \in A$.

Conversely, assume A has the property that for every directed set Y, $x = \bigvee Y \in Y_A^\uparrow$ implies $x \in A$. We want to show that for every ultrafilter $\mathcal{U}$ on P, $A \in \mathcal{U} \xrightarrow[L]{} x$ implies $x \in A$. Choose a directed set $Y \subseteq \mathcal{U}_\downarrow$ with $x = \bigvee Y = \bigvee \mathcal{U}_\downarrow$. For $y \in Y$ we get $\uparrow y \in \mathcal{U}$ and therefore $\uparrow y \cap A \in \mathcal{U}$. Hence, for every finite set F with $\uparrow y \cap A \subseteq \uparrow F$ for some $y \in Y$, we must have $\uparrow F \in \mathcal{U}$; as $\mathcal{U}$ is an ultrafilter, this implies $\uparrow z \in \mathcal{U}$, i.e., $z \in \mathcal{U}_\downarrow$ for some $z \in F$. But then, $z \le \bigvee \mathcal{U}_\downarrow = x$, whence $x \in \uparrow z \subseteq \uparrow F$. This proves $x \in Y_A^\uparrow$, and it follows that $x \in A$. ∎

Besides the above order-definable topologies, we shall also consider the following ones: the <u>Lawson topology</u> $\lambda(P) = \sigma(P) \vee \upsilon^\sim(P)$ and its dual $\lambda^\sim(P) = \lambda(P^\sim)$; the <u>Bi-Scott topology</u> $\Omega_\sigma(P) = \sigma(P) \vee \sigma^\sim(P)$; and, last but not least, the important <u>convex order topology</u> $\Omega_K(P)$ which has as a base the convex $\Omega(P)$-open sets (cf. Rennie [19], Lawson [17]). Observe that $\Omega_\iota(P) = \Omega_\iota(P^\sim)$, $\Omega_\sigma(P) = \Omega_\sigma(P^\sim)$, $\Omega_K(P) = \Omega_K(P^\sim)$, $\Omega(P) = \Omega(P^\sim)$, while the other topologies are not "self-dual." The hierarchy of these topologies is given in Diagram 2:

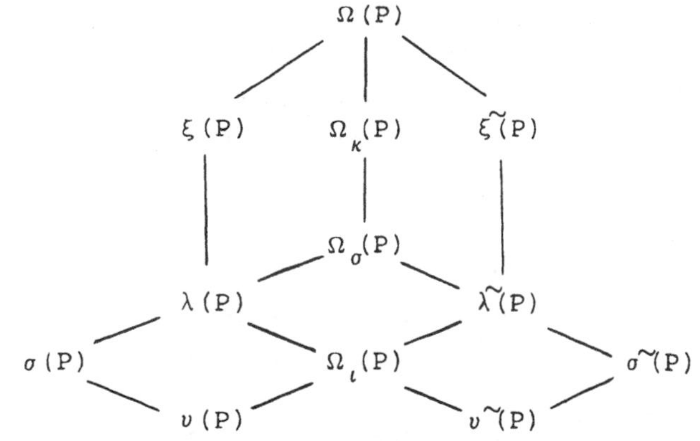

Diagram 2

A straightforward argument yields

1.3 LEMMA If P is weakly upper (resp., lower) continuous, then $U \in \Omega(P)$ implies $\uparrow U \in \sigma(P)$ [resp., $\downarrow U \in \tilde{\sigma}(P)$]. Hence, if P is both weakly upper and lower continuous, then $U \in \Omega(P)$ implies $\kappa U = \uparrow U \cap \downarrow U \in \Omega_\sigma(P)$, and consequently $\Omega_\kappa(P)$ coincides with $\Omega_\sigma(P)$.

The following facts are well known for complete lattices L:

(i) For any topology $\mathfrak{J}$ between $\sigma(L)$ and $\Omega(L)$, the $\mathfrak{J}$-open upper sets are precisely the σ-open sets.

(ii) For any topology $\mathfrak{J}$ between $\tilde{\upsilon}(L)$ and $\xi(L)$, the $\mathfrak{J}$-open lower sets are precisely the $\tilde{\upsilon}$-open sets (cf. [11,ch.III]).

While (i) remains true for arbitrary posets, this is not the case for (ii). E.g., the real plane $\mathbb{R}^2$ is a conditionally complete lattice where the lower half plane $\{(x,y) \in \mathbb{R}^2 ; x < 0\}$ is λ-open but not $\tilde{\upsilon}$-open (nor Ω_ι-open). $\lambda(\mathbb{R}^2) = \xi(\mathbb{R}^2) = \Omega_\sigma(\mathbb{R}^2) = \Omega_\kappa(\mathbb{R}^2)$ "$= \Omega(\mathbb{R}^2)$" is the euclidean topology, while $\Omega_\iota(\mathbb{R}^2)$ is strictly coarser. This and other examples show that generalizations from complete lattices to arbitrary lattices or posets have to be treated very cautiously. The following result, extending a well-known theorem of Frink [9], demonstrates that completeness is an essential ingredient in the theory of compact lattice topologies:

1.4 PROPOSITION Let L be an arbitrary lattice and $\mathfrak{J}$ a topology on L that is finer than the interval topology and coarser than the lim inf topology [e.g., $\mathfrak{J} = \Omega_\iota(L)$, $\lambda(L)$ or $\xi(L)$]. Then L is complete iff $\mathfrak{J}$ is quasicompact (cf. [11(III.3)]).

Proof: If L is complete, then every ultrafilter lim inf converges, and therefore $\xi(L)$ and every coarser topology is quasi-compact. On the other hand, if $\Omega_\iota(L)$ is quasi-compact, then every set $Y \subseteq L$ has a join since the system $\{[y,z] : y \in Y, z \in Y^\uparrow\}$ has nonempty intersection. (For $Y = \emptyset$, consider the system $\{\downarrow z : z \in \emptyset^\uparrow = L\}$.) ■

2. CONTINUITY OF ISOTONE MAPS

In this section, we investigate continuity properties of isotone maps with respect to the convergence relations and topologies introduced in Section 1. For complete (semi-)lattices some of the subsequent results already belong to the folklore of lattice theory (cf. Lawson [18], Scott [20], and the

Compendium [11]). However, the general poset approach reveals some interesting new aspects.

Given a function τ assigning to each poset P a topology $\tau(P)$, we shall say a map $f : P \to Q$ is τ-continuous if it is $\tau(P)$ - $\tau(Q)$-continuous, i.e., $f^{-1}(U) \in \tau(P)$ for all $U \in \tau(Q)$.

It is well known that the Scott-continuous (i.e., σ-continuous) maps between complete lattices are precisely those that preserve directed joins (cf. [11(II.2.1)]). This observation may be generalized to arbitrary posets (not only to up-complete ones).

2.1 PROPOSITION Let f be a map between posets P and Q. Furthermore, let $\mathcal{J}$ be a topology on P such that $\sigma(P) = \{U \in \mathcal{J} : U \text{ is an upper set}\}$, and let $\mathfrak{T}$ be any order-consistent topology on Q (i.e., $\upsilon(Q) \subseteq \mathfrak{T} \subseteq \sigma(Q)$; cf. [11(II.1.17)]). Then the following statements are equivalent:

(a) f preserves directed joins, i.e., for every directed subset Y of P possessing a join, $f(\bigvee Y) = \bigvee f(Y)$.

(b) f is continuous with respect to strong Scott convergence, i.e., $\mathfrak{F} \xrightarrow[\underline{S}]{} x$ in P implies $f(\mathfrak{F}) \xrightarrow[\underline{S}]{} f(x)$ in Q [where $f(\mathfrak{F})$ is the filter generated by the sets $f(G)$, $G \in \mathfrak{F}$].

(c) f is σ-continuous.

(d) f is isotone and continuous with respect to $\mathcal{J}$ and $\mathfrak{T}$.

(e) Inverse images of principal ideals under f are σ-closed.

Proof: (a) $\Rightarrow$ (b) : $\mathfrak{F} \xrightarrow[\underline{S}]{} x$ means $x \le \bigvee Y$ for some directed subset Y of $\mathfrak{F}_{\downarrow}$. As f is isotone, it follows that $f(Y)$ is a directed subset of $f(\mathfrak{F})_{\downarrow}$, and by hypothesis (a), $f(x) \le f(\bigvee Y) = \bigvee f(Y)$. Thus $f(\mathfrak{F}) \xrightarrow[\underline{S}]{} f(x)$ in Q.

(b) $\Rightarrow$ (c): This follows from a well-known general result on convergence relations which states that a continuous map between convergence spaces is also continuous as a map between their topological modifications.

(c) $\Rightarrow$ (d) $\Rightarrow$ (e): Clear. A σ-continuous map is isotone because inverse images of principal ideals are lower sets.

(e) $\Rightarrow$ (a): Suppose Y is a directed subset of P and $x = \bigvee Y$. Then $f(x)$ is an upper bound of $f(Y)$ because f is isotone. If z is any other upper bound of $f(Y)$ in Q, then $Y \subseteq f^{-1}(\downarrow z)$, and this is a σ-closed set. Hence, $x = \bigvee Y \in f^{-1}(\downarrow z)$, i.e., $f(x) \le z$. This proves $f(x) = \bigvee f(Y)$. ∎

In Proposition 2.1, we may choose, for example, $\mathcal{J} = \sigma(P), \lambda(P), \xi(P), \Omega_\sigma(P), \Omega_\kappa(P)$ or $\Omega(P)$, $\mathfrak{T} = \sigma(P)$ or $\upsilon(P)$. The "two-sided" analogue of 2.1 is

2.2 PROPOSITION Let $\tau \in \{\Omega_\sigma, \Omega_\kappa, \Omega\}$. Then for an isotone map $f : P \to Q$ the following statements are equivalent:

(a) f preserves up-directed joins and down-directed meets.

(b) f is continuous with respect to strong order convergence, i.e., $\mathfrak{F} \xrightarrow[\underline{O}]{} x$ in P implies $f(\mathfrak{F}) \xrightarrow[\underline{O}]{} f(x)$ in Q.

(c) f is τ-continuous.

(d) f is continuous with respect to topologies $\mathcal{J}$ and $\mathfrak{J}$, where $\tau(P) \subseteq \mathcal{J} \subseteq \Omega(P)$ and $\Omega_\iota(Q) \subseteq \mathfrak{J} \subseteq \tau(Q)$.

(e) Inverse images of intervals are τ-closed.

Proof: (a) $\Rightarrow$ (b): Clear by Proposition 2.1 and its dual.

(b) $\Rightarrow$ (c): As in Proposition 2.1, (b) implies Ω-continuity. But f is isotone, so inverse images of upper (lower, convex) sets are again upper (lower, convex) sets. Thus, Ω-continuity of f implies Ω_κ-continuity, which in turn implies Ω_σ-continuity.

(c) $\Rightarrow$ (d) $\Rightarrow$ (e) $\Rightarrow$ (a): See the proof of Proposition 2.1. ∎

We call an isotone map f <u>bicontinuous</u> if the equivalent conditions in Proposition 2.2 are fulfilled. Notice that the hypothesis that f is isotone is implied by (a) but not by (b), (c), or (d). For example, every map between <u>finite</u> posets is continuous with respect to all topologies occurring in Proposition 2.2 (because they are all discrete).

Observing that the Lawson topology and the lim inf topology are finer than the interval topology and coarser than the order topology, it is clear from Proposition 2.2 that λ-continuity or ξ-continuity, respectively, of an isotone map imply bicontinuity. However, one can prove the converse implication only under certain additional assumptions.

2.3 PROPOSITION Let $\tau \in \{\sigma, \lambda, \xi, \Omega_\sigma, \Omega_\kappa, \Omega\}$. Then, for an $\tilde{\upsilon}$-continuous map $f: P \to Q$, the following statements are equivalent:

(a) f preserves directed joins (see Proposition 2.1).

(b) f is bicontinuous (see Proposition 2.2).

(c) f is continuous with respect to strong lim inf convergence, i.e., $\mathfrak{F} \xrightarrow[\underline{L}]{} x$ in P implies $f(\mathfrak{F}) \xrightarrow[\underline{L}]{} f(x)$ in Q.

(d_τ) f is τ-continuous.

Proof: First, observe that $\tilde{\upsilon}$-continuity implies that inverse images of principal dual ideals are upper sets, whence f must be isotone.

(a) $\Rightarrow$ (d_σ): Proposition 2.1.

(b) ⇔ (d_τ) for $\tau \in \{\Omega_\sigma, \Omega_\kappa, \Omega\}$: Proposition 2.2.

(a) ⇒ (c): By Proposition 2.1, f is continuous with respect to strong Scott convergence, and the equivalence

$$\mathfrak{F} \xrightarrow[\underline{L}]{} x \Leftrightarrow \mathfrak{F} \xrightarrow[\underline{S}]{} x \text{ and } \mathfrak{F} \xrightarrow[\tilde{\upsilon}]{} x$$

ensures that f is also continuous with respect to strong lim inf convergence.

The implications (c) ⇒ (d_ξ) ⇒ (b) ⇒ (a) and (d_σ) ⇒ (d_λ) ⇒ (b) are clear by Propositions 2.1 and 2.2, so we have closed the implication circles. ∎

The following example shows that a bicontinuous map need not be λ-continuous nor ξ-continuous.

2.4 EXAMPLE Consider a map f between complete lattices L and M as indicated in Diagram 3.

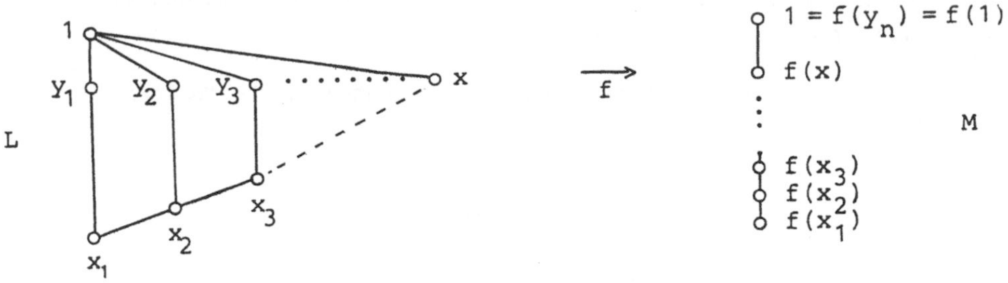

Diagram 3

Obviously, f preserves up-directed joins and down-directed meets. But {1} is λ-closed in M while $f^{-1}(\{1\}) = \{y_n : n \in \mathbb{N}\} \cup \{1\}$ is not even ξ-closed. Indeed, the Fréchet filter of the sequence (y_n) (strongly) lim inf converges to x whence x lies in the ξ-closure of $f^{-1}(\{1\})$.

It is easy to see that a ∧-homomorphism between complete ∧-semilattices preserves down-directed meets iff it is $\tilde{\upsilon}$-continuous. In particular, Proposition 2.3 states that such a ∧-homomorphism is bicontinuous iff it is τ-continuous, where $\tau \in \{\lambda, \xi, \Omega_\sigma, \Omega_\kappa, \Omega\}$ (cf. [18]).

Notice that Proposition 2.3 applies to adjoint maps (cf. [11(0.3)]). A lower adjoint is υ-continuous because inverse images of principal ideals are principal ideals. Dually, an upper adjoint is $\tilde{\upsilon}$-continuous. Furthermore, a lower adjoint preserves joins, an upper adjoint meets (to the extent they exist; cf. [11(0.3.3)]).

2.5 COROLLARY Let $f : P \to Q$ be a map between $\vee$-semilattices which has a lower adjoint $d : Q \to P$. Then each of the conditions stated in Proposition 2.3 is equivalent to the following:

(e) f preserves lim infs, i.e., $x = \underline{\lim}\, \mathfrak{F}$ in P implies $f(x) = \underline{\lim}\, f(\mathfrak{F})$ in Q.

Proof: In a $\vee$-semilattice, $x = \underline{\lim}\, \mathfrak{F} = \bigvee \mathfrak{F}_\downarrow$ implies that $\mathfrak{F}_\downarrow$ is directed, so it remains to verify that (e) holds when f preserves directed joins.

We already know that $x = \bigvee \mathfrak{F}_\downarrow$ implies $f(x) = \bigvee f(\mathfrak{F}_\downarrow)$; therefore it will be enough to prove the equation $\downarrow f(\mathfrak{F}_\downarrow) = f(\mathfrak{F})_\downarrow$. Using the lower adjoint $d : Q \to P$ of f, we obtain:

$$y \in \downarrow f(\mathfrak{F}_\downarrow) \Leftrightarrow y \leq f(x) \text{ for some } x \in \mathfrak{F}_\downarrow \Leftrightarrow d(y) \leq x \text{ for some } x \in \mathfrak{F}_\downarrow \Leftrightarrow$$

$$d(y) \in \mathfrak{F}_\downarrow \Leftrightarrow \uparrow d(y) \in \mathfrak{F} \Leftrightarrow f^{-1}(\uparrow y) \in \mathfrak{F} \Leftrightarrow \uparrow y \in f(\mathfrak{F}) \Leftrightarrow y \in f(\mathfrak{F})_\downarrow . \quad \blacksquare$$

The next example shows that Corollary 2.5 cannot be extended to arbitrary posets (or $\wedge$-semilattices).

2.6 EXAMPLE Consider the $\wedge$-semilattice S indicated by the following diagram:

S

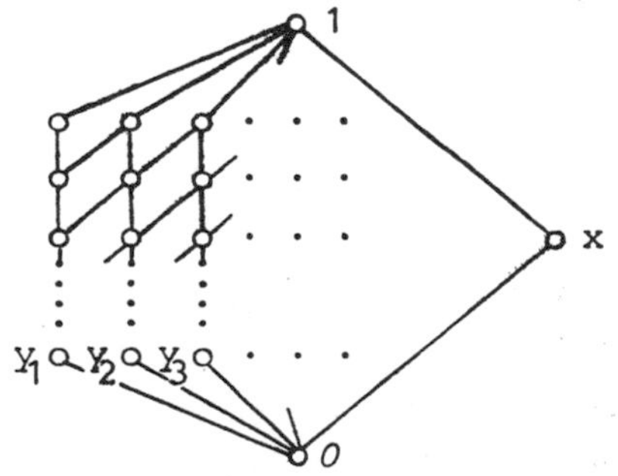

Diagram 4

The set $I = \{y_n : n \in \mathbb{N}\} \cup \{0\}$ is an ideal in the sense of Frink [10], but not directed. For the filter $\mathfrak{F}$ generated by the principal dual ideals $\uparrow y$ $(y \in I)$, one obtains $\mathfrak{F}_\downarrow = I$ (cf. [6]) and $1 = \bigvee \mathfrak{F}_\downarrow = \underline{\lim}\, \mathfrak{F}$. As S satisfies the ascending chain condition, the binary meet function $\wedge : S \times S \to S$ is bicontinuous and has a lower adjoint (see below), but it does not preserve lim infs because $\wedge(\underline{\lim}\, \mathcal{G}) = x \neq 0 = \underline{\lim}\, \wedge(\mathcal{G})$, where $\mathcal{G}$ denotes the filter on $S \times S$ generated by the sets $\{x\} \times F$ with $F \in \mathfrak{F}$.

The following fact is well known for complete lattices (cf. [11(II.2.9)]), and it extends without any alteration to arbitrary posets:

2.7 LEMMA A map $f : P_1 \times P_2 \to Q$ is σ-continuous (resp., bicontinuous) iff for each $x_1 \in P_1$ and each $x_2 \in P_2$ the maps

$$f_{x_1,\cdot} : P_2 \to Q,\ y_2 \mapsto f(x_1,y_2)$$
$$f_{\cdot,x_2} : P_1 \to Q,\ y_1 \mapsto f(y_1,x_2)$$

are σ-continuous (resp., bicontinuous).

Summarizing the results derived before, we are now in position to collect various topological characterizations of $\wedge$-continuous (semi)lattices. Observing that the binary meet $\wedge : S \times S \to S$ has a lower adjoint, namely, the diagonal map

$$d : S \to S \times S,\ x \mapsto (x,x)$$

we may apply 2.1-2.7 to this special situation and obtain (cf. [4(III)] and [18]):

2.8 THEOREM Let $\tau \in \{\sigma,\lambda,\xi,\Omega_\sigma,\Omega_\kappa,\Omega\}$. Then, for a $\wedge$-semilattice S, the following conditions are equivalent:

(a) S is $\wedge$-continuous.

(b) The binary meet operation $\wedge : S \times S \to S$ is τ-continuous.

(c) Meet is continuous with respect to strong Scott convergence: $\mathfrak{F} \xrightarrow[\underline{S}]{} x$ and $\mathfrak{G} \xrightarrow[\underline{S}]{} y$ implies $\mathfrak{F} \wedge \mathfrak{G} \xrightarrow[\underline{S}]{} x \wedge y$ (where $\mathfrak{F} \wedge \mathfrak{G}$ is the filter generated by the sets $F \wedge G$ with $F \in \mathfrak{F}$ and $G \in \mathfrak{G}$).

(d) Meet is continuous with respect to strong lim inf convergence.

(e) Meet is continuous with respect to strong order convergence.

(b′) The unary meet operations $\wedge_x : S \to S,\ y \mapsto x \wedge y$ $(x \in S)$ are τ-continuous (i.e., meet is "separately continuous" with respect to τ).

(c′),(d′),(e′) Same statements for the unary meet operations.

(f′) The maps $\wedge_x$ are $\mathcal{J}$-$\mathfrak{T}$-continuous, where $\mathcal{J}$ and $\mathfrak{T}$ are topologies such that $\upsilon(S) \subseteq \mathfrak{T} \subseteq \tau(S) \subseteq \mathcal{J} \subseteq \Omega(S)$.

2.9 COROLLARY Let $\tau \in \{\sigma,\lambda,\xi,\Omega_\sigma,\Omega_\kappa,\Omega\}$. If $(S,\tau(S))$ is a topological $\wedge$-semilattice (i.e., meet is "jointly continuous" with respect to τ), then each of the equivalent conditions in Theorem 2.8 is also satisfied. The converse implication holds whenever $\tau(S \times S)$ is the product topology $\tau(S) \times \tau(S)$.

The equation $\tau(S \times S) = \tau(S) \times \tau(S)$ proves to be independent of $\wedge$-continuity. Examples of $\wedge$-continuous (moreover, of Boolean) lattices where this equation fails for all relevant choices of τ because meet is not jointly continuous will be given in Section 4. Diagram 5 represents a (well-known) complete lattice L which is not $\wedge$-continuous although $\tau(L \times L) = \tau(L) \times \tau(L)$ for all choices of τ arising in Diagram 2.

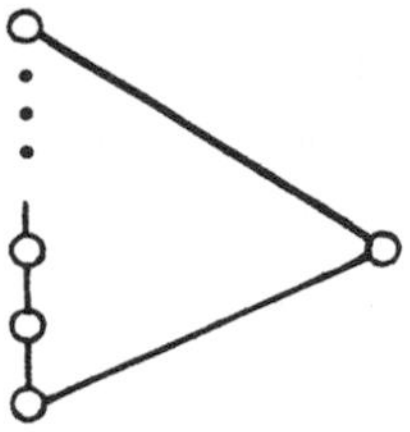

Diagram 5

Some interrelations between the topological properties of the lattice operations are collected in

2.10 PROPOSITION Consider the following conditions for a lattice L:

(a) $(L,\Omega(L))$ is a topological lattice.

(b) $(L,\Omega_\kappa(L))$ is a topological lattice.

(c) $(L,\Omega_\sigma(L))$ is a topological lattice.

(d) $(L,\sigma(L))$ and $(L,\sigma^\sim(L))$ are topological lattices.

(e) $(L,\sigma(L))$ is a topological $\wedge$-semilattice.

(f) $(L,\lambda(L))$ is a topological $\wedge$-semilattice.

Each of these conditions entails $\wedge$-continuity of L, and the following implications hold:

(a) $\Rightarrow$ (b) $\Leftrightarrow$ (c) $\Leftrightarrow$ (d) $\Rightarrow$ (e) $\Leftrightarrow$ (f).

Proof: By Corollary 2.9, each of the stated conditions is sufficient for $\wedge$-continuity, and by duality, (a), (b), (c) also imply $\vee$-continuity.

(a) $\Rightarrow$ (b): Given a convex neighborhood U of $x \wedge y$ in L, we find $\Omega(L)$-open sets V and W such that $x \in V$, $y \in W$, and $V \wedge W \subseteq U$. By $\vee$- and $\wedge$-continuity of L, it follows that the convex hulls κV and κW are $\Omega_\kappa(L)$-open neighborhoods of x and y, respectively, and $\kappa V \wedge \kappa W \subseteq \kappa(V \wedge W) \subseteq U$.

(b) $\Leftrightarrow$ (c): For $\vee$- and $\wedge$-continuous lattices L, one has $\Omega_\kappa(L) = \Omega_\sigma(L)$ (see Lemma 1.3).

(c) ⇔ (d): Analogously.

(d) ⇒ (e): Trivial.

(e) ⇒ (f): $(L,\upsilon^{\sim}(L))$ is always a topological $\wedge$-semilattice because $\wedge^{-1}(\uparrow y) = \uparrow y \times \uparrow y$ $(y \in L)$.

(f) ⇒ (e): Use the fact that L is $\wedge$-continuous [whence $U \in \lambda(L)$ implies $\uparrow U \in \sigma(L)$]. ∎

None of the conditions in Proposition 2.10 implies that $(L,\xi(L))$ is a topological $\wedge$-semilattice, as will be demonstrated at the end of this paper. But, on the other hand, it is easy to see that (e) is necessary for L to be a topological $\wedge$-semilattice with respect to $\xi(L)$.

3. DIRECT PRODUCTS AND TOPOLOGICAL GROUPOIDS

Given a family $((X_j,\mathfrak{T}_j): j \in J)$ of topological spaces and a topology $\mathfrak{T}$ on the cartesian product $X = \Pi_{j \in J} X_j$, one is often interested in the question whether $\mathfrak{T}$ is coarser, finer, or equal to the product topology $\Pi_{j \in J} \mathfrak{T}_j$ on X. It is well known that the inclusion

$$\prod_{j \in J} \mathfrak{T}_j \subseteq \mathfrak{T}$$

holds if and only if the <u>projections</u>

$$\pi_k : X \to X_k,\ x \mapsto x_k \qquad k \in J$$

are continuous with respect to $\mathfrak{T}$ and $\mathfrak{T}_k$ (one may even define the product topology by this property). But what about the inclusion $\mathfrak{T} \subseteq \Pi_{j \in J} \mathfrak{T}_j$? A necessary condition for this inclusion is that the family $((X_j,\mathfrak{T}_j) : j \in J)$ be $\mathfrak{T}$-<u>fibered</u>, i.e., for each $y \in X = \Pi_{j \in J} X_j$ and each $k \in J$, the injection

$$\eta_{y,k} : X_k \to X,\ v \mapsto x \qquad \text{where } x_j = \begin{cases} v & \text{for } j = k \\ y_j & \text{for } j \neq k \end{cases}$$

has to be continuous with respect to $\mathfrak{T}_k$ and $\mathfrak{T}$. Indeed, for each $U \in \mathfrak{T}$, the "fiber"

$$U(y,k) := \eta_{y,k}^{-1}(U)$$

must be $\mathfrak{T}_k$-open. To see this, choose $v \in U(y,k)$; then $x = \eta_{y,k}(v)$ is an element of U, so there exists a family of open sets $U_j \in \mathfrak{T}_j$ such that $x \in \Pi_{j \in J} U_j \subseteq U$, whence $v = \pi_k(x) \in U_k \subseteq U(y,k)$. Notice that continuity of the injections $\eta_{y,k}$ $(y \in X)$ implies that the projection π_k is open:

$$\pi_k(U) = \bigcup \{U(y,k) : y \in X\} \qquad \text{for } U \subseteq X$$

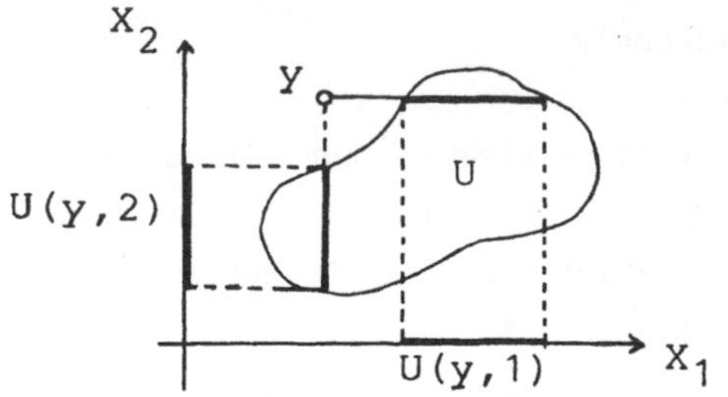

Diagram 6

Furthermore, it is evident that the inclusion $\mathfrak{J} \subseteq \Pi_{j \in J} \mathfrak{J}_j$ is impossible unless $\mathfrak{J}$ has the following "product property": For each $x \in U \in \mathfrak{J}$, there exists a finite index subset K of J such that

$$Q_K(x) := \{y \in X : y_j = x_j \text{ for } j \in K\} = \pi_K^{-1}(\pi_K(\{x\})) \subseteq U$$

Of course, this is not a restriction at all if J is finite (take K = J).

In general, the product property together with the continuity of the injections $\eta_{y,k}$ is necessary but not sufficient for the inclusion $\mathfrak{J} \subseteq \Pi_{j \in J} \mathfrak{J}_j$. As we shall see later on, both the product property and the continuity of the injections $\eta_{y,k}$ are satisfied if we are dealing with the special classes of topologies introduced in Section 1. Therefore, it will be useful to find additional conditions under which the inclusion $\mathfrak{J} \subseteq \Pi_{j \in J} \mathfrak{J}_j$ can be derived.

It turns out that the adequate setting for this problem is that of <u>topological groupoids</u>. By a <u>groupoid</u> we mean a set X together with a binary operation $\cdot$ (not necessarily associative) such that there exists a neutral element e (that is, $x \cdot e = e \cdot x = x$ for all $x \in X$). We shall not distinguish between X and the pair $(X, \cdot)$. Direct products of groupoids are defined in the obvious manner. A <u>topological groupoid</u> is a pair $(X, \mathfrak{J})$ where X is a groupoid and $\mathfrak{J}$ a topology on the underlying set such that the operation $\cdot$ is continuous with respect to $\mathfrak{J}$, i.e., for all $x, y \in X$ and all $U \in \mathfrak{J}$ with $x \cdot y \in U$ there exist $V, W \in \mathfrak{J}$ with $x \in V$ and $y \in W$ such that

$$V \cdot W = \{v \cdot w : v \in V, w \in W) \subseteq U$$

Given a finite family $(x_r : r = 1, \ldots, n)$ of elements in a groupoid X, we may define products inductively by setting

$$\bigodot_{r=1}^{0} x_r := e, \quad \bigodot_{r=1}^{m+1} x_r := \left(\bigodot_{r=1}^{m} x_r\right) \cdot x_{m+1} \qquad m < n$$

(Notice that this definition depends on the numbering of the elements x_r if the operation $\cdot$ is not associative or not commutative.) More generally, given any finite family $(x_k : k \in K)$ of elements in X and a bijection $\varphi : \{1,\ldots,n\} \to K$, define

$$\overset{\varphi}{\underset{k \in K}{\odot}} x_k := \overset{n}{\underset{r=1}{\odot}} x_{\varphi(r)}$$

(For empty K, $\odot^{\varphi}_{k \in K} x_k = e$.) Furthermore, for a finite family $(Y_k : k \in K)$ of subsets of X, set

$$\overset{\varphi}{\underset{k \in K}{\odot}} Y_k := \{ \overset{\varphi}{\underset{k \in K}{\odot}} x_k : x = (x_k) \in \prod_{k \in K} Y_k\}$$

In particular, $\odot^{\varphi}_{k \in \emptyset} Y_k = \{e\}$.

Now let $(X,\mathfrak{T})$ be a topological groupoid and $U \in \mathfrak{T}$. Then, by induction, it is easy to see that whenever $\odot^{\varphi}_{k \in K} x_k \in U$, there exist open sets $V_k \in \mathfrak{T}$ $(k \in K)$ such that $x_k \in V_k$ and $\odot^{\varphi}_{k \in K} V_k \subseteq U$.

The following fact belongs to the folklore of topological algebra:

3.1 LEMMA Let $(X_j : j \in J)$ be a family of groupoids and X their direct product. Furthermore, let there be given a family of topologies $\mathfrak{T}_j$ on X_j $(j \in J)$. Then $(X, \prod_{j \in J} \mathfrak{T}_j)$ is a topological groupoid iff each $(X_j,\mathfrak{T}_j)$ is a topological groupoid.

Now the main result can be formulated as follows:

3.2 THEOREM Let $(X_j : j \in J)$ be a family of groupoids and X their direct product. Furthermore, suppose that each X_j carries a topology $\mathfrak{T}_j$, and $\mathfrak{T}$ is a topology on X such that $(X,\mathfrak{T})$ is a topological groupoid. Then $\mathfrak{T}$ is contained in the product topology of the $\mathfrak{T}_j$ iff the family $((X_j,\mathfrak{T}_j) : j \in J)$ is $\mathfrak{T}$-fibered and $\mathfrak{T}$ has the product property.

Proof: The necessity of these conditions has already been established before. To show sufficiency, we use the following notation. For $x,y \in X$ and $K \subseteq J$, define $x \overset{*}{_K} y$ by

$$(x \overset{*}{_K} y)_j := \begin{cases} x_j & \text{for } j \in K \\ y_j & \text{for } j \in J\setminus K \end{cases}$$

For example, $x \overset{*}{_J} y = x$, $x \overset{*}{_\emptyset} y = y$, and $x \overset{*}{_K} y = y \overset{*}{_{J\setminus K}} x$.

Furthermore,

$$(x \underset{K}{*} e) \cdot (e \underset{K}{*} x) = x$$

where e is the neutral element of X. Note also that

$$x \underset{k}{*} y := x \underset{\{k\}}{*} y = \eta_{y,k}(x_k)$$

Now suppose the family $((X_j,\mathfrak{T}_j) : j \in J)$ is $\mathfrak{T}$-fibered, $\mathfrak{T}$ has the product property, and $(X,\mathfrak{T})$ is a topological groupoid. For $x \in U$ we wish to construct a product-open set $W \in \Pi_{j \in J} \mathfrak{T}_j$ with $x \in W \subseteq U$. As $x \cdot e = x \in U \in \mathfrak{T}$, we find $V,V' \in \mathfrak{T}$ with $x \in V$, $e \in V'$, and $V \cdot V' \subseteq U$. By the product property there exists a finite $K \subseteq J$ such that $Q_K(x) \subseteq V$ and $Q_K(e) \subseteq V'$. Fix an arbitrary bijection $\varphi : \{1,\ldots,n\} \to K$. Using once more the fact that $(X,\mathfrak{T})$ is a topological groupoid and that

$$\overset{\varphi}{\underset{k \in K}{\odot}} (x \underset{k}{*} e) = x \underset{K}{*} e \in Q_K(x) \subseteq V$$

we find open sets $V_k \in \mathfrak{T}$ $(k \in K)$ such that $x \underset{k}{*} e \in V_k$ and $\odot^{\varphi}_{k \in K} V_k \subseteq V$. For each $k \in K$, the fiber $W_k := V_k(e,k) = \eta_{e,k}^{-1}(V_k)$ is $\mathfrak{T}_k$-open and $x_k \in W_k$. Setting $W_j = X_j$ for $j \in J\backslash K$ and $W := \Pi_{j \in J} W_j$, we obtain $x \in W \in \Pi_{j \in J} \mathfrak{T}_j$.

Now consider an arbitrary $w \in W$. For $k \in K$ we have $w \underset{k}{*} e = \eta_{e,k}(w_k)$, and this is an element of V_k since $w_k \in W_k$. Thus

$$w \underset{K}{*} e = \overset{\varphi}{\underset{k \in K}{\odot}} (w \underset{k}{*} e) \in \overset{\varphi}{\underset{k \in K}{\odot}} V_k \subseteq V$$

On the other hand, $e \underset{K}{*} w \in Q_K(e) \subseteq V'$, and consequently $w = (w \underset{K}{*} e) \cdot (e \underset{K}{*} w) \in V \cdot V' \subseteq U$, as desired. ∎

3.3 COROLLARY Let $(X_j : j \in J)$ be a family of groupoids and X their direct product. Then for a family of topologies $\mathfrak{T}_j$ on X_j and a topology $\mathfrak{T}$ on X, the following two statements are equivalent:

(a) Each $(X_j,\mathfrak{T}_j)$ is a topological groupoid, and $\mathfrak{T} = \Pi_{j \in J} \mathfrak{T}_j$.

(b) $(X,\mathfrak{T})$ is a topological groupoid, $\mathfrak{T}$ has the product property, and the injections $\eta_{y,j} : (X_j,\mathfrak{T}_j) \to (X,\mathfrak{T})$ as well as the projections $\pi_j : (X,\mathfrak{T}) \to (X_j,\mathfrak{T}_j)$ are continuous.

Recall that for finite J the product property is trivially fulfilled for every topology $\mathfrak{T}$ on X.

The preceding general results will now apply to semilattices. Let τ be a function assigning to every partially ordered set P a certain topology $\tau(P)$. E.g., we may take for τ one of the choices $\upsilon, \sigma, \lambda, \xi$, etc. Then the following "product invariance" problem arises: Given a family $(P_j : j \in J)$ of posets, when is it true that the topology $\tau(\Pi_{j \in J} P_j)$ agrees with the product topology $\Pi_{j \in J} \tau(P_j)$?

Several aspects of this problem have already been discussed in [2], [4(I-III)], and [5]. Some of the basic facts and supplementary remarks will be given below.

Let P be the product of a family of posets P_j $(j \in J)$ and K a nonempty subset of J. Then we may define "partial products"

$$P_K = \prod_{j \in K} P_j$$

and "generalized projections"

$$\pi_K : P \to P_K, \ x \mapsto x|_K$$

3.4 PROPOSITION For $\tau \in \{\sigma, \tilde{\sigma}, \Omega_\sigma, \Omega_K, \Omega\}$ the generalized projections π_K are τ-continuous. In particular, this is true for the projections π_k, $k \in J$; i.e.,

$$\prod_{j \in J} \tau(P_j) \subseteq \tau(P)$$

If P has a least (resp., greatest) element, then the statement above also holds for $\tau \in \{\tilde{\upsilon}, \lambda, \xi\}$ [resp., $\tau \in \{\upsilon, \tilde{\lambda}, \tilde{\xi}\}$].

Proof: Generalized projections preserve arbitrary joins and meets. If P has a least element 0, then $\pi_K : P \to P_K$ has the lower adjoint

$$\psi_K : P_K \to P, \ y \mapsto x \quad \text{where } x_j = \begin{cases} y_j & \text{for } j \in K \\ 0_j & \text{for } j \in J \setminus K \end{cases}$$

Dually, if P has a greatest element, then π_K has an upper adjoint. Hence, the assertions follow from Propositions 2.1, 2.2, 2.3, and the corresponding dual statements. ■

If P has no least (resp., greatest) element, the projections π_k need not be $\tilde{\upsilon}$- (resp., υ-) continuous, as the example $P_k = \mathbb{R}$ shows. Here the inverse images of principal (dual) ideals are not $\tilde{\upsilon}$- (resp., υ-) closed. Indeed, the set $\pi_k^{-1}(\downarrow v) = \{x \in \mathbb{R}^J : x_k \leq v\}$ cannot be represented as an intersection of finite unions of principal ideals in $\mathbb{R}^J$ (when $|J| > 1$). On the other hand, it has been observed in [4(I)] that the inclusion $\tau(P) \subseteq \Pi_{j \in J} \tau(P_j)$ is always fulfilled for $\tau \in \{\upsilon, \upsilon^\sim, \Omega_\iota\}$, and equality holds whenever P is bounded.

3.5 COROLLARY For a product P of bounded posets P_j, the equations

$$\prod_{j \in J} \lambda(P_j) = \lambda(P) \qquad \prod_{j \in J} \sigma(P_j) = \sigma(P)$$

are equivalent.

It remains open whether the inclusion

$$\prod_{j \in J} \lambda(P_j) \subseteq \lambda(P)$$

is valid for arbitrary posets P_j. It is certainly true for finite products of chains (bounded or not).

Now consider the injections

$$\eta_{y,k} : P_k \to P \qquad y \in P,\ k \in J$$

introduced before. The next result ensures the "fibration property" for all topologies considered in Section 1.

3.6 PROPOSITION For $\tau \in \{\upsilon, \upsilon^\sim, \Omega_\iota, \sigma, \sigma^\sim, \lambda, \lambda^\sim, \xi, \xi^\sim, \Omega_\sigma, \Omega_K, \Omega\}$, each of the injections $\eta_{y,k}$ is τ-continuous. In particular, each of the projections π_k is τ-open.

Proof: Obviously, $\eta_{y,k}$ preserves all nonempty joins and meets. However, it may fail to have a lower or upper adjoint even if the posets P_j are complete lattices. But we have for $x \in P$

$$\eta_{y,k}^{-1}(\downarrow x) = \begin{cases} \downarrow x_k & \text{if } y_j \le x_j \text{ for all } j \ne k \\ \emptyset & \text{otherwise} \end{cases}$$

and dually for $\eta_{y,k}^{-1}(\uparrow x)$. This proves υ-, $\upsilon^\sim$-, and Ω_ι-continuity of $\eta_{y,k}$, and for the other choices of τ the assertion follows from Propositions 2.1, 2.2, 2.3, and their duals. ■

Using similar arguments, one can show that even the generalized projections π_K ($\emptyset \ne K \subseteq J$) are τ-open. This, together with Proposition 3.4, yields

3.7 COROLLARY $\pi_K(\tau(P)) = \tau(P_K)$ $(\emptyset \neq K \subseteq J)$

Next, let us demonstrate that the product property discussed at the beginning of this section is valid for a very large class of topologies on partially ordered sets.

3.8 PROPOSITION Let P be the direct product of an almost bounded family of posets P_j, $j \in J$; i.e., the number of indices j with unbounded P_j is finite. Then the order topology on P (and every coarser topology on P) has the product property.

Proof: For the bounded posets P_j, let 0_j and 1_j denote the least and greatest element of P_j, respectively. Then for $x \in P$, the set

$$Y := \{y \in \downarrow x : y_j = 0_j \text{ for almost all } j \in J\}$$

is up-directed and has join x, while the set

$$Z := \{z \in \uparrow x : z_j = 1_j \text{ for almost all } j \in J\}$$

is down-directed and has meet x. Hence $x \in U \in \Omega(P)$ implies $x \in [y,z] \subseteq U$ for certain elements $y \in Y$, $z \in Z$. Thus we find a finite set $K \subseteq J$ such that $y_j = 0_j$ and $z_j = 1_j$ for all $j \in J\setminus K$, and it follows that $Q_K(x) = \pi_K^{-1}(\pi_K(\{x\})) \subseteq [y,z] \subseteq U$. ∎

It should be mentioned that the product property may fail for $\Omega(P)$ [and even for $\sigma(P)$] if an infinite number of the posets P_j is unbounded. (Example: $\mathbb{R}^J$, $|J| > \aleph_0$).

Combining the results of Propositions 3.4, 3.6, and 3.8 with 3.3, we obtain immediately the following specialization:

3.9 THEOREM Let $(S_j : j \in J)$ be a family of bounded (∨- or ∧-) semilattices and S their direct product. Then for fixed $\tau \in \{\upsilon, \tilde{\upsilon}, \Omega_\iota, \sigma, \tilde{\sigma}, \lambda, \tilde{\lambda}, \xi, \tilde{\xi}, \Omega_\sigma, \Omega_K, \Omega\}$, the following two statements are equivalent:

(a_τ) Each $(S_j, \tau(S_j))$ is a topological semilattice, and $\tau(S) = \prod_{j \in J} \tau(S_j)$.

(b_τ) $(S, \tau(S))$ is a topological semilattice.

For $\tau \in \{\sigma, \tilde{\sigma}, \Omega_\sigma, \Omega_K, \Omega\}$, it suffices to postulate that $(S_j : j \in J)$ is an almost bounded family of unital (∨- or ∧-) semilattices. Using Corollary 2.9, we infer at once

3.10 COROLLARY Let S be a bounded $\wedge$-semilattice and $\tau \in \{\sigma, \lambda, \xi, \Omega_\sigma, \Omega_\kappa, \Omega\}$. Then $(S \times S, \tau(S \times S))$ is a topological $\wedge$-semilattice iff S is $\wedge$-continuous and $\tau(S \times S) = \tau(S) \times \tau(S)$.

Again, for $\tau \in \{\sigma, \Omega_\sigma, \Omega_\kappa, \Omega\}$ it suffices that S has a 1 (not necessarily a 0).

Since the binary join is a map that preserves arbitrary joins, it is always σ-continuous (Proposition 2.1; see also Scott [20]). As a consequence, we obtain

3.11 COROLLARY Let S be a $\vee$-semilattice with 0. Then $(S \times S, \sigma(S \times S))$ is a topological $\vee$-semilattice iff $\sigma(S \times S) = \sigma(S) \times \sigma(S)$.

The last two corollaries (3.10 only for $\tau = \sigma$) have also been mentioned by Isbell in [12], where he gives an example of a complete lattice L_0 with sober Scott topology but jointly discontinuous join and meet, and an example of a complete lattice L_1 that is a topology (in particular, $\wedge$-continuous) but has jointly discontinuous meet in the Scott topology. Examples of $\vee$- and $\wedge$-continuous (moreover, Boolean) lattices with jointly discontinuous join and meet are presented in Section 4.

The following extremely useful fact is due to Lawson [17]:

If L is a complete lattice and $\mathfrak{T}$ a locally convex regular topology on L such that $(L, \mathfrak{T})$ is a topological lattice, then $\mathfrak{T}$ must be the convex order topology $\Omega_\kappa(L)$.

Since local convexity and regularity are product-invariant properties, we infer from the preceding results:

3.12 COROLLARY Let $(L_j : j \in J)$ be a family of complete lattices and L their product. Then $(L, \Omega_\kappa(L))$ is a regular topological lattice iff each $(L_j, \Omega_\kappa(L_j))$ is a regular topological lattice. Furthermore, if one of these equivalent conditions holds, then $\Omega_\kappa(L) = \Pi_{j \in J}\, \Omega_\kappa(L_j)$, and L (resp., each L_j) is $\vee$- and $\wedge$-continuous.

For Boolean lattices, the regularity condition in Corollary 3.12 can be omitted because they may be regarded as topological rings in their convex order topology (the complement operation being continuous).

Our next theorem allows to extend "product theorems" for the Scott topologies from finite to infinite families of $\wedge$-continuous lattices.

3.13 THEOREM Let $(L_j : j \in J)$ be a family of bounded $\wedge$-continuous lattices such that for every finite subset K of J,

$$\sigma(\prod_{j \in K} L_j) = \prod_{j \in K} \sigma(L_j)$$

Then

$$\sigma(\prod_{j \in J} L_j) = \prod_{j \in J} \sigma(L_j)$$

(and conversely). Similarly for λ instead of σ.

Proof: Let $L = \Pi_{j \in J} L_j$, $L^0 = \{x \in L : \{j \in J : \{x_j \neq 0_j\}$ finite$\}$. For $u \in U \in \sigma(L)$, we have to find a product open set $W \in \Pi_{j \in J} \sigma(L_j)$ with $u \in W \subseteq U$. The set $Y = L^0 \cap {\downarrow}u$ is an ideal of L with $\bigvee Y = u$, so there exists a $y \in Y \cap U$. As L is $\wedge$-continuous, the unary meet operation $\wedge_y$: $x \mapsto x \wedge y$ is continuous in the Scott topology of L (see Theorem 2.8), and consequently

$$V := \{v \in L : v \wedge y \in U\} \in \sigma(L)$$

Clearly $y \leq u$, whence $u \in V$. By definition of Y, there exists a finite set $K \subseteq J$ such that $y_j = 0_j$ for $j \in J \setminus K$. Now $u \in V \in \sigma(L)$ implies $u_K \in \pi_K(V) \in \sigma(L_K) = \sigma(\Pi_{j \in K} L_j)$ (see Corollary 3.7). But by hypothesis, $\sigma(\Pi_{j \in K} L_j) = \Pi_{j \in K} \sigma(L_j)$, so there exist open sets $W_j \in \sigma(L_j)$ $(j \in K)$ such that $u_K \in \Pi_{j \in K} W_j \subseteq \pi_K(V)$. Setting $W_j = L_j$ for $j \in J \setminus K$, we obtain a product open set $W = \Pi_{j \in J} W_j \in \Pi_{j \in J} \sigma(L_j)$ containing u. For arbitrary $w \in W$, we have $w_K \in \pi_K(V)$, so we may choose a $v \in V$ with $v_K = w_K$. Then $v \wedge y \leq w$, and $v \in V$ implies $v \wedge y \in U$, whence $w \in U$ (this is the only place where we need the fact that U is an upper set). Accordingly, W is contained in U. For λ, apply Corollary 3.5. ∎

The boundedness condition in Theorem 3.13 is essential: For example, it can be shown that $\sigma(\mathbb{R}^J) = \sigma(\mathbb{R})^J$ if and only if J is countable (cf. [2] and [5]).

In order to transfer Theorem 3.13 from Scott topologies to bi-Scott topologies (resp., convex order topologies), we use Lemma 1.3 and obtain

3.14 PROPOSITION Let $(L_j : j \in J)$ be a family of $\vee$- and $\wedge$-continuous lattices, and let L denote their direct product. Then

(1) $\Omega_\sigma(L_j) = \Omega_K(L_j)$ $(j \in J)$ and $\Omega_\sigma(L) = \Omega_K(L)$

(2) $\Omega_\sigma(L) = \Pi_{j \in J} \Omega_\sigma(L_j)$ iff $\sigma(L) = \Pi_{j \in J} \sigma(L_j)$ and $\sigma^\sim(L) = \Pi_{j \in J} \sigma^\sim(L_j)$.

Now Theorem 3.13 is supplemented by

3.15 COROLLARY Let $(L_j : j \in J)$ be a family of bounded $\vee$- and $\wedge$-continuous lattices, and let $\tau \in \{\sigma, \tilde{\sigma}, \lambda, \tilde{\lambda}, \Omega_\sigma, \Omega_\kappa\}$. Then $\tau(\Pi_{j \in J} L_j) = \Pi_{j \in J} \tau(L_j)$ iff $\tau(\Pi_{j \in K} L_j) = \Pi_{j \in K} \tau(L_j)$ for each finite $K \subseteq J$.

Diagram 7 illustrates the main implications and equivalences so far derived. Of course, many questions concerning further implications remain open. L denotes a bounded lattice (not necessarily complete).

$(L \times L, \Omega(L \times L))$ topological lattice	$\Leftrightarrow$	$\Omega(L \times L) = \Omega(L) \times \Omega(L)$ L $\vee$-, $\wedge$-continuous	$\Rightarrow$	$(L, \Omega(L))$ topological lattice
$\Downarrow$		$\Downarrow$		$\Downarrow$
$(L \times L, \Omega_\kappa(L \times L))$ topological lattice	$\Leftrightarrow$	$\Omega_\kappa(L \times L) = \Omega_\kappa(L) \times \Omega_\kappa(L)$ L $\vee$-, $\wedge$-continuous	$\Rightarrow$	$(L, \Omega_\kappa(L))$ topological lattice
$\Updownarrow$		$\Updownarrow$		$\Updownarrow$
$(L \times L, \Omega_\sigma(L \times L))$ topological lattice	$\Leftrightarrow$	$\Omega_\sigma(L \times L) = \Omega_\sigma(L) \times \Omega_\sigma(L)$ L $\vee$-, $\wedge$-continuous	$\Rightarrow$	$(L, \Omega_\sigma(L))$ topological lattice
$\Downarrow$		$\Downarrow$		$\Downarrow$
$(L \times L, \lambda(L \times L))$ topological $\wedge$-semilattice	$\Leftrightarrow$	$\lambda(L \times L) = \lambda(L) \times \lambda(L)$ L $\wedge$-continuous	$\Rightarrow$	$(L, \lambda(L))$ topological $\wedge$-semilattice
$\Updownarrow$		$\Updownarrow$		$\Updownarrow$
$(L \times L, \sigma(L \times L))$ topological $\wedge$-semilattice	$\Leftrightarrow$	$\sigma(L \times L) = \sigma(L) \times \sigma(L)$ L $\wedge$-continuous	$\Rightarrow$	$(L, \sigma(L))$ topological $\wedge$-semilattice
$\Downarrow$		$\Downarrow$		
$(L \times L, \sigma(L \times L))$ topological $\vee$-semilattice	$\Leftrightarrow$	$\sigma(L \times L) = \sigma(L) \times \sigma(L)$	$\Rightarrow$	$(L, \sigma(L))$ topological $\vee$-semilattice

Diagram 7.

A sufficient condition for the equality

$$\prod_{j \in J} \tau(P_j) = \tau\left(\prod_{j \in J} P_j\right)$$

is given in

3.16 PROPOSITION Let $(P_j : j \in J)$ be a family of bounded continuous posets and P their direct product. Then P is also a continuous poset, and

$$\sigma(P) = \prod_{j \in J} \sigma(P_j) \qquad \text{(cf. [5])}$$

$$\lambda(P) = \prod_{j\in J} \lambda(P_j) = \xi(P) = \prod_{j\in J} \xi(P_j)$$

If not only the P_j but also their duals P_j are bounded continuous posets, then

$$\Omega(P) = \prod_{j\in J} \Omega(P_j) = \Omega_\kappa(P) = \prod_{j\in J} \Omega_\kappa(P_j) = \Omega_\sigma(P) = \prod_{j\in J} \Omega_\sigma(P_j)$$

Several hard difficulties arise if the boundedness condition is dropped. For details, the reader is referred to [2] and [5].

4. THE ROLE OF HAUSDORFF'S SEPARATION AXIOM

Under certain circumstances, the product invariance is strongly related to the Hausdorff separation property, as the following results show.

4.1 LEMMA For every poset P, the graph of the order relation $\le$ is closed in the lim inf topology $\xi(P \times P)$ and, a fortiori, in the order topology $\Omega(P \times P)$.

Proof: Let $\mathcal{U}$ be an ultrafilter on $P \times P$ containing the set $\le$ (considered as a subset of $P \times P$). If $\underline{\lim}\, \mathcal{U} = (x_1,x_2)$, then $x_j = \underline{\lim}\, \pi_j(\mathcal{U}) = \bigvee\bigcup\{\pi_j(F \cap \le)_\downarrow : F \in \mathcal{U}\}$, $j = 1,2$ (cf. [11(III.3.7)]), and the obvious inclusion $\pi_1(F \cap \le)_\downarrow \subseteq \pi_2(F \cap \le)_\downarrow$ for all $F \in \mathcal{U}$ yields $x_1 \le x_2$. ∎

4.2 COROLLARY If $\mathfrak{T}$ is a topology on P such that $\xi(P \times P) \subseteq \mathfrak{T} \times \mathfrak{T}$, then $(P,\mathfrak{T})$ is a pospace (i.e., $\le$ is closed in the product topology $\mathfrak{T} \times \mathfrak{T}$, and in particular, $\mathfrak{T}$ is Hausdorff (cf. [11(VI.1.4)]).

We apply this result to the order topology and to the lim inf topology itself.

4.3 COROLLARY If $\Omega(P \times P) = \Omega(P) \times \Omega(P)$, then $(P,\Omega(P))$ is a pospace, and $\Omega(P)$ is Hausdorff. Similarly, if $\xi(P \times P) = \xi(P) \times \xi(P)$, then $(P,\xi(P))$ is a pospace, and $\xi(P)$ is Hausdorff.

For complete lattices, one can say even more:

4.4 THEOREM For a complete lattice L the following statements are equivalent:

(a) L is a generalized continuous lattice (see [11(III.1.17)]).
(b) $\sigma(L)$ is a hypercontinuous lattice (see [11(III.3.22)]).
(c) $\lambda(L)$ is Hausdorff (and compact).
(d) $\xi(L)$ is Hausdorff (and compact).
(e) $(L,\lambda(L))$ is a pospace.

(f) $(L,\xi(L))$ is a pospace.

(g) $\xi(L \times L) = \xi(L) \times (L)$.

Each of these conditions implies

(h) $\sigma(L)$ is a continuous lattice.

(i) $\sigma(L \times L) = \sigma(L) \times \sigma(L)$.

(j) $\lambda(L \times L) = \lambda(L) \times \lambda(L)$.

(k) $\xi(L) = \lambda(L)$.

Proof: (a) $\Leftrightarrow$ (b) : [11(VII.4.12)].

(a) $\Leftrightarrow$ (c): [11(III.3.15)].

(c) $\Leftrightarrow$ (d) $\Leftrightarrow$ (e) $\Leftrightarrow$ (f) $\Rightarrow$ (k): [11(III.3.7 and 3.9; VI.1.4)].

(b) $\Rightarrow$ (h) $\Rightarrow$ (i): [11(III.3.22 and II.4.11)].

(i) $\Leftrightarrow$ (j): Corollary 3.5.

(e) $\Rightarrow$ (g): By the (proven) implications (e) $\Rightarrow$ (b) $\Rightarrow$ (j) and (k), $\lambda(L \times L) = \lambda(L) \times \lambda(L)$ is Hausdorff, whence $\xi(L \times L) = \lambda(L \times L) = \lambda(L) \times \lambda(L) = \xi(L) \times \xi(L)$.

(g) $\Rightarrow$ (f): Corollary 4.3. ∎

It should be mentioned that none of the conditions (h), (i), (j) implies the equivalent conditions (a) to (g). In [11(VI.4.5)] there is given an example of a unital compact topological $\wedge$-semilattice W without a basis of subsemilattices. By [11(VII.4.4)], W is a meet-continuous complete lattice and $\sigma(W)$ is a continuous lattice [whence $\sigma(W \times W) = \sigma(W) \times \sigma(W)$ and $\lambda(W \times W) = \lambda(W) \times \lambda(W)$]. On the other hand, by the fundamental theorem of compact semilattices [11(VI.3.4)], W cannot be a continuous lattice; and by [11(III.2.14)] it cannot even be a generalized continuous lattice.

Another example of a complete lattice L where $\sigma(L \times L) = \sigma(L) \times \sigma(L)$ and $\lambda(L \times L) = \lambda(L) \times \lambda(L)$ but $\xi(L \times L) \neq \xi(L) \times \xi(L)$ is the complete Boolean lattice of all Lebesgue-measurable sets modulo null sets (see the end of this paper). However, it seems to be open whether, for this lattice, $\sigma(L)$ is continuous.

Let us note two further consequences of Corollaries 4.3 and 3.10:

4.5 COROLLARY Let $\tau = \xi$ or $\tau = \Omega$. If S is a bounded $\wedge$-semilattice, such that the square $(S \times S, \tau(S \times S))$ is a topological $\wedge$-semilattice, then $\tau(S \times S)$ must be Hausdorff and coincide with $\tau(S) \times \tau(S)$.

This together with Theorem 4.4 yields a new characterization of (complete) continuous lattices:

4.6 COROLLARY A complete lattice L is continuous iff its square $L \times L$ is a topological $\wedge$-semilattice with respect to the lim inf topology.

(Recall that a $\wedge$-continuous and generalized continuous lattice is already continuous; cf. [11(III.2.14)].) We do not know whether L must be continuous whenever $(L,\xi(L))$ is a topological $\wedge$-semilattice. However, we have a related result for the order topology on Boolean lattices. We start with a more general result on pseudocomplemented semilattices.

4.7 PROPOSITION Let S be a $\wedge$-semilattice with pseudocomplementation ψ : $S \to S$, $x \mapsto x^*$ (i.e., $x \wedge y = 0 \Leftrightarrow y \leq x^*$). Let $\mathcal{S}$ and $\mathfrak{T}$ be topologies on S such that meet is jointly continuous with respect to $\mathfrak{T}$, and ψ is $\mathcal{S}$-$\mathfrak{T}$-continuous. Then the following three statements are equivalent:

(a) $\{0\}$ is closed in $\mathfrak{T}$.

(b) For each $x \neq 0$, there is a $\mathfrak{T}$-open upper set U and an $\mathcal{S}$-open lower set V such that $x \in U$, $0 \in V$, and $U \cap V = \emptyset$.

(c) Any two elements with different pseudocomplements can be separated by a $\mathfrak{T}$-open upper set and an $\mathcal{S}$-open lower set, respectively.

Proof: (a) $\Rightarrow$ (c): Consider two elements x,y with $x^* \not\leq y^*$. Then $x \wedge x^* = 0$ but $z = y \wedge x^* \neq 0$. Hence $S \setminus \{0\}$ is a $\mathfrak{T}$-open neighborhood of z, and we find $\mathfrak{T}$-open sets V,W such that $y \in V$, $x^* \in W$, and $V \wedge W \subseteq S\setminus\{0\}$. Moreover, we may assume $V = \uparrow V$ $(= \bigcup\{\wedge_s^{-1}(V) : s \in S\})$ and $W = \uparrow W$. Now, by continuity of ψ, $U = \psi^{-1}(W)$ is an $\mathcal{S}$-open lower set containing x. (Notice that $v \leq u \in U$ implies $v^* \geq u^* \in W$, whence $v^* \in \uparrow W = W$ and so $v \in U$.) Furthermore, $U \cap V = \emptyset$ since $u \in U \cap V$ would imply $0 = u \wedge u^* \in V \wedge W \subseteq S\setminus\{0\}$.

(c) $\Rightarrow$ (b) $\Rightarrow$ (a): Obvious. ■

This proposition has several interesting consequences. First, the case $\mathcal{S} = \mathfrak{T}$ leads to

4.8 THEOREM Let $(S,\mathfrak{T})$ be a topological pseudocomplemented semilattice in which $\{0\}$ is closed. Then any two elements with different pseudocomplements can be separated by open upper and lower sets, respectively.

4.9 COROLLARY A topological Boolean lattice with continuous complementation (i.e., a topological Boolean algebra) must be Hausdorff (moreover, a pospace) if $\{0\}$ is closed.

Of course, via Boolean rings, this also follows from the fact that a topological group in which the neutral element is closed must be Hausdorff (and even regular).

Second, taking for $\mathcal{J}$ the order topology, we obtain

4.10 THEOREM Let $\mathfrak{J}$ be a topology on a Boolean lattice L such that $\{0\}$ is closed and $\mathfrak{J} \subseteq \Omega(L)$. If $(L,\mathfrak{J})$ is a topological $\wedge$-semilattice, then $(L,\Omega(L))$ must be a pospace (in particular, Hausdorff).

4.11 COROLLARY If L is a Boolean lattice with non-Hausdorff order topology, then there is no topology between $\upsilon(L)$ and $\Omega(L)$ making the binary meet jointly continuous.

As Flachsmeyer [7] has shown, all infinite free Boolean lattices as well as their MacNeille completions are examples of Boolean lattices whose order topologies are not Hausdorff. The first example of this kind, namely, the complete lattice G of all regular open sets of the real line, is due to Floyd [8]. Recently, Isbell [13] has observed (using the above methods) that there is no T_0 topology on G making binary meet and infinitary join continuous. More precisely, we have that for a topology $\mathfrak{J}$ on a σ-complete lattice L the following two statements are equivalent:

(a) Increasing sequences converge to their join, each $\mathfrak{J}$-open set is an upper set, and the binary join is continuous.

(b) The countable join operation $\bigvee : L^{\omega} \to L$ is continuous.

Proof: For the implication (b) $\Rightarrow$ (a) see Isbell [13]. The converse can be checked as follows: Choose $(x_n) \in L^{\omega}$ with $\bigvee_{n \in \omega} x_n = x \in U \in \mathfrak{J}$. Then (y_n) with $y_n := x_0 \vee x_1 \vee \cdots \vee x_n$ is an increasing sequence in L with $\bigvee_{n \in \omega} y_n = x$. Hence there exists an index m such that $y_m \in U$. Now choose $U_0, \ldots, U_m \in \mathfrak{J}$ with $x_j \in U_j$ and $U_0 \vee \cdots \vee U_m \subseteq U$. As $U = \uparrow U$, we have a fortiori $\bigvee_{n \in \omega} U_n \subseteq U$, where $U_n = L$ for $n > m$, and $\Pi_{n \in \omega} U_n$ is a product open neighborhood of (x_n). ■

Now, if L is a Boolean lattice satisfying the countable chain condition (e.g., a free Boolean lattice or its completion), then the join of any ideal $I \subseteq L$ is reached by an increasing sequence in I. Hence a T_0-topology on L satisfying the equivalent conditions (a) and (b) must be coarser than the Scott topology on L, and $\{0\}$ must be closed. But by Theorem 4.10 no such topology $\mathfrak{J}$ can exist when $\Omega(L)$ fails to be Hausdorff, and from Proposition 2.10 we infer that none of the topologies $\sigma(L),\lambda(L),\xi(L),\Omega_{\kappa}(L) = \Omega_{\sigma}(L),\Omega(L)$ can make the binary meet jointly continuous. In particular, in this situation $\tau(L \times L) \neq \tau(L) \times \tau(L)$ for $\tau \in \{\sigma,\lambda,\xi,\Omega_{\kappa},\Omega\}$ (cf. Lawson [17]).

Another interesting example of a complete atomless Boolean lattice is the lattice L of all Lebesgue-measurable subsets of the unit interval (or of the real line) modulo null sets. If we denote by ℓ the usual Lebesgue measure, then the Nikodym metric d_ℓ on L defined by $d_\ell(x,y) = \ell(x \Delta y)$ [where $x \Delta y$ means symmetric difference, i.e., $x \Delta y = (x' \wedge y) \vee (x \wedge y')$] induces the order topology on L (cf. [7]). Now, since ℓ is subadditive [i.e., $\ell(x \vee y) \leq \ell(x) + \ell(y)$] and monotone, we have that $K_\epsilon(x) \wedge K_\epsilon(y) \subseteq K_{2\epsilon}(x \wedge y)$ for all $x,y \in L$ and $\epsilon > 0$ [where $K_\epsilon(x) = \{z \in L : \ell(x \Delta z) < \epsilon\}$], which implies that $(L,\Omega(L))$ is a topological lattice. Notice that the complementation map $'$ of a Boolean lattice is always continuous with respect to the order topology. Obviously $L \cong L^2$, so $(L^2,\Omega(L^2))$ is a topological lattice, too.

Now observe that L is a complete $\wedge$-continuous lattice, but not a continuous lattice (since L is atomless) and so L cannot be Hausdorff in its Lawson topology (cf. [11(III)]). Thus, for $\tau \in \{\sigma,\sigma^\sim,\lambda,\lambda^\sim,\Omega_\sigma,\Omega_K,\Omega\}$ we have $\tau(L^2) = \tau(L) \times \tau(L)$ but $\xi(L^2) \neq \xi(L) \times \xi(L)$, and $(L,\xi(L))$ cannot be a topological (semi)lattice. Notice that the topologies $\Omega(L)$, $\Omega_\sigma(L)$, and $\Omega_K(L)$ coincide in this example.

Summarizing some of the previous results, we obtain the following chain of implications for a Boolean lattice L:

$$\Omega(L \times L) = \Omega(L) \times \Omega(L) \Leftrightarrow (L^2,\Omega(L^2)) \text{ topological lattice}$$
$$\Downarrow$$
$$(L,\Omega(L)) \text{ topological lattice}$$
$$\Downarrow$$
$$(L,\sigma(L)) \text{ topological lattice} \Leftrightarrow (L,\Omega_\sigma(L)) \text{ topological lattice}$$
$$\Downarrow$$
$$(L,\Omega(L)) \text{ pospace}$$
$$\Downarrow$$
$$\Omega(L)\ T_2$$

It is certainly an interesting and nontrivial problem to check which of these implications may be inverted.

REFERENCES

1. P. Crawley and R. P. Dilworth, Algebraic Theory of Lattices. Prentice-Hall Inc., Englewood Cliffs, N.J. (1973).

2. H. Dobbertin and M. Erné, Intrinsic topologies on lattices. Preprint, Math. Inst. Techn. Univ. Hannover 123 (1981).

3. M. Erné, Order-topological lattices. Glasgow Math. J. 21 (1980), 57-68.

4. M. Erné, Topologies on products of partially ordered sets I, II, III. Algebra Universalis 11 (1980), 295-311, 312-319; 13 (1981), 1-23.

5. M. Erné, Scott convergence and Scott topology in partially ordered sets II, in Lecture Notes Math. 871, Springer, Berlin-Heidelberg-New York (1981), 61-96.

6. M. Erné and S. Weck, Order convergence in lattices. Rocky Mountain J. Math. 10 (1980), 805-818.

7. J. Flachsmeyer, Einige topologische Fragen in der Theorie der Booleschen Algebren. Arch. Math. 16 (1965), 25-33.

8. E. E. Floyd, Boolean algebras with pathological order topologies. Pacific J. Math. 5 (1955), 687-689.

9. O. Frink, Topology in lattices. Trans. AMS 51 (1942), 569-582.

10. O. Frink, Ideals in partially ordered sets. Amer. Math. Monthly 61 (1954), 223-234.

11. G. Gierz, K. H. Hofmann, K. Keimel, J. D. Lawson, M. Mislove, D. S. Scott, A Compendium of Continuous Lattices. Springer-Verlag, Berlin-Heidelberg-New York (1980).

12. J. Isbell, Discontinuity of meets and joins. (This volume, Chapter 8.)

13. J. Isbell, A frame with no admissible topology. Math. Proc. Cambridge Phil. Soc. 94 (1983), 447-448.

14. D. C. Kent, Convergence functions and their related topologies. Fund. Math. 54 (1964), 125-133.

15. D. C. Kent, On the order topology in a lattice. Illinois J. Math. 10 (1966), 90-96.

16. D. C. Kent, The interval topology and order convergence as dual convergence structures. Amer. Math. Monthly 74 (1967), 426-427, 1231.

17. J. D. Lawson, Intrinsic topologies in topological lattices and semilattices. Pacific J. Math. 32 (1970), 459-466.

18. J. D. Lawson, Intrinsic lattice and semilattice topologies. Proc. Lattice Theory Conf., U. Houston (1973), 206-260.

19. B. C. Rennie, Lattices. Proc. London Math. Soc., II. Ser. 52 (1951), 386-400.

20. D. S. Scott, Continuous lattices, in Lecture Notes Math. 274. Springer, Berlin-Heidelberg-New York (1972), 97-136.

21. A. J. Ward, On relations between certain intrinsic topologies in partially ordered sets. Proc. Cambridge Philos. Soc. 51 (1955), 254-261.

22. S. Weck, Scott convergence and Scott topology in partially ordered sets I, in Lecture Notes Math. 871, Springer-Verlag, Berlin-Heidelberg-New York (1981), 372-383.

3

Natural Topologies, Essential Extensions, Reductive Lattices, and Congruence Extension

GERHARD GIERZ and ALBERT STRALKA

University of California
Riverside, California

In this paper we discuss some recent work on distributive lattices where we show that the seemingly disparate collection of concepts which forms its title is interrelated, albeit often in a very complicated fashion. We began with the simple question: Does every lattice have a best intrinsic topology? This best topology we call the natural topology for the lattice. As with most research projects, we have not yet been able to answer our initial question. In fact, we feel that only now are we able to make significant progress on the question. Nevertheless, we have established several partial solutions and along the way we have developed a good bit of theory involving essential extensions. So much of the work has to do with essential extensions that the natural topology problem has been forced into the background.

1. ESSENTIAL EXTENSIONS AND THE INJECTIVE HULL OF A DISTRIBUTIVE LATTICE

In this section we discuss the concept of essential extensions of distributive lattices, which is totally algebraic or even category theoretical in nature. On the surface, there is nothing to indicate that topology or topological lattices are involved. However, upon closer inspection we shall see in later sections that essential extensions are intrinsically topological and they have considerable importance in the area of topological algebra.

Let $\mathcal{D}$ be the category of all distributive lattices together with all lattice homomorphisms as morphisms. In this category (as we could do in every category) we may define the notion of injective objects:

1.1 DEFINITION A distributive lattice P is called *injective* if whenever L is a sublattice of M and φ: L → P is a lattice homomorphism, then φ may be extended to a lattice homomorphism ψ: M → P.

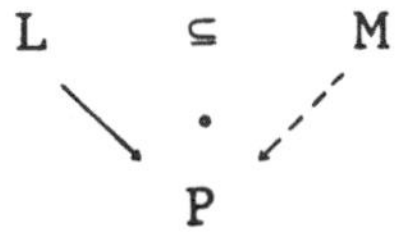

It is well known that the category of distributive lattices has enough injectives, i.e., every distributive lattice may be embedded into an injective lattice. Moreover, the injective distributive lattices are exactly the complete Boolean algebras (see Banaschewski and Bruns [1967,1968] for details). Furthermore, for every distributive lattice L there is (up to isomorphism) a unique smallest complete Boolean algebra in which L can be embedded. This Boolean algebra is called the *injective hull* of L.

Injective hulls are closely tied to the idea of essential extensions (again see Banaschewski and Bruns [1967,1968]). Let L be a sublattice of M. If every nontrivial congruence θ on M intersects L × L in a subset strictly larger than the diagonal of L, we say that M is an *essential extension* of L, or that the embedding of L into M is *essential*. Another way to express this is:

1.2 DEFINITION Let L be a sublattice of M. If every lattice homomorphism φ: M → N is injective whenever its restriction $\varphi|L$ to L is injective, we call M an essential extension of L.

Every lattice L has a unique maximal essential extension and this is the injective hull of L. (This is true in every equational class of algebras having enough injectives; see Banaschewski [1970] for details.)

In the rest of this section we explain how to construct the injective hull of a lattice. One way to do this is to embed L into a Boolean algebra $B_1(L)$ and then take $B_0(L)$ to be the Boolean subalgebra of $B_1(L)$ which L generates. A theorem of Hashimoto [1952] then states that L and $B_0(L)$ have exactly the same congruences. Thus, $B_0(L)$ will be an essential extension of L. Glivenko [1929] has discussed how to find the smallest complete Boolean algebra B(L) containing $B_0(L)$. The Boolean algebra B(L) will be the injective hull of L. One way of constructing B(L) from $B_0(L)$ is to take the MacNeille completion of $B_0(L)$. (The construction mentioned in this paragraph is due to Banaschewski and Bruns [1968].)

Going carefully through the work of Hashimoto [1952] and Glivenko [1929], one may discover two facts:

(1) The smallest Boolean algebra $B_0(L)$ containing L may be recovered as the set of all compact congruences on L, provided L has 0 and 1.

(2) Pseudocomplements play an important role when one tries to complete $B_0(L)$.

With these two facts in mind, we suggest the following construction for the injective hull of L. In this explanation, we will omit all the proofs, which can be found in Gierz and Stralka [1982a].

As always, let L be a distributive lattice. The congruence lattice of L will be denoted by $\Theta(L)$. It is well known that $\Theta(L)$ is an algebraic lattice. In particular, $\Theta(L)$ is Brouwerian, i.e., for every congruence $\theta \in \Theta(L)$ and every family $(\theta_i)_{i\in I} \subseteq \Theta(L)$, the following equation holds:

$$\theta \cap \sup_{i\in I} \theta_i = \sup_{i\in I} \theta \cap \theta_i$$

In every Brouwerian lattice we have pseudocomplements, i.e., for every congruence $\theta \in \Theta(L)$, there is a largest congruence $\theta^\perp \in \Theta(L)$ such that $\theta \cap \theta^\perp = \Delta$ (as usual, Δ denotes the smallest and ∇ denotes the largest congruence on L). The congruence $\theta^\perp$ may also be defined by

$$\theta^\perp := \sup\{\psi \in \Theta(L) : \psi \cap \theta = \Delta\}$$

The following results are folklore and go back to the beginning of lattice theory:

(i) $\theta \subseteq \psi$ implies $\psi^\perp \subseteq \theta^\perp$.

(ii) $\theta \subseteq \theta^{\perp\perp}$.

(iii) $\theta^{\perp\perp\perp} = \theta^\perp$.

(iv) $(\theta \vee \psi)^\perp = \theta^\perp \cap^\perp$; more generally, $(\sup_{i\in I} \theta_i)^\perp = \bigcap_{i\in I} \theta_i^\perp$.

Furthermore, the mapping $\theta \mapsto \theta^{\perp\perp} : \Theta(L) \to \Theta(L)$ is a hull operator on $\Theta(L)$ and the subset

$$\Theta^*(L) := \{\theta^\perp : \theta \in \Theta(L)\}$$

is a complete Boolean algebra. If $\theta, \psi \in \Theta^*(L)$, then the infimum of θ and ψ in $\Theta^*(L)$ agrees with the infimum taken in $\Theta(L)$, whereas the supremum of θ and ψ in $\Theta^*(L)$ is given by

$$\theta \cup \psi = (\theta \vee \psi)^{\perp\perp}$$

For every $a \in L$ we let θ_a be the smallest congruence collapsing $\downarrow a$. Note that θ_a is given by

$$\theta_a = \{(x,y) \in L \times L : x \vee a = y \vee a\}$$

i.e., θ_a is the kernel of the lattice homomorphism $x \mapsto x \vee a : L \to \uparrow a$. Dually, we let

$$\theta^a = \{(x,y) \in L \times L : x \wedge a = y \wedge a\}$$

be the smallest congruence collapsing $\uparrow a$. Clearly, the fact that θ_a and θ^a are lattice congruences depends heavily on the distributivity of L. Furthermore, θ_a and θ^a are complements of each other in $\Theta(L)$, i.e., we have

$$\theta_a \cap \theta^a = \Delta \qquad \theta_a \vee \theta^a = \nabla$$

This puts θ_a and θ^a automatically into $\Theta^*(L)$.

Now assume that $a,b \in L$ are comparable, say $a \le b$. Then there exists a smallest congruence relation identifying a and b, and this congruence is given by

$$\theta(a,b) = \theta_b \cap \theta^a$$

Since $\Theta^*(L)$ is closed in $\Theta(L)$ under infima, congruences of the form $\theta(a,b)$ belong to $\Theta^*(L)$, too. What are their complements in $\Theta^*(L)$? If we define

$$\pi_{a,b} := \theta(a,b)^{\perp}$$

then De Morgan's law implies

$$\pi_{a,b} = \theta_b^{\perp} \sqcup (\theta^a)^{\perp} = \theta^b \sqcup \theta_a$$

Since both θ^b and θ_a admit complements in $\Theta(L)$, an easy calculation shows that the suprema of θ^b and θ_a in $\Theta(L)$ and $\Theta^*(L)$ agree; hence we have

$$\pi_{a,b} = \theta_a \vee \theta^b$$

It is now an easy exercise in lattice theory to verify the equation

$$\pi_{a,b} = \{(x,y) \in L \times L : (a \vee x) \wedge b = (a \vee y) \wedge b\}$$

i.e., $\pi_{a,b}$ is the kernel of the projection $x \mapsto (a \vee x) \wedge b : L \to [a,b]$ of L onto the interval [a,b].

These last few properties as well as the next proposition may be found in Hashimoto [1952].

1.3 PROPOSITION If L is a distributive lattice, then the map $a \mapsto \theta_a : L \to \Theta(L)$ is an injective lattice homomorphism.

Using again the facts that the suprema of θ_a and θ_b in both $\Theta(L)$ and $\Theta^*(L)$ agree, and θ_a (and θ_b) already admit complements in $\Theta(L)$, we conclude from Proposition 1.3:

1.4 PROPOSITION The map $a \mapsto \theta_a : L \to \Theta^*(L)$ is an injective lattice homomorphism.

At this point, let us pause for a moment and see what happens if L has 0 and 1. In this case, $\theta_a = \theta(0,a)$ is the smallest congruence identifying 0 and a. Dually, $\theta^a = \theta(a,1)$ is the smallest congruence identifying a and 1, and therefore all the congruences θ_a, θ^a, $\theta(a,b)$, and $\pi(a,b) = \theta_a \vee \theta^b$ are compact congruences. In fact, Hashimoto [1952] has shown that for a bounded distributive lattice, the compact congruences $\Theta_c(L)$ form a Boolean sublattice of $\Theta(L)$ and this Boolean sublattice is generated by L when L is embedded into $\Theta_c(L)$ in the way described in Proposition 1.3. How do we complete a Boolean algebra B? Glivenko [1929] suggests the following method: Let Id(B) be the ideal lattice of B. If $I \in \mathrm{Id}(B)$, we let

$$I^{\perp} = \{x \in B : x \wedge y = 0 \text{ for all } y \in I\}$$

Then $I^{\perp}$ is the largest ideal intersecting I in $\{0\}$, i.e., $I^{\perp}$ is indeed the pseudocomplement of I. Again, we let

$$\mathrm{Id}^*(B) = \{I^{\perp} : I \in \mathrm{Id}(B)\}$$

Then $\mathrm{Id}^*(B)$ is a complete Boolean algebra. If we embed B into $\mathrm{Id}^*(B)$ via the map $x \mapsto \downarrow x$, it turns out that $\mathrm{Id}^*(B)$ is the Dedekind-MacNeille completion of B.

Now let $B = \Theta_c(L)$ in this construction. Since every algebraic lattice is the ideal lattice of the sup semilattice of the compact elements, we find that $\mathrm{Id}(\Theta_c(L)) \cong \Theta(L)$ and hence $\mathrm{Id}^*(\Theta_c(L)) \cong \Theta^*(L)$. Then, recalling the construction of Banaschewski and Bruns mentioned earlier, it should be clear that $\Theta^*(L)$ is the injective hull of L. This may also be verified directly, even for distributive lattices without 0 and 1 (see Gierz and Stralka [1982a] for details).

1.5 THEOREM If L is a distributive lattice, then the embedding $a \mapsto \theta_a : L \to \Theta^*(L)$ is essential and $\Theta^*(L)$ is the injective hull of L.

We now give a description of all the elements in $\Theta^*(L)$. Let us start with any congruence $\theta \in \Theta(L)$. Then θ is the supremum of compact congruences; explicitly,

$$\theta = \sup\{\theta(a,b) : (a,b) \in \theta\}$$

Taking complements, we obtain

$$\theta^\perp = \inf\{\theta(a,b)^\perp : (a,b) \in \theta\} = \bigcap\{\pi_{a,b} : (a,b) \in \theta\}$$

Thus, every congruence $\psi \in \Theta^*(L) = \{\theta^\perp : \theta \in \Theta(L)\}$ is an intersection of congruences of the form $\pi_{a,b}$. Since all the $\pi_{a,b}$s belong to $\Theta^*(L)$, which is closed in $\Theta(L)$ under intersections, we obtain:

1.6 THEOREM A congruence $\theta \in \Theta(L)$ belongs to $\Theta^*(L)$ iff θ is an intersection of congruences of the form $\pi_{a,b}$, $a,b \in L$, $a \le b$.

We conclude this section with some examples.

1.7 EXAMPLES (i) If $L = [0,1]$ is the unit interval, then $\Theta^*(L)$ is isomorphic with the Boolean algebra $\mathcal{O}_{reg}([0,1])$ of regular open sets of $[0,1]$.
(ii) Let $L = [0,1] \times [0,1]$ be the unit square. The chain $C = \{0\} \times [0,1] \cup [0,1] \times \{1\}$ is isomorphic with the unit interval and the embedding of C into $[0,1] \times [0,1]$ is essential. Hence the square is an essential extension of the unit interval. Therefore, both the unit interval and the square have the same maximal essential extension and we conclude that $\Theta^*([0,1] \times [0,1]) \cong \mathcal{O}_{reg}([0,1])$. Similarly, we see that $\Theta^*([0,1]^\gamma) = \mathcal{O}_{reg}([0,1])$ for every $\gamma \le \omega$.
(iii) Let L be an algebraic lattice. Then $\Theta^*(L)$ is isomorphic with the Boolean algebra of all subsets of a certain set X. Indeed, in this case $\Theta^*(L)$ has enough atoms to separate points. The atoms are exactly the congruences of the form $\theta(a,b)$, where a is a lower neighbor of b.

Next, we give an example of a congruence θ on the unit interval that does not belong to $\Theta^*([0,1])$.
(iv) Let $C \subseteq [0,1]$ be the Cantor chain and let $\gamma_0 : C \twoheadrightarrow [0,1]$ be the gap-closing map. We may extend γ_0 to a monotone map $\gamma : [0,1] \to [0,1]$. Note that γ is a lattice homomorphism. Let Γ be the kernel congruence of γ. Then Γ is a closed congruence on $[0,1]$ and we have

$(x,y) \in \Gamma$ iff $x = y$ or x and y belong to the closure of the same component of $[0,1]\setminus C$

Now suppose that $x,y \in [0,1]$ with $x < y$. We will show that $\theta(x,y) \cap \Gamma \neq \Delta$. Indeed, since $[0,1]\setminus C$ is dense in $[0,1]$, there is some component A of $[0,1]\setminus C$ that intersects the open interval $]x,y[$. Choose $x < a < b < y$, where $a,b \in A$. Then $(a,b) \in \theta(x,y) \cap \Gamma$. We may conclude that $\Gamma^{\perp} = \Delta$ and thus $\Gamma^{\perp\perp} = \nabla$. But $\Gamma \neq \Delta$ so $\Gamma \neq \Gamma^{\perp\perp}$. Thus $\Gamma \notin \Theta^*([0,1])$.

Our final example deals with the following problem: The maximal essential extension of a distributive lattice L is obtained by first enriching L with relative complements and then taking the Dedekind-MacNeille completion. Is this true for every essential extension V of L, i.e., is V obtained from L by first throwing in some (not all!) relative complements and then going through a completion process? It is not hard to verify that every V obtained from L in this way is indeed an essential extension of L, once the completion process is defined in an appropriate way. One obvious way to define this "appropriate" completion of V is to take the smallest complete sublattice of $B(V) = \Theta^*(V)$ containing V. The following example shows that $[0,1] \times [0,1]$ is the essential extension of a closed sublattice which is closed under relative complements in $[0,1] \times [0,1]$, refuting the above conjecture.

EXAMPLE (v) As in example (iv), let $\gamma:[0,1] \to [0,1]$ be the extension of the gap-closing map of the Cantor chain onto the unit interval and let

$$L \subseteq [0,1] \times [0,1]$$

be the kernel congruence associated with γ, i.e.,

$$L = \{(x,y) : \gamma(x) = \gamma(y)\}$$

It is left to the reader to verify that L is complete and closed under relative complements in $[0,1] \times [0,1]$, i.e., we have:

If $a,b,x,y \in [0,1] \times [0,1]$ satisfy $x \wedge y = a$, $x \vee y = b$, and if $a,b,x \in L$, then $y \in L$

Moreover, $[0,1] \times [0,1]$ is an essential extension of L. Indeed, let θ be a congruence on $[0,1] \times [0,1]$ that is nontrivial. Then there are $x,y,a \in [0,1]$, $x < y$, such that either $(x,a)\theta(y,a)$ or $(a,x)\theta(a,y)$. By symmetry, we only have to consider the first case. Using the same reasoning as in example (iv), we can find elements $r,s \in [0,1]$ such that $x \leq r < s \leq y$ and $\gamma(r) = \gamma(s)$. Since congruence classes are order convex, we conclude that $(r,a)\theta(s,a)$. We now consider two cases:

(a) $a \leq r$. Then $(r,r) = (r,a)\vee(0,r)\theta(s,a)\vee(0,r) = (s,r)$.

(b) $r \leq a$. Then $(r,r) = (r,a)\wedge(1,r)\theta(s,a)\wedge(1,r) = (s,r)$.

Hence, in any case we have $(r,r)\theta(s,r)$. By construction, $(r,r),(s,r) \in L$ and therefore $\theta \cap L \times L \neq \Delta$.

2. INFINITE DISTRIBUTIVITY, ESSENTIAL EXTENSIONS, AND A COMPLETION FOR DISTRIBUTIVE LATTICES

It is a well-known and unfortunate fact that the Dedekind-MacNeille completion of a distributive lattice is no longer distributive in general. One reason for this may be found in the fact that the Dedekind-MacNeille completion does not pay any attention to meet continuity or join continuity. In this section we shall construct a completion of a distributive lattice which is not only complete (as the term completion indicates) but also is meet continuous and join continuous. Of course, for an ambitious program like this, we have to pay a price. In this setting, we have to give up the idea that our completion respects all existing suprema and infima. On the other hand, when "completing" the open unit square $]0,1[\times]0,1[$ in our sense, we will obtain the closed square $[0,1] \times [0,1]$, whereas the Dedekind-MacNeille completion leads to an undesirable object like $]0,1[\times]0,1[\cup \{(0,0),(1,1)\}$.

Recall that a complete lattice L is meet continuous if for every directed subset D of L and every element $x \in L$, $x \wedge \sup D = \sup\{x \wedge d : d \in D\}$. Join continuity is defined dually. Algebraic lattices and continuous lattices (in the sense of Scott) are meet continuous. Although meet continuity is purely a completeness property, when the adjective directed is dropped from the definition, this is no longer the case since distributivity would be implied. A complete distributive lattice that is both meet and join continuous will be called infinitely distributive. The category of infinitely distributive lattices and maps between such lattices that preserve arbitrary infima and arbitrary suprema will be denoted by $\mathcal{ID}$. Note that if $\varphi: L \to M$ is an $\mathcal{ID}$-morphism, then the associated congruence is closed in $L \times L$ with respect to arbitrary infima and arbitrary suprema. Such congruences will be called complete congruences. The lattice of complete congruences on L will be denoted by $\mathcal{C}\Theta(L)$.

The next proposition may be found in Gierz and Stralka [1982a].

2.1 PROPOSITION Let L be a distributive lattice. The canonical embedding $e: L \to \Theta^*(L)$ will preserve arbitrary infima if and only if L is join continuous. Thus, by symmetry, L is infinitely distributive if and only if e is an $\mathcal{ID}$ embedding.

Hence, it seems to be reasonable to define a completion of L in the following way:

2.2 DEFINITION Let L be a distributive lattice. The infinitely distributive hull L^* of L is the smallest complete sublattice V of $B(L) = \Theta^*(L)$ containing L [or, more precisely, containing the image of L via the imbedding $a \to \theta_a : L \to \Theta^*(L)$].

If M is a completely distributive lattice that is connected in the interval topology and if L is an open sublattice of M, then L^* agrees with the topological closure of L in M (see Gierz and Stralka [1982c]).

2.3 PROPOSITION L^* is an essential extension of L; as a matter of fact, L^* is the smallest infinitely distributive essential extension L admits.

Proof: Since $\Theta^*(L)$ is a complete Boolean algebra, it is infinitely distributive and so is the complete sublattice L^* of $\Theta^*(L)$. Next, let θ be a nontrivial congruence on L^*. Since distributive lattices have the congruence extension property, θ may be extended to a congruence ψ on $B(L) = \Theta^*(L)$, which has to be nontrivial. Because $\Theta^*(L)$ is an essential extension of L, we conclude $\Delta \neq \psi \cap L \times L = \theta \cap L \times L$. Finally, let V be an infinitely distributive lattice and assume that α: $L \to V$ is an essential imbedding. Since $\Theta^*(L)$ is the maximal essential extension of L, V may be embedded into V by a map β: $V \to \Theta^*(L)$ such that $\beta \bullet \alpha(x) = \theta_x$ for every $x \in L$. By an easy argument we see that β is an essential embedding and hence $\Theta^*(L)$, being a complete Boolean algebra, has to be the injective hull of V, i.e., $\Theta^*(L) \cong \Theta^*(V)$. Now Proposition 2.1 yields that β preserves all suprema and infima. The last assertion of Proposition 2.3 becomes obvious, since the image of V contains (the image of) L.

As the example of the open square enriched with (0,0) and (1,1) shows, the embedding $L \hookrightarrow L^*$ does not preserve all infima and suprema, although quite a number of them are preserved. Let us identify them.

2.4 DEFINITION Let D be a directed subset of L. We say that D admits a topological supremum if

(i) $x = \sup D$ exists.

(ii) For every $y \in L$, $\sup(y \wedge D)$ exists and $x \wedge y = \sup(y \wedge D)$.

Topological infima are defined dually.

Making minor adjustments in the proof of Proposition 2.1 as it may be found in Gierz and Stralka [1982a], we can show:

2.5 PROPOSITION Let $D \subseteq L$ be directed and assume that sup D exists in L. Then D admits a topological supremum iff the embedding $L \hookrightarrow L^*$ preserves sup D.

The role of topological suprema becomes more impressive once we define a suitable topology on L, which will agree with the usual topology on $]0,1[\times]0,1[$. For the moment, however, we will be content with the information gathered so far.

3. ESSENTIAL EXTENSIONS AND CONGRUENCE EXTENSION

Let L and V be distributive lattices and assume that L is a sublattice of V. Given a congruence θ on L, we always can extend θ to a congruence ψ on V. This elementary fact may be found in every textbook on lattice theory.

Now let us assume that L and V are (Hausdorff) topological lattices and that θ is a closed congruence on L. Under which conditions can θ be extended to a closed congruence on V? Earlier discussions of this problem may be found in Clinkenbeard [1981] and Stralka [1971,1981].

Let us start our discussion with the case where L is a 0-dimensional compact topological lattice. We will also restrict the lattices V which contain L as a sublattice to the same class. Hence we are asking: Given a 0-dimensional compact topological lattice V and a closed congruence θ on a closed sublattice L of V, can we extend θ to a closed congruence on V? Now remember that 0-dimensional compact lattices are always algebraic and meet and join continuous. Hence, given such an L, its maximal essential extension $\Theta^*(L)$ is a power set lattice by Example 1.7(iii) and the embedding $L \hookrightarrow \Theta^*(L)$ is continuous when $\Theta^*(L) \cong 2^X$ carries its usual compact 0-dimensional topology by Proposition 2.1. Hence we may use $\Theta^*(L)$ as a test object in this case.

3.1 THEOREM Let L be a 0-dimensional compact topological distributive lattice and let θ be a (closed) congruence on L. Then the following statements are equivalent:

(i) $\theta \in \Theta^*(L)$.

(ii) θ may be extended to a closed congruence on $\Theta^*(L)$, where L is considered as a closed sublattice of $\Theta^*(L)$ via the embedding $a \mapsto \theta_a$.

(iii) θ may be extended to a closed congruence of every 0-dimensional compact distributive lattice V which contains L as a closed sublattice.

(For the proof see Gierz and Stralka [1982a].)

We should point out here that congruences $\theta \in \Theta^*(L)$ are automatically closed on every T_2 topological lattice L, since they are the intersections of congruences of the form $\pi_{a,b}$, and a congruence of form $\pi_{a,b}$ is closed since it is the kernel congruence of the continuous map $x \mapsto (a \vee x) \wedge b$.

Using all the information we have already gathered together, the proof of Theorem 3.1 turns out to be not too complicated. The situation changes, however, if we leave the class of 0-dimensional lattices. As the example of the unit interval shows, the maximal essential extension of a lattice does not admit any reasonable Hausdorff lattice topology. Thus, in the general case, $\Theta^*(L)$ is no longer appropriate as a test object. Therefore, the proof of an analog of Theorem 3.1 turns out to be more complicated (see Gierz and Stralka [1982a]).

3.2 THEOREM Let L be a distributive Hausdorff topological lattice and let θ be a congruence on L. Then $\theta \in \Theta^*(L)$ iff θ is closed and, for every continuous one-to-one map $\varphi: L \to V$ into a distributive Hausdorff topological lattice V, there is a closed congruence ψ such that $\psi \cap L \times L = \theta$.

In the last equation, we identified L with its image.

3.3 COROLLARY Let L be a compact distributive lattice. Then $\theta \in \Theta^*L$ iff θ is closed and θ can be extended to a closed congruence on every compact distributive lattice V which contains L as a closed sublattice.

4. METRIZABILITY CONDITIONS, CONTINUOUS LATTICES, AND ESSENTIAL EXTENSIONS

A second use of essential extensions in topological lattices may be found in Gierz, Lawson, and Stralka [1982]. Let us briefly review these results in our context.

We will call a lattice L <u>essentially metrizable</u> if it is the essential extension of a countable sublattice. The following result actually is a result in universal topological algebra, i.e., this result may be stated for every compact topological algebra having only a countable number of operations acting on it.

4.1 THEOREM Let L be an essentially metrizable compact topological lattice. Then L is (topologically) metrizable.

The obvious question now arising is: Is every metrizable compact topological lattice essentially metrizable? We shall see that essential metrizability is an algebraic condition leading to continuous lattices. First, however, we need a definition.

4.2 DEFINITION A complete lattice L is called topologically incompatible if every Scott open neighborhood of 1 contains a filter having 0 as its infimum. A complete lattice is called topologically compatible if it contains no topologically incompatible intervals.

Examples of topologically compatible lattices are continuous lattices, or, more generally, compact topological semilattices. Topologically incompatible lattices may be found among complete Boolean algebras. In particular, the Boolean algebra of regular open sets of the unit interval is incompatible (see Floyd [1955]).

4.3 THEOREM Let the lattice L be complete, distributive, meet continuous, join continuous, essentially metrizable, and topologically compatible. Then L is a continuous lattice; as a matter of fact, L is completely distributive.

The last statement of Theorem 4.3 may appear to be a bit surprising. On the surface, the assumption of L being topologically compatible is not self-dual. However, we obtain a self-dual result. With this background, we are now able to show

4.4 THEOREM Let L be a distributive compact topological lattice. If L is essentially metrizable, then L is completely distributive.

4.5 THEOREM Let L be a distributive compact topological lattice which is metrizable. Then the following statements are equivalent:

(i) L is completely distributive.
(ii) L is essentially metrizable.

5. NATURAL TOPOLOGIES ON DISTRIBUTIVE LATTICES

The third use of essential extensions is based on the following ideas: Given two distributive compact topological lattices L and V, assume that V is an essential extension of L. It follows easily from the results in

Section 2 (notably from Proposition 2.1), that the embedding of L into V preserves arbitrary suprema and infima. In particular, the embedding of L into V is topological. Furthermore, we shall see later on that every lattice L of finite breadth admits an essential extension that is a direct product $\Pi_{i\in I} C_i$ of complete chains. Hence L may be endowed with the topology induced from the interval topology on $\Pi_{i\in I} C_i$. This topology, which we will call the H-topology on L, does not depend on the particular choice of an essential embedding of L into a product of complete chains. (For the verification of these and all further unproved statements, we refer to Gierz and Stralka [1982b,c].) It has the desirable property that a directed set that admits a topological supremum (see Definition 2.4) indeed converges to its supremum.

The impatient reader need no longer wait for an abstract and general definition of the H-topology.

5.1 DEFINITION Let L be a distributive lattice. The H-topology on L is the topology having the sets of the form

$$\{x \in L : a \vee x \geq b\} \qquad \{x \in L : b \wedge x \leq a\}$$

where $a,b \in L$, $a \leq b$, as a subbase for its closed sets.

It is obvious that the H-topology on an infinitely distributive lattice is the interval topology.

Further evidence for the importance of the H-topology is given by the following example: Let L be an open or a closed sublattice of a cube I^X. Then the H-topology on L is the topology induced from I^X. Moreover,

5.2 PROPOSITION Let L,V be distributive lattices and let $\varphi: L \hookrightarrow V$ be an essential embedding. Then φ is a topological embedding for the H-topologies.

5.3 PROPOSITION The translations $x \mapsto x \vee a : L \to L$ and $x \mapsto x \wedge a : L \to L$ are continuous for the H-topology.

5.4 PROPOSITION Let $(L_i)_{i\in I}$ be a family of distributive lattices. Then the H-topology on the cartesian product ΠL_i is the product topology of the H-topologies on the L_i, $i \in I$.

5.5 PROPOSITION Let $\alpha: L \twoheadrightarrow V$ be a surjective lattice homomorphism; then α is a continuous map for the H-topologies, provided that $\ker \alpha \in \Theta^*(L)$.

Thus, these properties of the H-topology as well as the examples preceding Definition 5.1 suggest that in the H-topology we found the natural topology on a distributive lattice. (More discussion concerning the notion of a "natural topology" can be found in Gierz and Stralka [1983c].) Of course, in these considerations we do not pay any attention to one-sided topologies, like CL-topologies on continuous lattices. These one-sided topologies may very well be natural for L as an $\wedge$-semilattice, but not for L as a distributive lattice.

We conclude this paper with a discussion of when the H-topology is Hausdorff. An answer to this question was first given by Ball [1982], although he never stated his result explicitly, as in the following (for a more direct proof, see Gierz and Stralka [1982c]):

5.6 THEOREM The H-topology on a distributive lattice L is Hausdorff iff L admits a completely distributive essential extension.

The characterization given in Theorem 5.6 is somewhat "external" and we would like to have an internal condition that yields the Hausdorff property for the H-topology. One way to do this utilizes the following:

5.7 DEFINITION Let L be a distributive lattice. Then L is called _reductive_ if whenever $a,b \in L$ are given, $a < b$, there are elements $c,d \in L$ such that $a \leq c < d \leq b$ and the interval $[c,d]$ is a chain.

Examples of reductive lattices are algebraic lattices, $[0,1]^X$, lattices of finite breadth, and lattices of the form $\mathcal{O}(X)$, X a T_1-space. Not reductive are the free distributive lattice on countably many generators, complete atomless Boolean algebras, and the lattice of closed lower ends of the unit square.

The following result is taken from Gierz and Stralka [1982b]:

5.8 THEOREM A lattice is reductive iff it has an essential extension which is a product of (complete) chains.

Now Proposition 5.2 yields:

5.9 THEOREM The H-topology on a reductive lattice is T_2.

REFERENCES

Banaschewski, B., Injectivity and essential extensions in equational classes of algebras, in _Proc. of Conf. on Universal Algebras_, Queen's Univ., Kingston, Ontario (1970), pp. 131-147.

Banaschewski, B. and G. Bruns, Categorical characterization of the MacNeille completion, Arch. Math. _18_ (1967), 369-377.

Banaschewski, B., and G. Bruns, Injective hulls in the category of distributive lattices, J. für die reine und angewandte Math. _232_ (1968), 102-109.

Ball, Richard, Distributive Cauchy lattice, Algebra Universalis, to appear.

Clinkenbeard, D. S., The lattice of closed congruences on a topological lattice, Trans. Amer. Math. Soc. 263 (1981), 457-467.

Floyd, E. E., Boolean algebras with pathological order topologies, Pacific J. Math. 5 (1955), 687-689.

Glivenko, V., Sur quelques points de la logique de M. Brouwer, Bull. Acad. des Sci. de Belgique 15 (1929), 183-188.

Gierz, G., J. D. Lawson, and A. R. Stralka, Metrizability conditions for completely distributive lattice, Canad. J. Math., to appear.

Gierz, G. and A. Stralka, The injective hull of a distributive lattice and congruence extension, preprint (1982a).

Gierz, G. and A. Stralka, Reductive lattices and essential embeddings, preprint (1982b).

Gierz, G. and A. Stralka, Natural topologies on distributive lattices, preprint (1982c).

Hashimoto, J., Ideal theory for lattices, Math. Japan 2 (1952), 149-186.

Stralka, A. R., The congruence extension property for compact topological lattices, Pacific J. Math. 38 (1971), 795-802.

Stralka, A. R., Fundamental congruences on Lawson semilattices, in Springer Lecture Notes in Math. 871, Springer-Verlag, Berlin-Heidelberg-New York (1981), pp. 348-359.

4
The Fell Compactification Revisited

RUDOLF-E. HOFFMANN
Universität Bremen
Bremen, Federal Republic of Germany

Fell [1962] considers, for a topological space X, a certain topology on the complete lattice $\underline{A}(X)$ of all closed subsets of X (ordered by the inclusion relation) for which the sets

$$U(C;V_1,\ldots,V_n) := \{A \in \underline{A}(X) \mid A \cap C = \emptyset,\ A \cap V_i \neq \emptyset \text{ for } i = 1,\ldots n\}$$

with C quasi-compact and V_i open in X, $n \in \mathbb{N} \cup \{0\}$, form an open basis.

As noted in <u>A Compendium of Continuous Lattices</u> [1980, pp. 151-152], the <u>Fell topology</u> is, for a locally quasi-compact space X, the <u>Lawson topology</u> λ of the lattice $\underline{O}(X)$ of open subsets of X (ordered by inclusion) transferred to $\underline{A}(X)$ via the bijection $\underline{O}(X) \to \underline{A}(X)$, $V \mapsto X-V$, where a space X is said to be <u>locally quasi-compact</u> iff every point has a neighborhood basis consisting of quasi-compact (but not necessarily open) subsets.†

The <u>Fell compactification</u> $\underline{H}(X)$ of a locally quasi-compact space X is the closure of

$$\{\mathrm{cl}\{x\} \mid x \in X\}$$

in $\underline{A}(X)$ with regard to the Fell topology. By the <u>Compendium</u> [VI-3.4(1) and VI-1.14], $(\underline{O}(X),\lambda_{\underline{O}(X)})$ is a <u>compact partially ordered (po) space</u> (in the sense of Nachbin [1965], the <u>Compendium</u> [VI-1.1]), hence so is $\underline{H}(X)$ in its inclusion order (reversing the order is no problem in a pospace).

For a locally compact, noncompact Hausdorff space X, $\underline{H}(X)$ is the <u>Alexandrov one-point compactification</u> of X with $\emptyset$ adjoined as a new point

†It has been observed by Flachsmeyer [1964] that, for a locally compact Hausdorff space X, the Fell topology induced on $\underline{A}(X)-\{\emptyset\}$ coincides with the "lbc-topology" of Mrówka [1958].

(Fell [1962, p. 475]), considered as a compact pospace in which $\emptyset \leq \{x\}$ for every $x \in X$ is the only nontrivial occurrence of $\leq$. Thus, in the setting of locally quasi-compact spaces, the Fell compactification may be viewed as a substitute for the Alexandrov one-point compactification. (In other contexts, of course, different substitutes can be adequate, cf. e.g., Hoffmann [1979a, sec. 3] and Wilansky [1967].) Fell [1961, sec. 2] has provided, in a special case, an interpretation of his construction in functional-analytic terms.

Hofmann and Lawson [1978] have given, for a distributive continuous lattice L, various characterizations of the closure of the set consisting of the greatest element 1 of L and all meet-prime elements of L with regard to the Lawson topology of L. By the celebrated theorem of Hofmann and Lawson that the distributive continuous lattices L are – up to an isomorphism – precisely the lattices $\underline{O}(X)$ of open sets (ordered by the inclusion relation) of locally quasi-compact (T_0-)spaces X, these results are, as has been observed by Hoffmann [1982a], intimately related to Fell's construction $\underline{H}(X)$, since X can be chosen as a sober space. In that case, X is uniquely determined by L (up to a homeomorphism) and can be canonically represented by the set of meet-prime elements of L (endowed with a topology): The points of $\underline{H}(X)$ correspond, via the obvious anti-isomorphism $\underline{A}(X) \to \underline{O}(X) \cong L$, to the pseudo-meet-prime elements of L, i.e., the suprema of the prime ideals ($\neq$ L) of L.†

Using this latter observation, we have noted (Hoffmann, 1982b), with benefit of discussions with K. H. Hofmann, that the points of the Fell compactification $\underline{H}(X)$ of a locally quasi-compact T_0-space X are contained in an extension $X \hookrightarrow \gamma X$ of the space X studied in Hoffmann [1979b].‡ This extension $X \hookrightarrow \gamma X$, defined for arbitrary T_0-spaces X, is an equivalent representation of the greatest essential extension $X \hookrightarrow \lambda X$ of the T_0-space X discovered by Banaschewski [1977, sec. 2]. Thus, for a locally quasi-compact T_0-space X, the corestriction of the extension $X \hookrightarrow \gamma X$ to the points of the Fell compactification $\underline{H}(X)$ gives an extension $X \hookrightarrow \psi X$ with a new

†It has to be noted, however, that $1 \in L$ is (in the definition used by Hoffmann [1982a] never meet prime and that it need not be contained in the closure of the set of meet-prime elements of L with regard to the Lawson topology. To this extent the definitions of Hoffmann [1982a] differ from those of Hofmann and Lawson [1978].

‡This observation can be also based upon Fell's result [1962, p. 475] that the points of $\underline{H}(X)$ are the convergence sets of the "primitive" nets of X; cf. Hoffmann [1979b, p. 419; 1982a, 3.13)].

topology on ψX such that X is contained in ψX as a subspace) which is an essential extension of X.

Whereas in Hoffmann [1982a, sec. 3] the extension $\psi X \hookrightarrow \gamma X$ for an arbitrary (T_0-)space X has been studied, we shall investigate here the extension $X \hookrightarrow \psi X$ for locally quasi-compact T_0-spaces X or, slightly more generally†, for those T_0-spaces X for which $\underline{O}(X)$ is a continuous lattice.

ψX is a sober space, and $\underline{O}(\psi X)$ is a continuous lattice, in which the greatest element is compact and the way below relation $\ll$ is multiplicative, i.e.,

$$U \ll V \text{ and } U \ll W \text{ imply } U \ll V \cap W$$

or, equivalently (by a result of Hofmann and Mislove [1981]),

X is a quasi-compact, locally quasi-compact, super-sober space.

The canonical embedding $\psi X \hookrightarrow \psi(\psi X)$ is a homeomorphism.

The space ψX corresponds to the compact ordered space $\underline{H}(X)$ via an isomorphism between the category of compact ordered spaces and isotone continuous maps and the category of quasi-compact, locally quasi-compact, super-sober spaces and "perfect" continuous maps (where perfect means that the preimage map between the lattices of open sets preserves the way below relation), which is (a slight modification of) the isomorphism described in the Compendium [1980, VII-3]. Moreover, both $\underline{H}(\cdot)$ and $\psi(\cdot)$ can be extended to functors defined on the category of locally quasi-compact sober spaces and continuous perfect mappings.

For the proofs, we develop a program which is of interest in itself, since it exhibits an intriguing interaction between two of the basic constructions for continuous posets: the dual (Lawson, 1977,1979; Hoffmann, 1981b) and the injective hull (in the category $\underline{T}_0$ of T_0-spaces and continuous maps; cf. Hoffmann [1981a, 3.14]).

For a distributive continuous lattice L, let D(L) denote the "dual" of L, consisting of those filters (= down-directed upper sets) of L which

---

† Isbell [1974] and Hofmann and Lawson [1978] have provided an example of a T_0-space X for which $\underline{O}(X)$ is a continuous lattice but X fails to be locally quasi-compact. Note that a sober space X is locally quasi-compact if and only if $\underline{O}(X)$ is a continuous lattice.

are open in the Scott topology σ_L of L. This is, with regard to the inclusion order, a continuous 1,$\wedge$-semilattice. Then we form the injective hull

$$(D(L),\sigma_{D(L)}) \hookrightarrow (I(D(L)),\sigma_{I(D(L))})$$

in the category $\underline{T}_0$, using the fact established in Hoffmann [1981a, 3.14][†] that the continuous posets in their Scott topology have an injective hull in $\underline{T}_0$, i.e., their greatest essential extension space (in the sense of Banaschewski [1977]) is an injective T_0-space (i.e., by Scott's result [1972, 2.12], a continuous lattice in its Scott topology).

Let us explain briefly why we expect that $\mathbb{D}(L)$, or rather $D\mathbb{D}(L)$ is related to our problem, viz.,

$$D\mathbb{D}(L) \cong \underline{O}(\Psi X)$$

if $L = \underline{O}(X)$ is a distributive continuous lattice.

First, note that we had seen in Hoffmann [1982a] (slightly extending a result of Hofmann and Lawson [1978]) that the canonical embedding $X \hookrightarrow \Psi X$ for a locally quasi-compact sober space X is a homeomorphism iff the way below relation of $\underline{O}(X)$ is multiplicative and the greatest element of $\underline{O}(X)$ is compact.

Second, recall that, by a result of Lawson [1977], the dual D(S) of a continuous 1,$\wedge$-semilattice S is a continuous lattice iff 1 is compact in S and the way below relation $\ll$ of S is multiplicative. Thus, for a continuous lattice L, D(L) has these properties [since $DD(L) \cong L$].

Now there is some hope that these properties are preserved under the injective hull

$$D(L) \hookrightarrow \mathbb{D}(L)$$

since an injective hull of a continuous poset preserves arbitrary infima (to the extent they exist) and the way below relation (Hoffmann [1982b, 1.1]).

[†]The stronger assertion that every sober space with an injective hull in $\underline{T}_0$ is a continuous poset in its Scott topology is false. (Indeed, the second implication of Banaschewski [1977, cor. 2, p. 240], upon which this claim was based, is false.) Counterexamples have been found recently by Hofmann and Mislove [1982]: sober spaces with an injective hull in $\underline{T}_0$ which fail to be continuous posets. Also, there is a continuous poset with ascending chain condition carrying a compatible sober topology which has an injective hull in $\underline{T}_0$, but differs from the Scott topology (cf. Hoffmann [1982b]).

Thus the greatest element of $\mathbb{D}(L)$ must be compact, and also, a "torso" of multiplicativity of the way below relation in $\mathbb{D}(L)$ is present.

As one may suspect, the proof of the full multiplicativity of $\ll$ in $\mathbb{D}(L)$ requires the use of a suitable representation. The one we need seems not to have been explicitly used before: $\mathrm{Filt}_\sigma L$, the complete lattice of all those filters of L that are generated by Scott-open subsets of L, ordered by inclusion, seems to be a reasonable candidate for $\mathbb{D}(L)$, since it is a "natural" enlargement of D(L). Indeed, the extension

$$D(L) \hookrightarrow \mathrm{Filt}_\sigma L$$

is, via the intrinsic characterization of the injective hull recently provided in Hoffmann [1982b], equivalent to the injective hull $D(L) \hookrightarrow \mathbb{D}(L)$. The treatment of $\mathrm{Filt}_\sigma L$ in this paper also owes much to the skillful techniques introduced by Lawson [1977] for D(L) (cf. Lawson [1979]).

The distributivity of $\mathrm{Filt}_\sigma L$ will be deduced from the inclusion $\mathrm{Filt}_\sigma L \hookrightarrow \mathrm{Filt}\, L$ (= all filters of L, ordered by inclusion). The latter is distributive iff L is.

0 BASIC CONCEPTS

0.1 For an arbitrary partially ordered set (poset) $(P,\le)$, we have

$$x \ll y \qquad (x \text{ is \underline{way below} } y)$$

iff whenever $y \le \sup D$ (the supremum of D) for some nonempty, <u>up-directed</u> subset D (i.e., $a,b \in D$ implies $a,b \le c$ for some $c \in D$) of P, then $x \le d$ for some $c \in D$ (cf. Scott [1972 (p. 110)]). We note the following properties of $\ll$:

$s \le t,\ t \ll x,\ x \le y$ imply $s \ll y$.

$x \ll y$ implies $x \le y$.

$x \ll y$ and $y \ll z$ imply $x \ll z$ for $s,t,x,y,z \in P$.

0.2 A poset $(P,\le)$ is said to be a <u>continuous poset</u> iff

(i) P is <u>up-complete</u>, i.e., for every up-directed subset D of $(P,\le)$, sup D exists.

(ii) For every $x \in P$, $\{y \in P \mid y \ll x\}$ is nonempty and up-directed, and

$$\sup\{y \in P \mid y \ll x\} = x.$$

Note that, in a continuous poset P, the way below relation has the following <u>interpolation property</u>: If $x \ll y$ in P, then $x \ll z$ and $z \ll y$ for some $z \in P$ (cf. Markowsky [1977, 2.5]).

A <u>continuous lattice</u> L is a continuous poset which is a complete lattice (or, equivalently, a 0,∨-semilattice).

The notion of a continuous lattice is due to Scott [1970,1972]; that of a continuous poset is a natural generalization of it in the realm of up-complete posets introduced by Smyth [1977] and Markowsky [1977]. See also Hoffmann [1981a,1979c,1981b], Lawson [1977,1979], Wilson [1977].

0.3 For an arbitrary poset $(P,\leq)$, a subset M is said to be open in the <u>Scott topology</u> (= Scott-open) (Scott [1972, p. 101]), iff

(i) M is an "upper set," i.e., $x \leq y$, $x \in M$, $y \in P$ imply $y \in M$.

(ii) Whenever $\sup D \in M$ for a nonempty, up-directed subset D of M, then $D \cap M \neq \emptyset$.

The Scott topology on a poset P is designated by σ_P.

For up-complete posets P,Q, a map $f\colon (P,\sigma_P) \to (Q,\sigma_Q)$ is continuous iff f preserves suprema of nonempty up-directed subsets, i.e., $f(\sup D) = \sup(f[D])$ for every nonempty up-directed subset D of P (cf. Wyler [1981, 3.5]).

For a continuous poset P, the sets of the form

$$\Uparrow x := \{y \in P \mid x \ll y\}$$

with x ranging through P, form an open basis of the Scott topology σ_P. It results that, in a continuous poset P, $x \ll y$ iff

$$y \in U \subseteq \uparrow x := \{z \in P \mid x \leq z\}$$

for some Scott-open subset U of P (cf. Markowsky [1977, 3.2]).

A subset U of a continuous poset P is Scott-open iff it is an upper set of P and for every $y \in U$ there is some $x \in U$ with $x \ll y$.

0.4 Every topology τ on a set M induces a preorder (= quasi-order), i.e., a transitive and reflexive relation on this set:

$$x \leq y \text{ iff } x \in \operatorname{cl}\{y\} \qquad x,y \in M$$

This preorder is called the <u>specialization preorder</u> (Artin, Grothendieck, and Verdier [1972, IV, 4.2.2]). It is antisymmetric (i.e., a partial order) iff (M,τ) is T_0. The <u>compatible</u> topologies on a preordered set are those which induce the given preorder.

On every preordered set $(P,\leq)$ there is a weakest compatible topology, the <u>weak topology</u> for which the sets

$$\downarrow x := \{y \in P \mid y \leq x\}$$

with x ranging through P, form a subbasis for the closed sets. (See Hoffmann [1979b, sec. 2] for references.) In the Compendium [II-1.16], this topology is called the "upper topology" $\upsilon(P)$ of P. The weak topology on $(P,\leq)^{op}$ (i.e., the "opposite" $(P,\leq^*)$ of $(P,\leq)$, where $x \leq^* y$ iff $y \leq x$ will be designated by ω_P (= the "lower topology" of $(P,\leq)$ in the Compendium [III-1.1]).

Note that the Scott topology on a poset is compatible.

The preorder induced on a subspace is the induced preorder, i.e., a topological embedding induces an order embedding. [For preordered sets P and Q, a map $e: P \to Q$ is an order embedding (= order extension) iff e is one-to-one and $x \leq y$ in P is equivalent to $e(x) \leq e(y)$ in Q.]

0.5 A subset F of a poset $(P,\leq)$ is said to be a filter of P or in P iff

(i) F is an upper set.

(ii) F is nonempty and down-directed, i.e., for every $x,y \in F$ there is some $z \in F$ with $z \leq x$ and $z \leq y$.

A filter F is said to be proper iff $F \neq P$; otherwise F is improper. A filter F on a set M is a filter in the complete lattice $\underline{P}(M)$ of all subsets of M, ordered by the inclusion relation. An open filter of a topological space X is a filter in the lattice $\underline{O}(X)$ of open subsets of X.

For 1,$\wedge$-semilattices P, condition (ii) can be replaced by: $1 \in F$ and $x \wedge y \in F$ whenever $x,y \in F$.

A subset J of a poset P is an ideal iff it is a filter in P^{op}.

0.6 The following observation, due to Lawson [1977] (cf. the Compendium [p. 84]), is crucial for the duality of continuous posets:

If $x \ll y$ in a continuous poset P, then, by the interpolation property,

$$x \ll \cdots \ll y_n \ll \cdots \ll y_1 \ll y_0 = y$$

for some $y_1, y_2, \cdots \in P$. Thus

$$F = \bigcup\{\uparrow y_n \mid n \in \mathbb{N}\}$$

is a Scott-open filter of P with

$$y \in F \subseteq \Uparrow x$$

Thus every Scott-open subset of a continuous poset P is the union of Scott-open filters.

The dual D(P) of a continuous poset P is the set of all Scott-open filters of P, ordered by inclusion. The poset D(P) is a continuous poset:

For $F,G \in D(P)$, $F \ll G$ in $D(P)$ iff $F \subseteq \uparrow x \subseteq G$ for some $x \in P$. The natural map

$$\epsilon_P : P \to DD(P), \qquad x \mapsto \{G \in D(P) | x \in G\}$$

is an isomorphism. The inverse

$$\mu_P : DD(P) \to P$$

of ϵ_P assigns to $\underline{F} \in DD(P)$ the supremum, in P, of the nonempty up-directed subset

$$\{x \in P | G \subseteq \uparrow x \text{ for some } G \in \underline{F}\}$$

The dual of a continuous poset P is uniquely determined up to an isomorphism by the fact that it is a continuous poset whose lattice of Scott-open sets is anti-isomorphic to σ_P. The duality theory for continuous posets has been developed in Lawson [1979] and Hoffmann [1981b] (with fore-runners in Lawson [1977] and Hoffmann [1979d]).

0.7 A topological space X is called sober (Artin, Grothendieck, and Verdier [1972 (IV,4.21)]; Bruns [1962 (II, condition (1) on p. 17)]†) iff every non-empty, irreducible, closed subspace A of X has a unique "generic" point x, i.e., a point x with $cl\{x\} = A$. (A subspace A is irreducible iff it is not the union of two proper closed subsets; "sober" is strictly between T_0 and T_2, it does not imply nor is it implied by T_1.)

The category Sob of sober spaces and continuous maps is a full reflective subcategory of the category Top of all topological spaces and continuous maps. For a space X, let ${}^sX$ be the space of all nonempty irreducible closed subsets of X with open sets

$${}^sO := \{C \in {}^sX | C \cap O \neq \emptyset\}$$

with O ranging through the lattice $\underline{O}(X)$ of all open subsets of X. Then the mapping

$$\tilde{s}_X : X \to {}^sX, \ x \mapsto cl\{x\} \qquad \text{(the closure of x in X)}$$

is the Sob-reflection morphism (Artin, Grothendieck, and Verdier [1972 (IV, 4.2.1)]). This mapping $\tilde{s}_X$ is one-to-one iff it is an embedding iff X is a T_0-space; $\tilde{s}_X$ is bijective iff $\tilde{s}_X$ is a homeomorphism (onto) iff X is sober. Further, note that the lattice homomorphism

† Further historical information is given in Hoffmann [1977, pp. 365-366].

$$\underline{O}(\tilde{s}_X) : \underline{O}(^sX) \to \underline{O}(X)$$

induced by $\tilde{s}_X$ is an isomorphism.

0.8 For T_0-spaces X and Y, a continuous map $f:X \to Y$ is called an <u>essential extension</u> in the category $\underline{T}_0$ of T_0-spaces and continuous maps iff

(i) $f:X \to Y$ is an embedding (= extension).

(ii) Whenever $gf: X \to Z$ is an embedding for some continuous map $g: Y \to Z$, then g is an embedding.

Banaschewski [1977, prop. 2 (p. 237] has shown that every T_0-space X has an <u>essential hull</u>, viz., a unique greatest essential extension $\lambda_X:X \hookrightarrow \lambda X$ (i.e., whenever $f:X \hookrightarrow Y$ is an essential extension, then $hf = \lambda_X$ for some embedding $h:Y \hookrightarrow X$). Banaschewski's space λX is a subspace of the filter space $\Phi(X)$, the algebraic lattice of all (proper or improper) open filters of X (ordered by the inclusion relation) endowed with the Scott topology.

A T_0-space X is said to be <u>essentially complete</u> iff $\lambda_X: X \hookrightarrow \lambda X$ is a homeomorphism, i.e., iff X does not admit any nontrivial essential extension. Every essentially complete T_0-space is sober (Hoffmann [1979b, 0.1]). For further information see Hoffmann [1979b], in particular, sections 1 and 2.

This theme will be pursued further in Section 4 below.

0.9 One of the major insights at the root of the theory of continuous lattices is a result of Scott's [1972, 2.12]: The continuous lattices endowed with their Scott topology are precisely the <u>injective T_0-spaces</u>, i.e., the injective objects X in the category $\underline{T}_0$ of T_0-spaces and continuous maps with regard to the class of all (topological) embeddings, i.e., whenever $e: Y \hookrightarrow Z$ is an embedding for T_0-spaces Y and Z and $f:Y \to Z$ is a continuous map, then there is a continuous map $g:Z \to X$ rendering

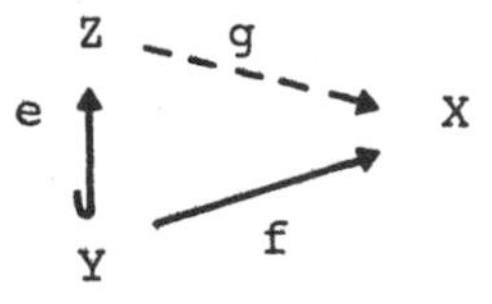

commutative.

Every injective T_0-space is essentially complete, hence, a fortiori, sober.

0.10 Banaschewski [1977, sec. 2] showed that in the category $\underline{T}_0$ of T_0-spaces and continuous maps every space X satisfying

(*) Whenever $x \in V \in \underline{O}(X)$, then there is an open neighborhood W of x in X such that $V \cap \bigcap \{cl\{z\} \mid z \in W\} \neq \emptyset$ has an injective hull in $\underline{T}_0$

i.e., λX is an injective T_0-space (Banaschewski [1981 (cor. 2, p. 240)]), i.e., λX is, by Scott's result [1972, 2.12], a continuous lattice in its Scott topology. In Hoffmann [1981a, 3.14], it is established that the sober spaces satisfying (*) are precisely the continuous posets endowed with their Scott topology. From this analysis, for a continuous poset P, the greatest essential extension of (P,σ_P) in $\underline{T}_0$ is an injective hull; hence it has the form

$$(P,\sigma_P) \hookrightarrow (I(P),\sigma_{I(P)})$$

for some continuous lattice I(P) which is, up to an isomorphism, uniquely determined. By an abuse of language, the order extension $P \hookrightarrow I(P)$ may be called the injective hull of P (Hoffmann [1981a,1982b].[†]

The continuous posets endowed with their Scott topology are also known as the projective sober spaces (Hoffmann [1981a, 2.19]).

0.11 The Lawson topology (or λ-topology, or $\underline{CL}$-topology) of a (continuous) poset $(P,\leq)$ is the weakest topology on P finer than both the Scott topology of $(P,\leq)$ and the weak topology of $(P,\leq)^{op}$ (cf. the Compendium [III-1.5]). It is designated by λ_P.

The Lawson topology of a continuous lattice is compact Hausdorff (the Compendium [III-1.10]).

1. MEET-PRIME, PSEUDO-MEET-PRIME, AND QUASI-MEET-PRIME ELEMENTS. LOCALLY QUASI-COMPACT STRONGLY SOBER SPACES

This is a survey of known results.[‡] After reviewing the (classical) theory of meet-prime elements, we discuss the notion of a pseudo-meet-prime element and that of a quasi-meet-prime element. The latter notion is — in a continuous lattice — essentially an equivalent description of what has been called

[†]See footnote on p. 60.

[‡]I am indebted to K. H. Hofmann for discussions on some of the material in this Section.

a "weakly prime element" by Hofmann and Lawson [1976-1977 (1.7, p. 313)]. The notion of a pseudo-meet-prime element is essentially due to Keimel and Mislove [1976] and Hofmann and Lawson [1978]. "Essentially" indicates the following difference: Whereas these authors assume that the unit element 1 of a complete lattice L is always "prime," we insist here that 1 is never meet prime. It is, in a sense (which will become precise later), a consequence of this modification that 1 can be, but need not be, pseudo-meet-prime or quasi-meet-prime. Thus the results of these authors need a slight adaptation to the present definitions.

The results of Hofmann and Lawson [1976-1977, 1978] and Keimel and Mislove [1976] appear in the Compendium [in particular, in I-3.23 to I-3.27 and V-3]. In Hoffmann [1982a, sec. 3], some of the modifications we need here have been derived from the results in the Compendium. Here we give direct proofs. The ideas involved are not new, but we have reduced the number of "auxiliary concepts" introduced in the Compendium and we have tried to single out the precise hypotheses actually needed for the lemmas into which the proofs are decomposed.

We conclude with a theorem, largely due to Hofmann and Mislove [1981], on strongly sober locally quasi-compact spaces.

1.1 Let L be a complete lattice. An element $p \in L$ is said to be meet prime iff for every finite subset F of L, $\inf F \le p$ implies $x \le p$ for some $x \in F$.

Note that $1 = \inf \emptyset$ is not meet prime (in contrast to the definition of a "prime" element given in the Compendium [1980 (I-3.11)]).

A theorem due to Büchi [1952] and Papert [1959] (and other authors) says that a complete lattice L is isomorphic to the lattice $\underline{O}(X)$ of open subsets of a topological space X iff every element x of L is the infimum of a family of meet-prime elements.

The set of meet-prime elements of a complete lattice L will be endowed with the trace of the weak topology of L^{op}: The resulting space will be designated by Spec^*L. This notation is a compromise between the notation used in the Compendium [1980] and Hofmann and Lawson [1978], and that of Hoffmann [1979b,1982a].[†] The specialization partial order of Spec^*L is inverse to the order induced by L.

[†] In the Compendium and Hofmann and Lawson [1978], "Spec L" is used to designate Spec*L, whereas in Hoffmann [1979b,1982a] "Spec L" (or $\vee$-Spec L) designates the set of join-prime elements of L, endowed with the trace of the weak topology of L.

A subset M of Spec^*L is closed iff

$$M = \mathrm{Spec}^*L \cap \uparrow x$$

for some $x \in L$.

For every complete lattice L, Spec^*L is a sober space (Hoffmann [1979b, 3.5], Compendium (V-4.4)]), and we have

$$\underline{O}(\mathrm{Spec}^*L) \cong L$$

iff every element of L is a meet of meet-prime elements.

Note that, for a space X, there is a homeomorphism

$$^sX \to \mathrm{Spec}^*\underline{O}(X)$$

since the elements of ^sX are the join-prime elements of the lattice $\underline{A}(X)$ of all closed subsets of X ordered by inclusion which correspond – via the anti-isomorphism $\underline{A}(X) \to \underline{O}(X)$, $A \mapsto X - A$ – to the meet-prime elements of $\underline{O}(X)$.

1.2 An ideal J in a 0,∨-semilattice is said to be a prime ideal iff J is a meet-prime element in the complete lattice Id L (ordered by inclusion) consisting of all ideals of L.

An ideal J in a lattice L with 0 and 1 is a prime ideal iff

(i) $x \wedge y \in J$ always implies $x \in J$ or $y \in J$.

(ii) $1 \notin J$, i.e., $J \neq L$.

A prime filter in a lattice L with 0 and 1 is a meet-prime element of Filt L := $\mathrm{Id}(L^{op})$. Note that, by the preceding criterion (and its dual), $M \subseteq L$ is a prime filter iff $L - M$ is a prime ideal (cf. the Compendium [I-3.16]).

Now, an element x of a lattice L with 0 and 1 is meet prime iff $\downarrow x$ is a prime ideal.

1.3 (i) An element a of a complete lattice is said to be pseudo-meet-prime iff there exists a prime ideal J such that a = sup J.

(ii) An element $b \in L$ is quasi-meet-prime iff, whenever $\inf F \ll b$ for a finite subset F of L, there is some $x \in F$ with $x \leq b$. (Note that $F = \emptyset$ is not excluded.)

The set of pseudo-meet-prime elements of L and the set of quasi-meet-prime elements of L will be designated by ψ^*L and κ^*L, respectively.

Later (in Section 4), we will endow ψ^*L with a topology. The resulting space will be also denoted by ψ^*L. The notation ψL will be reserved to designate both the set and the space $\psi^*(L^{op})$ of all pseudo-join-prime elements of L. Note that in this section ψ^*L is always the set of pseudo-meet-prime elements of L without any topology.

1.4 LEMMA (a) Every meet-prime element is pseudo-meet-prime. (b) Every pseudo-meet-prime element is quasi-meet-prime.

Proof: (a) $x \in L$ is meet prime iff $\downarrow x$ is a prime ideal. Clearly, $x = \sup \downarrow x$. (Cf. the Compendium [p. 75].)

(b) Cf. the Compendium [I-3.24, "(1) implies (2)"]. (Referring to the continuity of L is clearly unnecessary.) ∎

We now have the following inclusions

$$\mathrm{Spec}^*L \subseteq \psi^*L \subseteq K^*L$$

of sets.

1.5 REMARKS Let L be a complete lattice.

(1) An element p is quasi-meet-prime in L iff the filter generated by $L - \downarrow p$ does not meet $\Downarrow p$.

(2) In the unit square $[0,1] \times [0,1]$ (a distributive continuous lattice) the element $(1,1/2)$ is meet prime, hence quasi-meet-prime and $(1,0) \wedge (0,1) \ll (1,1/2)$, but neither $(1,0) \ll (1,1/2)$ nor $(0,1) \ll (1,1/2)$, since $(1,1)$ is the supremum of the ideal $[0,1) \times [0,1)$. Thus the definition of a quasi-meet-prime element b cannot be strengthened to $\inf F \ll b$ with $F \subseteq L$ finite implies $x \ll b$ for some $x \in F$. Cf. the Compendium [V-(remark after 3.4, p. 248)].

(3) If the greatest element 1 is compact in L, i.e., $1 \ll 1$, then 1 fails to be quasi-meet-prime.

(4) If, in addition, L is distributive and if 1 is not pseudo-meet-prime, then 1 is compact.

Proof: (1) The filter generated by $L - \downarrow p$ is

$$\{y \in L \mid \inf F \leq y \text{ for some finite } F \subseteq L - \downarrow p\}$$

Now our assertion is immediate from (the contraposition) of the definition of quasi-meet-prime.

(3) If 1 is compact, then $\inf \emptyset \ll 1$, hence 1 fails to be quasi-meet-prime.

(4) If 1 not $\ll 1$, then there exists an ideal J of L such that $\sup J = 1$, but $1 \notin J$. By a standard argument for distributive lattices (cf. the Compendium [I-3.19]), there is a prime ideal P with $J \subseteq P$, but $1 \notin P$. Clearly, $1 = \sup J \leq \sup P \leq 1$, hence $1 = \sup P$ is pseudo-meet-prime. ∎

Note that Remark 1.5(1) corrects a slight inaccuracy of the Compendium [I-3.24] where it is overlooked that a filter F, by definition (the Compendium [O-1.3]), contains inf $\emptyset = 1$.

Parts (2) and (3) of the following Lemma 1.6 are modifications of the Compendium [I-3.27, "(2) iff (3)", and I-3.24, "(1) iff (2)", respectively]. Part (1) extends the additional remark in the Compendium [I-3.27] set free from the hypothesis of continuity for L.

1.6 LEMMA Let L be a complete lattice: (1) Suppose, in addition, that L is distributive. If $\mathrm{Spec}^*L = \psi^*L$, then 1 is compact in L, i.e., $1 \ll 1$, and the way below relation $\ll$ in L is multiplicative, i.e., $x \ll y$ and $x \ll z$ imply $x \ll y \wedge z$ for all $x,y,z \in L$. (2) Suppose L is a continuous lattice, $1 \in L$ is compact, and $\ll$ is multiplicative. Then $\mathrm{Spec}^*L = \kappa^*L$, i.e., every quasi-meet-prime element p of L is meet prime. (3) If L is a distributive continuous lattice, then $\psi^*L = \kappa^*L$, i.e., every quasi-meet-prime element of L is pseudo-meet-prime.

Proof: (1) (a) Assume $\ll$ is not multiplicative; hence there are elements $a,x,y \in L$ with $a \ll x$ and $a \ll y$, but a not $\ll x \wedge y$. Thus there is an ideal J with $x \wedge y \leq \sup J$, but $a \notin J$. By a standard argument for distributive lattices (cf. the Compendium [I-3.19]), there is a prime ideal P with $J \subseteq P$, but $a \notin P$. Now let $p = \sup P$. Clearly, $x \wedge y \leq \sup J \leq \sup P = p$. However, $x \leq p$ would give $a \ll x \leq \sup P$, hence $a \in P$, contradicting the choice of P. Thus $x \nleq p$ and, analogously, $y \nleq p$. As a consequence, the pseudo-meet-prime element p fails to be meet prime, a contradiction.

(b) If 1 not $\ll 1$, then 1 is pseudo-meet-prime by Remark 1.5(4), but not meet prime, a contradiction to the hypothesis.

(2) Suppose $x \wedge y \leq p$ with $x,y \in L$. Then, by continuity of L, $x \leq p$ or $y \leq p$, as in the Compendium [I-3.27, "(2) implies (3)"]. Since 1 is compact, it fails to be quasi-meet-prime [by Remark 1.5(3)], hence $1 \neq p$. Thus p is meet prime.

(3) See the Compendium [I-3.24, "(3) implies (1)"]. ∎

Recall that λ denotes the Lawson- or λ-topology of a complete lattice L (cf. Section O.11 above).

1.7 LEMMA Let L be a complete lattice. (1) If every element of L is a meet (= infimum) of meet-prime elements, then ψ^*L is contained in the λ-closure of Spec^*L in L.

(2) Suppose that the way below relation $\ll$ has the interpolation property, i.e., $x \ll y$ implies $x \ll z$ and $z \ll y$ for some $z \in L$; then κ^*L is λ-closed in L.

Proof: (1) Suppose $a \in L$ is pseudo-meet-prime, i.e., $a = \sup P$ for a prime ideal P of L and $a \in U - (\uparrow x_1 \cup \cdots \cup \uparrow x_n)$ for some Scott-open subset U of L and $x_1, \ldots, x_n \in L(n \in \mathbb{N} \cup \{0\})$; note that these sets are the standard basic Lawson-open neighborhoods of a in L. Since U is Scott-open and P is an ideal with $\sup P \in U$, there is some $q \in P$ with $q \in U$. Assume now that $\inf\{x_1, \ldots, x_n\} \leq q$; then $x_i \in P$ for some $i \in \{1, \ldots, n\}$, since P is prime. Thus $x_i \leq \sup P = a$, contradicting our hypothesis that $a \in U - (\uparrow x_1 \cup \cdots \cup \uparrow x_n)$. Thus we have $x := \inf\{x_1, \ldots, x_n\} \not\leq q$.

By hypothesis, $q = \inf\{p_k | k \in K\}$ for meet-prime elements p_k of L. There is some $k_0 \in K$ such that $x \not\leq p_{k_0}$ (otherwise $x \leq \inf\{p_k | k \in K\} = q$, contradicting the above). It results that $p_{k_0} \notin \uparrow x_1 \cup \cdots \cup \uparrow x_n$. Since $q \in U$ and $q \leq p_k$, we have $p_k \in U$. Thus, as required,

$$p_{k_0} \in U - (\uparrow x_1 \cup \cdots \cup \uparrow x_n)$$

(2) Let $p \in L$ denote a limit of a net $(p_j)_{j \in J}$ of quasi-meet-prime elements of L in the λ-topology, and let

$$x := \inf\{x_1, \ldots, x_n\} \ll p$$

for some $x_1, \ldots, x_n \in L$. We have to show that $x_i \leq p$ for some $i \in \{1, \ldots, n\}$. Suppose, on the contrary, that $x_i \not\leq p$ for every $i \in \{1, \ldots, n\}$. Then $U := \Uparrow x - (\uparrow x_1 \cup \cdots \cup \uparrow x_n)$ contains p. Since $\ll$ interpolates, $\Uparrow x$ is Scott-open, hence U is Lawson-open. Thus there is some $j \in J$ with $p_j \in U$. However, $\inf\{x_1, \ldots, x_n\} \ll p_j$ implies $x_i \leq p_j$ for some $i \in \{1, \ldots, n\}$, i.e., $p_j \in \uparrow x_i$, contradicting $p_j \in U$. Thus $x_i \leq p$ for some $i \in \{1, \ldots, n\}$. This shows that p is quasi-meet-prime. ■

1.8 REMARK It is immediate from Lemmas 1.6(2) and 1.7(2) that in a continuous lattice L (not necessarily distributive) with 1 compact in L and $\ll$ multiplicative, Spec^*L is closed with regard to the λ-topology of L. (Cf. Hofmann and Lawson [1978, 6.8].)

1.9 PROPOSITION For a distributive continuous lattice L, $\psi^*L = \kappa^*L$ is the closure of Spec^*L with regard to the Lawson topology λ on L.

Proof: By Lemma 1.6(3), $\psi^*L = \kappa^*L$. Since by the Compendium [I-3.7], every element of a distributive continuous lattice L is the meet of meet-prime elements, we have

$$\mathrm{Spec}^*L \subseteq K^*L = \psi^*L \subseteq \mathrm{cl}_\lambda(\mathrm{Spec}^*L)$$

by Lemma 1.7(1). Since in a continuous lattice L the way below relation interpolates (the Compendium [I-1.18]), K^*L is λ-closed by Lemma 1.7(2). Thus $K^*L = \mathrm{cl}_\lambda\ (\mathrm{Spec}^*L)$. ■

1.10 COROLLARY For a distributive continuous lattice L, Spec^*L is λ-closed iff 1 is compact in L and the way below relation $\ll$ is multiplicative.

Proof: The first implication is established in Remark 1.8. Suppose Spec^*L is λ-closed; then, by Proposition 1.9, $\mathrm{Spec}^*L = \psi^*L$. Now Lemma 1.6(1) applies. ■

The preceding results modify (and sharpen) analogous results in the Compendium [V-3].

1.11 By a celebrated theorem of Hofmann and Lawson [1978], every distributive continuous lattice is isomorphic to the lattice $\underline{O}(X)$ of open subsets, ordered by inclusion, of a topological space X, and X can be chosen as a locally quasi-compact sober space, namely, $X = \mathrm{Spec}^*L$. Furthermore, if X is a locally quasi-compact space, then $\underline{O}(X)$ is a distributive continuous lattice. Thus every result on distributive continuous lattices may be viewed as a result on locally quasi-compact (sober) spaces, and – for sober spaces – conversely.

For a space X, let

$$\psi^*X := \psi^*\underline{O}(X) \qquad \psi X := \psi\underline{A}(X)$$

Note that there is a canonical mapping

$$\psi_X\colon X \to \psi X \qquad x \mapsto \mathrm{cl}\{x\}$$

the composite of $\tilde{s}_X\colon X \to {}^sX$ (1.1) and the inclusion ${}^sX \to \psi X$. This mapping is one-to-one iff X is a T_0-space.

Note that, for a locally quasi-compact (sober) space X, ψX consists, by Proposition 1.9, of the same points as the Fell compactification $\underline{H}(X)$ of X. (As noted in Hoffmann [1982a, 3.13], this observation extends to nonsober locally quasi-compact spaces.)

For a filter $\underline{F}$ on a space X, let

$$\mathrm{conv}\ \underline{F} = \{x \in X \mid \underline{O}(x) \subseteq \underline{F}\}$$

denote the convergence set of $\underline{F}$, where $\underline{O}(x)$ denotes the open neighborhood filter of x in X. Note that the improper filter $\underline{P}(X) := \{M | M \subseteq X\}$ is here not excluded.

1.12 DEFINITION A space X is said to be strongly sober iff for every ultrafilter $\underline{U}$ on X, conv $\underline{U}$ has a unique generic point, i.e.,

$$\text{conv } \underline{U} = \text{cl}\{x\}$$

for a unique element x of X.

The notion of strong sobriety is a slight modification of the notion of super sobriety (Compendium [VII-1.10, p. 310]; strongly sober = super sober + quasi-compact).

Every essentially complete T_0-space is strongly sober, since, by Hoffmann [1979b, 3.11], a space X is an essentially complete T_0-space iff, for every filter $\underline{F}$ on X, conv $\underline{F}$ has a unique generic point. Every strongly sober space is sober (the Compendium [VII-1.11]), since, by Hoffmann [1977, 1.9], a space X is sober iff conv $\underline{U}$ has a unique generic point for all those ultrafilters $\underline{U}$ which enjoy the property that conv $\underline{U} \in \underline{U}$ ("irreducible" ultrafilters, Hoffmann [1977, 1.4]). Further, note that the strongly sober T_1-spaces are precisely the compact Hausdorff spaces.

The following result is (up to slight modifications of the statement) due to Hofmann and Mislove [1981, 4.8].

1.13 THEOREM Let X be a locally quasi-compact sober space; then the following conditions are equivalent:

(1) The way below relation of the (continuous) lattice $\underline{O}(X)$ is multiplicative, and the unit element of $\underline{O}(X)$ is compact (i.e., X is a quasi-compact space).

(2) $\text{Spec}^*\underline{O}(X)$ is λ-closed in $\underline{O}(X)$.

(3) X is strongly sober.

(4) X is quasi-compact and the intersection of two quasi-compact saturated subsets is quasi-compact. (A subset of a space is saturated iff it is the intersection of its open neighborhoods – cf. the Compendium [V-5.2, p. 258].)

The equivalence of (1) and (2) is established in Corollary 1.10. In view of Proposition 1.9, we may add the following equivalent condition:

(5) The canonical mapping $\psi_X: X \to \psi X$ is bijective.

It has been observed in Hoffmann [1982a, 3.12] that this is equivalent to:

(6) For every open prime filter $\underline{F}$ of X, conv $\underline{F}$ has a unique generic point.

1.14 It may be worth pointing out that strong sobriety per se does not imply local quasi-compactness: The essential hull X^* of a Hausdorff space X (i.e., $X \cup \{0,1\}$, where 0 and 1 are adjoined as a smallest and a largest point, respectively; cf. Hoffmann [1979b, sec. 5]) is, of course, essentially complete, hence strongly sober, but X is an intersection of an open and a closed set of X^*; hence X^* cannot be locally quasi-compact unless X is locally compact.

1.15 REMARK Simmons [1982] has shown that the strongly sober locally quasi-compact spaces form the Eilenberg-Moore algebras for a "triple" (or "monad") $\Psi = \langle P,\eta,\mu\rangle$ on the category $\underline{T}_0$ of T_0-spaces and continuous maps. The functor part P of this monad assigns to a T_0-space X its space of open prime filters [i.e., meet-prime elements in the complete lattice, ordered by inclusion, of all filters of the lattice $\underline{O}(X)$ of open sets of X], with the topology inherited from the space of all open filters of X (cf. Banaschewski [1977, sec. 1]). This space P(X) is, via a result of Schröder [1977], homeomorphic to the extension space of X induced by the open finite decomposition spectrum ("Zerlegungsspektrum") of X introduced by Flachsmeyer [1961, p. 264]. The Ψ-homomorphisms are those continuous maps $f: X \to Y$ which enjoy the property that the induced map $\underline{O}(f): \underline{O}(Y) \to \underline{O}(X)$ (assigning to an open subset of Y its inverse image) preserves the way below relation.

It may be noted that, as a consequence of this result, the category of locally quasi-compact strongly sober spaces and those continuous maps whose inverse image map preserves the way below relation is complete [i.e., has (projective) limits for all diagrams indexed over categories which are "small" with regard to the given universe] and that these limits can be constructed in the category $\underline{T}_0$. (An Eilenberg-Moore category over a complete category is complete and limits are "constructed" in the underlying category – cf. MacLane [1971, VI-2, exercise 2]; Schubert [1972, 21.3.9].) An argument different from this results from the isomorphism described in Section 6 below.

1.16 REMARK Suppose X and Y are locally quasi-compact sober spaces. In the Compendium [1980, V-5.14, 5.15] it is shown that for a continuous map $f: X \to Y$ the following conditions are equivalent:

(i) $\underline{O}(f): \underline{O}(Y) \to \underline{O}(X)$ preserves the way below relation.

(ii) The inverse image of every saturated quasi-compact subset of Y is quasi-compact (and saturated) in X.

For locally compact Hausdorff spaces X and Y the mappings enjoying property (ii) are known as the <u>perfect maps</u> (note that in a T_1-space every subset is saturated). This name may be extended to the setting of locally quasi-compact sober spaces.

1.17 REMARK It has been observed in Hoffmann [1982a, 3.8] that the very definition of $\psi X := \psi(\underline{A}(X))$ implies that the points of ψX are the convergence sets of the open prime filters of the space X. For a locally quasi-compact space X, the points of ψX also can be characterized as the convergence sets of the primitive nets; this results – via the observation in Section 1.11 above – from Fell [1962, p. 475]. (A net is said to be primitive iff every adherence point of the net is also a limit point. By the topological equivalence between nets and filters, observed in Bruns and Schmidt [1955], primitive nets can be replaced by primitive filters.) I do not know whether every member of ψX is a convergence set of an ultrafilter (under the proviso that X is a locally quasi-compact sober space).

1.18 REMARK It has been observed in Hoffmann [1982a, sec. 3] (in the notes added) that the greatest essential extension of a strongly sober (not necessarily locally quasi-compact) space coincides, on the level of the specialization orders, with the MacNeille completion. Thus a space X is essentially complete if (and only if) X is strongly sober and a complete lattice in its specialization order. As a consequence, by Hoffmann [1979b, 1.5], the Scott topology on a complete lattice makes $\vee$ continuous iff it is strongly sober.

Also note that every spectral space (i.e., every prime spectrum of a commutative, associative ring with 1 (cf. Hochster [1969]) is strongly sober and locally quasi-compact. Indeed, the strongly sober locally quasi-compact spaces are precisely the retracts, in $\underline{T}_0$, of the spectral spaces (Johnstone [1981, sec. 2], Simmons [1982]; cf. also Banaschewski [1981, prop. 2]).

2. THE INJECTIVE HULL OF D(L) FOR A CONTINUOUS 1,∧-SEMILATTICE L

In this section, L denotes a continuous 1,∧-semilattice L. However, some of the results are based on the assumption that L is a <u>distributive continuous lattice</u>.

2.0 For a continuous 1,$\wedge$-semilattice L, let

$$\mathrm{Filt}_\sigma L = \{\text{all filters in } L \text{ which are generated by a Scott-open subset of } L\}$$

Recall that, in a 1,$\wedge$-semilattice L, the filter generated by a subset M of L (i.e., the smallest filter of L containing M) is the set

$$\varphi(M) = \{x \in L \mid \inf F \leq x \text{ for some finite set } F \subseteq M\}$$

If L is a distributive lattice and M is an upper set, then

(*) $\varphi(M) = \{x \in L \mid \inf F = x \text{ for some finite set } F \subseteq M\}$

2.1 For a continuous 1,$\wedge$-semilattice L, we have

(i) $D(L) \subseteq \mathrm{Filt}_\sigma L$

where D(L) denotes the dual of L (consisting of all Scott-open filters of L) – cf. 0.6.

(ii) $\mathrm{Filt}_\sigma L$ is a complete lattice, since

$$\varphi(\bigcup_i M_i) = \varphi(\bigcup_i \varphi(M_i))$$

is the supremum of $(\varphi(M_i))_{i\in I}$ in $\mathrm{Filt}_\sigma L$ for every family $(M_i)_{i\in I}$ of Scott-open sets M_i of L.

(iii) Since every Scott-open set is the union of Scott-open filters (cf. 0.6), D(L) is join-dense in $\mathrm{Filt}_\sigma L$, i.e., every member of $\mathrm{Filt}_\sigma L$ is a supremum of a family of members of D(L) in $\mathrm{Filt}_\sigma L$.

2.2 LEMMA For a continuous 1,$\wedge$-semilattice L, and $F,G \in \mathrm{Filt}_\sigma L$, the following are equivalent:

(i) $F \ll G$ in $\mathrm{Filt}_\sigma L$.

(ii) $F \subseteq \uparrow x \subseteq G$ for some $x \in L$.

Proof: It is readily clear that, for $F_i \in \mathrm{Filt}_\sigma L$, $\bigcup_i F_i$ is the supremum of $\{F_i \mid i \in I\}$ in $\mathrm{Filt}_\sigma L$ if this family is nonempty and up-directed. Thus "(ii) implies (i)" is evident.

To prove that (i) implies (ii), let $F = \varphi(K)$ and $G = \varphi(M)$ for $K,M \in \sigma_L$ (= the Scott topology of L). For every $x \in M$ choose some $x' \in M$ with $x' \ll x$ in L (by 0.3), and some Scott-open filter F_x in L such that

$$x \in F_x \subseteq \uparrow x'$$

(by Lawson's argument; cf. 0.6). Then

$$M = \bigcup\{F_x | x \in M\}$$

Hence,

$$\varphi(M) = \sup\{F_x | x \in M\}$$

where sup denotes the supremum in $\mathrm{Filt}_\sigma L$. Since

$$F_x \subseteq \uparrow x' \subseteq M \subseteq \varphi(M)$$

we have $F_x \ll \varphi(M)$ by "(ii) implies (i)"; hence, by hypothesis (i), there are $x_1, \ldots, x_n \in M$ $(n \geq 0)$ with

$$\varphi(K) \subseteq \sup\{F_{x_1}, \ldots, F_{x_n}\}$$

Clearly,

$$\begin{aligned}\sup\{F_{x_1}, \ldots, F_{x_n}\} &= \varphi(F_{x_1} \cup \cdots \cup F_{x_n}) \\ &\subseteq \varphi(\uparrow x_1' \cup \cdots \cup \uparrow x_n') \\ &\subseteq \uparrow \inf\{x_1', \ldots, x_n'\}\end{aligned}$$

Since $x_1', \ldots, x_n' \in M$, we infer that

$$y := \inf\{x_1', \ldots, x_n'\} \in \varphi(M)$$

Thus

$$\varphi(K) \subseteq \uparrow y \subseteq \varphi(M)$$

as claimed. ∎

2.3 PROPOSITION For a continuous 1,∧-semilattice L, $\mathrm{Filt}_\sigma L$ is a continuous lattice.

Proof: For $M \in \sigma_L$ we have

$$\varphi(M) = \sup\{F_x | x \in M\}$$

where (as in the proof of Lemma 2.2) $x \in F_x \subseteq \uparrow x'$ for some $F_x \in D(L)$ and some $x' \in M$. Hence, by Lemma 2.2, $F_x \ll \varphi(M)$ in $\mathrm{Filt}_\sigma L$ for all $x \in M$. ∎

2.4 The dual D(L) of a continuous 1,∧-semilattice L is a continuous 1,∧-semilattice.

To show that

$$D(L) \hookrightarrow \mathrm{Filt}_\sigma L$$

is the (an) injective hull in $\underline{T}_0$ (with regard to the respective Scott topologies), the following characterization of the injective hull of a continuous 1,$\wedge$-semilattice will be used:

Suppose S is a continuous 1,$\wedge$-semilattice. An order embedding f: S $\to$ K into a continuous lattice K is an injective hull iff the following conditions are satisfied:

(1) f: S $\to$ K preserves suprema of nonempty up-directed subsets.
(2) f: S $\to$ K preserves the way below relation.
(3) f[S] is (topologically) dense in (K,λ_K).
(4) f[S] is join dense in K, i.e., every member of K is a supremum, in K, of elements of f[S].

This is established in Hoffmann [1982b].

2.5 THEOREM For a continuous 1,$\wedge$-semilattice L,

$$D(L) \hookrightarrow \mathrm{Filt}_\sigma L$$

is an injective hull.

Proof: We have already seen that $\mathrm{Filt}_\sigma L$ is a continuous lattice (2.3). For condition (4), see 2.1(iii). Now (1) and (2) are immediate consequences of the explicit description of suprema of nonempty up-directed subsets (= set-theoretic unions) and of the way below relation in D(L) and $\mathrm{Filt}_\sigma L$, respectively (cf. 0.6 and 2.2).

Thus it remains to show that D(L) is (topologically) dense in $\mathrm{Filt}_\sigma L$ with regard to the Lawson topology.

Suppose

$$V = \Uparrow F - (\uparrow G_1 \cup \cdots \cup \uparrow G_n)$$

(where, for once, $\Uparrow$ and $\uparrow$ have to be interpreted in $\mathrm{Filt}_\sigma L$) is nonempty for $F, G_1, \ldots, G_n \in \mathrm{Filt}_\sigma L$ $(n \geq 0)$, i.e.,

$$F \subseteq \uparrow x \subseteq H \qquad G_i \nsubseteq H \qquad i = 1, \ldots, n$$

for some $H \in \mathrm{Filt}_\sigma L$ and some $x \in L$.

For every i = 1, ..., n, there is some $x_i \in L$ with $x_i \ll x$ in L and

$$G_i \nsubseteq \uparrow x_i$$

(otherwise $G_i \subseteq \bigcap\{\uparrow y | y \in L, y \ll x\} = \uparrow x \subseteq H$, since L is continuous). Thus, for $z = \sup\{x_1, \ldots, x_n\}$, we have

$$z \ll x \qquad G_i \not\subseteq \uparrow z \qquad \text{for every } i$$

There is a Scott-open filter M of L such that

$$x \in M \subseteq \Uparrow z$$

Clearly, $F \subseteq \uparrow x \subseteq M$ and $G_i \not\subseteq M$ $(i = 1, \ldots, n)$; hence $M \in V$, as we want.

Since these sets V form an open basis of the Lawson topology of $\mathrm{Filt}_\sigma L$, this proves that D(L) is dense in $\mathrm{Filt}_\sigma L$ with regard to the Lawson topology. ∎

2.6 For a 1,∧-semilattice, let Filt L denote the complete (algebraic) lattice of all filters of L. Note that the meet (= infimum) in Filt L is the set-theoretic intersection.

For a continuous 1,∧-semilattice L, we have

(a) $\mathrm{Filt}_\sigma L$ is stable in Filt L under arbitrary joins.

(b) The order embedding $\mathrm{Filt}_\sigma L \hookrightarrow \mathrm{Filt}\, L$ preserves and reflects the way below relation, i.e., for $F,G \in \mathrm{Filt}_\sigma L$, $F \ll G$ in $\mathrm{Filt}_\sigma L$ iff $F \ll G$ in Filt L.

Proof: (a) is clear from 2.1(ii), since the formula given there describes the suprema in Filt L, when M_i is interpreted as an arbitrary subset of L.

(b) In the algebraic lattice Filt L we have $G \ll F$ (for $G,F \in$ Filt L) iff

$$G \subseteq \uparrow x \subseteq F$$

for some $x \in L$ (since the principal filters $\uparrow x$ are the compact elements of Filt L). Now Lemma 2.2 applies. ∎

2.7 LEMMA For a distributive (!) continuous lattice L, $\mathrm{Filt}_\sigma L$ is stable in Filt L under finite meets.

Proof: Clearly, L is the greatest element of both $\mathrm{Filt}_\sigma L$ and Filt L. Suppose K,M are Scott-open subsets of L. We prove that

$$\varphi(K) \cap \varphi(M) \subseteq \varphi(K \cap M)$$

(The other inclusion is evident.)

If $x \in \varphi(K) \cap \varphi(M)$, then, by 2.0(*), there are $k_1, \ldots, k_r \in K$, and $m_1, \ldots, m_n \in M$ $(r,n \in \mathbb{N} \cup \{0\})$ with

$$x = \inf\{k_1, \ldots, k_r\} = \inf\{m_1, \ldots, m_n\}$$

It results that

$$k_i \vee m_j \in K \cap M$$

and, by distributivity of L,

$$x = \inf\{k_i \vee m_j \mid i \in \{1, \ldots, r\} \text{ and } j \in \{1, \ldots, n\}\}$$

This implies that

$$x \in \varphi(K \cap M)$$

as claimed. ∎

2.8 The dual D(P) of a continuous poset P is a continuous lattice if and only if P is a 1,$\wedge$-semilattice, the way below relation $\ll$ of P is multiplicative (i.e., $x \ll y$ and $x \ll z$ for $x,y,z \in P$ imply $x \wedge y \ll z$), and the greatest element 1 of P is compact (i.e., $1 \ll 1$) (Lawson [1977; 1979, 9.6], Hoffmann [1981b, 3.13]).

2.9 LEMMA For a continuous lattice L, we have:

(a) The greatest element L of $\mathrm{Filt}_\sigma L$ is compact.

(b) If, in addition, L is distributive, then the way below relation of $\mathrm{Filt}_\sigma L$ is multiplicative.

Proof: (a) Evidently, $L = \uparrow 0$; hence $L \ll L$, by Lemma 2.2.

(b) If $G \ll F_1$ and $G \ll F_2$ in $\mathrm{Filt}_\sigma L$, then

$$G \subseteq \uparrow x_1 \subseteq F_1 \qquad G \subseteq \uparrow x_2 \subseteq F_2$$

Hence,

$$G \subseteq \uparrow(x_1 \vee x_2) \subseteq F_1 \cap F_2$$

so

$$G \ll F_1 \cap F_2$$

(Recall that, by Lemma 2.7, $F_1 \cap F_2$ is the meet of F_1 and F_2 in $\mathrm{Filt}_\sigma L$.) ∎

It is well known that a lattice L is distributive iff the lattice Filt L of all filters of L is distributive (Birkhoff [1967, p. 114]). The following observation is now immediate from 2.3, 2.6(a), and 2.7.

2.10 LEMMA For a distributive continuous lattice L, $\mathrm{Filt}_\sigma L$ is a distributive continuous lattice.

2.11 LEMMA[†] Let K be a distributive continuous lattice such that $1 \in K$ is compact and the way below relation is multiplicative. Then D(K) is a distributive continuous lattice.

Proof: For $F,G,H \in D(K)$ we clearly have

$$(F \wedge G) \vee (F \wedge H) \subseteq F \wedge (G \vee H)$$

where $\wedge$ and $\vee$ denote the binary infimum (= intersection $\cap$, by Lemma 2.7) and the binary supremum in D(K), respectively.

For the proof of distributivity of D(K), suppose $x \in F \wedge (G \vee H) = F \cap (G \vee H)$; hence $x \in F$ and $x \in G \vee H$. Since

$$G \vee H = \{y \in L \mid g \wedge h = y \text{ for some } g \in G \text{ and some } h \in H\}$$

(by distributivity of K, multiplicativity of $\ll$, and compactness of 1), $x = g \wedge h$ for some $g \in G$, $h \in H$; hence

$$g \in F \cap G \qquad h \in F \cap H$$

so

$$x = g \wedge h \in (F \wedge G) \vee (F \wedge H)$$

as we want. ∎

2.12 THEOREM For a distributive continuous lattice L, $D(\mathrm{Filt}_\sigma L)$ is a distributive continuous lattice.

Proof: Immediate from Lemmas 2.10 and 2.11. ∎

2.13 REMARK For a continuous 1,$\wedge$-semilattice L, the order embedding

$$D(L) \hookrightarrow \mathrm{Filt}_\sigma L$$

preserves finite meets (as every join-dense order embedding does) and nonempty up-directed joins. Thus it induces a Scott-continuous 1,$\wedge$-homomorphism

$$D\,\mathrm{Filt}_\sigma L \to DD(L)$$

assigning to a Scott-open filter Φ of $\mathrm{Filt}_\sigma L$ the Scott-open filter $\Phi \cap D(L)$

[†] A proof of this observation is also immediate from the representation of D(K) in terms of quasi-compact saturated subsets of a locally quasi-compact space (Hofmann and Mislove [1981, sec. 2]).

of D(L). Via the canonical isomorphism μ_L: DD(L) → L (cf. 0.6) this induces a morphism (to be studied further in Section 5.3)

$$D\ \mathrm{Filt}_\sigma L \to L$$

2.14 REMARKS

(a) For a continuous 1,∧-semilattice L, D Filt_σ(L) is an "idempotent" construction in the sense that the induced morphism (cf. Remark 2.13)

$$D\ \mathrm{Filt}_\sigma(D\ \mathrm{Filt}_\sigma L) \to D\ \mathrm{Filt}_\sigma L$$

is an isomorphism.

This is readily clear from the observation that, for a continuous 1,∧-semilattice K such that D(K) is complete (i.e., 1 is compact and ≪ is multiplicative in K), Filt_σK coincides with D(K) (in other words, the filter generated by a Scott-open subset of K is itself Scott-open).

The idempotency of D Filt_σ• can also be based upon Theorem 2.5. This argument may be sketched in the following way:

$$\begin{aligned}(D\ \mathrm{Filt}_\sigma)^2(L) &\cong (DID)^2(L) \cong DID^2ID(L)\\ &\cong DI^2D(L) \quad \cong DID(L) \cong D\ \mathrm{Filt}_\sigma L\end{aligned}$$

since $D^2(P) \cong P$ and $I^2(P) \cong I(P)$ for every continuous poset P [where $P \hookrightarrow I(P)$ denotes an injective hull of P].

(b) For a distributive continuous lattice L,

$$(\mathrm{Filt}_\sigma)^3 L \cong \mathrm{Filt}_\sigma L$$

since, by Theorem 2.12, D Filt_σL is complete [i.e., IDID(L) ≅ DID(L)]; hence

$$\begin{aligned}(\mathrm{Filt}_\sigma)^3(L) &\cong (ID)^3(L) \cong IDIDID(L) \cong ID^2ID(L)\\ &\cong I^2D(L) \quad \cong ID(L) \quad\quad \cong \mathrm{Filt}_\sigma L\end{aligned}$$

2.15 REMARK For a distributive continuous lattice L,

$$\psi^* D\ \mathrm{Filt}_\sigma L = \mathrm{Spec}^*\ D\ \mathrm{Filt}_\sigma L$$

by 2.8, 2.12, and 1.6(2).

3. THE MEET-PRIME ELEMENTS OF D Filt_σL CORRESPOND TO THE QUASI-MEET-PRIME ELEMENTS OF L FOR A DISTRIBUTIVE CONTINUOUS LATTICE L

In this section, L denotes a continuous lattice, D(L) its dual (0.6), and Filt_σL denotes the continuous lattice defined in 2.0. The hypothesis of distributivity for L can be deferred until the final step in the proof of Proposition 3.9.

3.1 LEMMA Suppose L is a continuous lattice. An element $F \in D(L)$ is meet prime in $D(L)$ iff

(1) $0 \notin F$.

(2) Whenever $x \vee y \in F$ for some $x,y \in L$, then $x \in F$ or $y \in F$.

Proof: (a) Suppose (1) and (2) are satisfied. Then $F \neq L$, by (1). Suppose $G \cap H \subseteq F$ for some $G,H \in D(L)$. Assume, on the contrary, $G \not\subseteq F$ and $H \not\subseteq F$, i.e., $x \in G$, $y \in H$, and $x,y \notin F$ for some $x,y \in L$. Then $x \vee y \in G \cap H \subseteq F$; hence, by (2), $x \in F$ or $y \in F$, a contradiction. Thus F is meet-prime in $D(L)$.

(b) Suppose F is meet-prime in $D(L)$. Then $F \neq L$; hence $0 \notin F$. Suppose $x \vee y \in F$ for some $x,y \in L$. Since

$$x = \sup \twoheaddownarrow x \qquad y = \sup \twoheaddownarrow y$$

in the continuous lattice L, we have

$$x \vee y = \sup D$$

for $D := \{s \vee t \mid s,t \in L,\ s \ll x,\ t \ll y\}$. Since D is nonempty and up-directed, and since F is Scott-open, there are $s \ll x$ and $t \ll y$ with $s \vee t \in F$. Since $s \ll x$ and $t \ll y$, there are Scott-open filters F_x and F_y with

$$x \in F_x \subseteq \twoheaduparrow s \qquad y \in F_y \subseteq \twoheaduparrow t$$

hence

$$F_x \cap F_y \subseteq \twoheaduparrow s \cap \twoheaduparrow t = \twoheaduparrow (s \vee t) \subseteq F$$

As a consequence,

$$F_x \subseteq F \qquad \text{or} \qquad F_y \subseteq F$$

since F is meet prime (by hypothesis); hence $x \in F$ or $y \in F$, as claimed. ■

3.2 Suppose $\underline{G} \in \underline{P}\,\mathrm{Filt}_\sigma L$, the power set of $\mathrm{Filt}_\sigma L$, for a continuous lattice L. Let

$$\Delta(\underline{G}) := \{x \in L \mid x \text{ is a lower bound of some } F \in \underline{G}\}$$

and

$$\chi(\underline{G}) := \sup \Delta(\underline{G})$$

It is easy to see that $\chi\colon \underline{P}\,\mathrm{Filt}_\sigma L \to L$ is an isotone map.

3.3 LEMMA For a continuous lattice L and $\underline{G} \in D\,\mathrm{Filt}_\sigma L$, $\Delta(\underline{G})$ is a nonempty and up-directed lower set of L.

Proof: We shall write Δ instead of $\Delta(\underline{G})$, for brevity. If $a,b \in \Delta$, then $F \subseteq \uparrow a$ and $G \subseteq \uparrow b$ for some $F,G \in \underline{G}$. Since $\underline{G}$ is a filter, there is some $H \in \underline{G}$ with $H \subseteq F$ and $H \subseteq G$; hence

$$H \subseteq \uparrow a \cap \uparrow b = \uparrow(a \vee b)$$

hence $a \vee b \in \Delta$. Consequently, Δ is a nonempty $(0 \in \Delta)$ and up-directed lower set of L. ∎

For a continuous lattice L and $x \in L$, let

$$H_x := \{F \in \mathrm{Filt}_\sigma L \mid x \in F\}$$

$$K_x := \{F \in \mathrm{Filt}_\sigma L \mid x \in \mathrm{int}\,F\}$$

$$= \{F \in \mathrm{Filt}_\sigma L \mid x' \ll x \text{ for some } x' \in F\}$$

(where int denotes the <u>interior operator</u> of the Scott topology). We also write $H(x)$ instead of H_x and $K(x)$ instead of K_x.

3.4 LEMMA For a continuous lattice L and $x \in L$,

$$K_x := \{F \in \mathrm{Filt}_\sigma L \mid x \in \mathrm{int}\,F\}$$

is an element of $D\,\mathrm{Filt}_\sigma L$.

Proof: (a) Clearly, K_x is an upper set of $\mathrm{Filt}_\sigma L$.

(b) If $\{F_i\}_{i\in I}$ is a nonempty up-directed family of members of $\mathrm{Filt}_\sigma L$ and

$$\sup\{F_i \mid i \in I\} = \bigcup\{F_i \mid i \in I\} \in K_x$$

then there is some $x' \in \bigcup\{F_i \mid i \in I\}$ with $x' \ll x$. Consequently, $x' \in F_i$ for some $i \in I$; hence

$$x \in \mathrm{int}\,F_i = \{y \in F_i \mid y' \ll y \text{ for some } y' \in F_i\}$$

It results that $F_i \in K_x$ for some $i \in I$.

By (a) and (b), K_x is a Scott-open subset of $\mathrm{Filt}_\sigma L$.

(c) In order to see that K_x is a filter in the lattice $\mathrm{Filt}_\sigma L$, note first that $L \in K_x$. If $F,G \in K_x$, then $\mathrm{int}\,F \cap \mathrm{int}\,G$ is a Scott-open neighborhood of x in L; hence there is a Scott-open filter H of L with

$$x \in H \subseteq \mathrm{int}\,F \cap \mathrm{int}\,G$$

since the Scott-open filters form a basis for the open sets of the Scott topology of L. Clearly, we have $H \in K_x$ and $H \subseteq F,G$. ∎

3.5 LEMMA For a distributive (!) continuous lattice L and $x \in L$,

$$H_x = \{F \in \mathrm{Filt}_\sigma L \mid x \in F\}$$

is an element of D $\mathrm{Filt}_\sigma L$.

Proof: Reasoning similar to the proof of Lemma 3.4 yields that H_x is a Scott-open subset of $\mathrm{Filt}_\sigma L$. Part (c) of this proof can be replaced by the observation that, for a distributive continuous lattice L, the meet of a finite number of members of $\mathrm{Filt}_\sigma L$ is, by Lemma 2.7, their set-theoretic intersection. ∎

3.6 REMARK For a meet-prime element p of a continuous lattice L we have $H_p = K_p$.

3.7 LEMMA For every element x of a continuous lattice L,

$$x = \chi(H_x) = \chi(K_x)$$

Proof: Since $K_x \subseteq H_x$, we have $\Delta(K_x) \subseteq \Delta(H_x)$; hence

$$\chi(K_x) = \sup \Delta(K_x) \leq \sup \Delta(H_x) = \chi(H_x)$$

Since $y \leq x$ for every $y \in \Delta(H_x)$, we can infer $\chi(H_x) \leq x$. Suppose $z \in L$ with $z \ll x$; then there is some Scott-open filter F in L with

$$x \in F \subseteq \Uparrow z$$

hence z is a lower bound of $F \in K_x$. Thus $z \leq \chi(K_x)$ for every $z \in L$ with $z \ll x$. As a consequence, $x \leq \chi(K_x)$ since $x = \sup\{z \in L \mid z \ll x\}$. ∎

3.8 LEMMA Suppose $\underline{G} \in$ D $\mathrm{Filt}_\sigma L$ for a continuous lattice L. Then

$$K_x \subseteq \underline{G} \subseteq H_x$$

for $x := \chi(\underline{G})$.

Proof: We write Δ instead of $\Delta(\underline{G})$.

(i) Suppose $x \in \mathrm{int}\, F$ for some $F \in \mathrm{Filt}_\sigma L$, i.e., $x' \ll x$ for some $x' \in F$. Since $x = \sup \Delta$, and Δ is a nonempty up-directed lower set of L (by Lemma 3.3), we infer that $x' \in \Delta$; hence there is some $G \in \underline{G}$ such that

$$G \subseteq \uparrow x' \subseteq F$$

Consequently, $F \in \underline{G}$, as claimed.

(ii) Suppose $H \in \underline{G}$. Since $\underline{G}$ is a Scott open subset of $\mathrm{Filt}_\sigma L$, $H' \ll H$ in $\mathrm{Filt}_\sigma L$ for some $H' \in \underline{G}$, i.e., by Lemma 2.2,

$$H' \subseteq \uparrow y \subseteq H$$

for some $y \in L$; hence $y \in \Delta$. Consequently, $y \leq x$; hence $x \in H$, as claimed. ∎

3.9 PROPOSITION Suppose L is a distributive continuous lattice. A member $\underline{P}$ of $D\,\mathrm{Filt}_\sigma L$ is meet prime in $D\,\mathrm{Filt}_\sigma L$ iff

$$\underline{P} = K_x = \{S \in \mathrm{Filt}_\sigma L \mid x \in \mathrm{int}_\sigma S\}$$

for a quasi-meet-prime element x of L. This element x is uniquely determined: $x = \chi(\underline{P})$.

Proof: (a) Suppose $x \in L$ is quasi-meet-prime. Let

$$K_x := \{S \in \mathrm{Filt}_\sigma L \mid x' \in S \text{ for some } x' \ll x\}$$

We verify the conditions (i) and (ii) of Lemma 3.1 for K_x: Suppose $F \vee G \in K_x$. Then there is some $x' \in F \vee G$ with $x' \ll x$; hence

$$\inf\{f_1, \ldots, f_n, g_1, \ldots, g_m\} \leq x'$$

for some $f_1, \ldots, f_n \in \mathrm{int}\, F$ and some $g_1, \ldots, g_m \in \mathrm{int}\, G$ with natural numbers $n, m \geq 0$.

Since x is quasi-meet-prime in L, there is some natural number i with $1 \leq i \leq n$ or $1 \leq i \leq m$ such that

$$f_i \leq x \qquad \text{or} \qquad g_i \leq x$$

Hence $x \in \mathrm{int}\, F$ or $x \in \mathrm{int}\, G$; hence $F \in K_x$ or $G \in K_x$.

It remains to show that $\{1\} = \varphi(\emptyset)$ is not an element of K_x. Suppose, on the contrary, that $\{1\} \in K_x$; hence $x \in \{1\}$ and there is some $x' \ll x$ with $x' \in \{1\}$, i.e., $1 \ll 1$. However, 1 fails to be quasi-meet-prime if it is compact [by Remark 1.5(3)].

(b) Now suppose that $\underline{P}$ is a meet-prime element of $D\,\mathrm{Filt}_\sigma L$, and let $x := \chi(\underline{P})$.

(1) Since we already know from Lemma 3.8 that $K_x \subseteq \underline{P}$, suppose that $F \in \underline{P}$. Since $\underline{P}$ is a Scott-open subset of $\mathrm{Filt}_\sigma L$, there is some $F' \in \underline{P}$ with $F' \ll F$ in $\mathrm{Filt}_\sigma L$, i.e., (by Lemma 2.2),

$$F' \subseteq \uparrow y \subseteq F$$

for some $y \in L$. Consequently, $y \in \Delta(\underline{P})$; hence $y \le x$. Since $y \in F$, there are $y_i \in \text{int } F$ $(i = 1, \ldots, n)$ with $n \ge 0$ such that

$$\inf\{y_1, \ldots, y_n\} \le y$$

Since $y_i \in \text{int } F$, there are $y_i', y_i'' \in F$ with $y_i'' \ll y_i' \ll y_i$ $(i = 1, \ldots, n)$. (The existence of the y_i' is guaranteed by the interpolation property of $\ll$.) Thus there are Scott-open filters G_i in L such that

$$y_i \in G_i \subseteq \Uparrow y_i' \subseteq F$$

Since $\inf\{y_1, \ldots, y_n\} \le y$, we conclude that

$$y \in G_1 \vee \cdots \vee G_n$$

where $\vee$ denotes the join (= supremum) in $\text{Filt}_\sigma L$.

Since $F' \subseteq \uparrow y$ and $F' \in \underline{P}$, we can infer that $F' \subseteq \uparrow y \subseteq G_1 \vee \cdots \vee G_n$; hence

$$G_1 \vee \cdots \vee G_n \in \underline{P}$$

Since $\underline{P}$ is meet prime in D $\text{Filt}_\sigma L$, we conclude, by Lemma 3.1, that $G_i \in \underline{P}$ for some $i \in \{1, \ldots, n\}$. Consequently, $y_i' \le x$, since y_i' is a lower bound of $G_i \in \underline{P}$. It results that $y_i'' \ll x$. Since $y_i'' \in F$, we infer that $x \in \text{int } F$, as claimed.

In all, this says that $K_x = \underline{P}$.

(2) We infer from $\underline{P} = K_y$ that

$$y = \chi(K_y) = \chi(\underline{P}) = \chi(K_x) = x$$

hence x is uniquely determined.

(3) To show that $x = \chi(\underline{P})$ is quasi-meet-prime, suppose that

$$\inf\{y_1, \ldots, y_n\} \ll x$$

for some $y_1, \ldots, y_n \in L$ and a natural number $n \ge 0$. Then

$$H(y_1) \cap \cdots \cap H(y_n) = H(\inf\{y_1, \ldots, y_n\}) \subseteq K_x \subseteq \underline{P}$$

since every $F \in \text{Filt}_\sigma L$ containing $\inf\{y_1, \ldots, y_n\}$ contains x as an inner point.

Since L is, by hypothesis, a <u>distributive</u> lattice,

$H(z) \in$ D $\text{Filt}_\sigma L$

for every $z \in L$, by Lemma 3.5. Consequently (in view of Lemma 2.7), $H(y_i) \subseteq \underline{P}$ for some $i \in \{1, \ldots, n\}$, since $\underline{P}$ is meet prime in $D\ \mathrm{Filt}_\sigma L$. Since χ is an isotone map $\underline{P}\ \mathrm{Filt}_\sigma L \to L$, we infer

$$y_i = \chi(H(y_i)) \leq \chi(\underline{P}) = x$$

as we want. ∎

4. THE TRACE OF THE Γ^*-TOPOLOGY ON ψ^*L

In Hoffmann [1979b, sec. 3] I have given another representation $\gamma_X: X \hookrightarrow \gamma X$ of the essential hull $\lambda_X: X \hookrightarrow \lambda X$ of a T_0-space X, discovered by Banaschewski [1977]. The elements of γX are the <u>convergence sets</u> of X, i.e., those (closed) subsets M of X such that either M = X or there exists an ordinary proper filter (or, equivalently, a net) on X which converges precisely to the points of M.

Also, in Hoffmann [1979b, 3.4(a)] the notion of a γ-element of a lattice L is defined, and it is shown there that the γ-elements of the (complete) lattice $\underline{A}(X)$ of all closed subsets (ordered by inclusion) of a T_0-space X are precisely the convergence sets of X (Hoffmann [1979b, 3.9; 1982a, 2.8]). Furthermore, the Γ-topology is introduced on L such that the set of γ-elements of L endowed with the trace of this topology constitutes a space γL which, for the lattice $L = \underline{A}(X)$ of closed subsets of a T_0-space X, coincides with γX.

In Hoffmann [1982a] an extensive study of the space γL is made, and it is observed there that every pseudo-join-prime element (Section 1.3) of a complete lattice L is a γ-element (Hoffmann [1982a,3.4(2)]).

Since in the present paper the basic notion is that of a distributive continuous lattice, i.e., a continuous lattice which is, by a celebrated theorem of Hofmann and Lawson [1978], representable as the lattice $\underline{O}(X)$ of open subsets (ordered by inclusion) of some T_0-space X, it is convenient here to adapt the definition of the Γ-topology and of a γ-element so as to apply to the lattice $\underline{O}(X)$ rather than $\underline{A}(X)$, i.e., in a sense, to dualize:

4.1 For a complete lattice L and $a,b \in L$ we write

$$a \vdash b$$

iff, whenever $\inf F \leq a$ for a finite subset F of L (where $F = \emptyset$ is not excluded) then $x \leq b$ for some $x \in F$.

This is the dual of the relation $\dashv$ of Hoffmann [1979b, sec. 3; 1982a]. It "relativizes" the notion of a meet-prime element p of a complete lattice

L in the same way as the way below relation relativizes the notion of a compact element, viz., $p \in L$ is meet prime in L iff $p \vdash p$. Thus $\vdash$ may be read as relatively meet prime below.

4.2 It is observed in Hoffmann [1982a, 1.1] that $\inf F \vdash a$ for a finite subset F of a complete lattice L implies $x \vdash a$ for some $x \in F$. It results that the sets

$$\Gamma^*(a) := \{x \in L \mid a \vdash x\}$$

(with a ranging through L) form a basis for the closed sets of a topology on (the underlying set of) L, which will be referred to as the Γ^*-topology of L (i.e., the Γ-topology of L^{op} (cf. Hoffmann [1979b, 3.2; 1982a, 1.2]).

4.3 An element p of a complete lattice L is said to be a γ^*-element (i.e, γ-element of L^{op}) iff it enjoys one of the following conditions (1), (2), and (3), which are pairwise equivalent (Hoffmann [1982a, 1.5, 2.7]):

(1) $p = \sup\{x \in L \mid x \vdash p\}$.

(2) $p = \sup(L - F)$ for some filter (i.e., nonempty, down-directed lower set) F of L.

(3) $\uparrow p = \{y \in L \mid p \leq y\}$ is closed in the Γ^*-topology of L.

It results from (3) that the trace of the Γ^*-topology of L on the set of all γ^*-elements defines a topological T_0-space γ^*L whose associated specialization partial order is inverse to (the trace of) the order of L.

4.4 For a T_0-space X there is an embedding

$$\gamma_X^*: X \hookrightarrow \gamma^*X$$

into the space $\gamma^*X := \gamma^*(\underline{O}(X))$, given by

$$x \mapsto X - \mathrm{cl}\{x\}$$

Obviously, $\gamma_X^*: X \hookrightarrow \gamma^*X$ is (an equivalent representation of) the greatest essential extension $\gamma_X: X \to \gamma X$ of the T_0-space X.

4.5 DEFINITION (a) For a complete lattice L, let ψ^*L denote (both the set and) the space of all pseudo-meet-prime elements of L enodwed with the trace of the Γ^*-topology of L.

(b) For a T_0-space X, let $\psi^*X := \psi^*(\underline{O}(X))$, and let

$$\psi_X^*: X \hookrightarrow \psi^*X$$

denote the embedding $x \mapsto X - \mathrm{cl}\{x\}$ to be referred to as the "ψ^*-extension" of X.

4.6 It is immediate from 4.3(2) that every pseudo-meet-prime element of a complete lattice is a γ^*-element (cf. Hoffmann [1982a, 3.4(2)]), since the complement of a prime ideal is a filter; hence ψ^*X is a subspace of γ^*X. Thus we can infer (from Banaschewski [1977, lemma 2, p. 235]).

4.7 PROPOSITION For a T_0-space X, ψ^*_X: $X \hookrightarrow \psi^*X$ is an essential extension.

4.8 REMARK It is convenient also to topologize the set ψL of all join-prime elements of a complete lattice L with the trace of the Γ-topology of L.

For a T_0-space X, we obtain an essential extension, the ψ-extension of X,

$$\psi_X: X \hookrightarrow \psi X := \psi \underline{A}(X)$$

corestricting the extension γ_X: $X \hookrightarrow \gamma X$.

The points of ψX are, as observed in 1.11, precisely the points of the Fell compactification $\underline{H}(X)$ of X if X is a locally quasi-compact space. The elements of ψX are the complements in X of the members of ψ^*X. Obviously, every result on the ψ^*-extension has an analogue for the ψ-extension.

4.9 THEOREM For a distributive continuous lattice L, the mapping

$$K: \psi^*L \to \mathrm{Spec}^*D\ \mathrm{Filt}_\sigma L$$

with

$$K(x) := \{F \in \mathrm{Filt}_\sigma L \mid x \in \mathrm{int}_\sigma F\}$$

is a homeomorphism.

Proof: In Proposition 3.9, it is shown that, for a distributive continuous lattice L, K(x) is a member of $\mathrm{Spec}^*D\ \mathrm{Filt}_\sigma L$, i.e., a meet-prime element of $D\ \mathrm{Filt}_\sigma L$, if and only if x is a quasi-meet-prime element of L. Moreover, this element $x \in L$ is uniquely determined. By 1.6(3), an element $x \in L$ is quasi-meet-prime iff x is pseudo-meet-prime. Thus

$$K: \psi^*L \to \mathrm{Spec}^*D\ \mathrm{Filt}_\sigma L$$

is a bijection.

It remains to show that this mapping K is continuous and that every (basic) closed subset of ψ^*L is the inverse image of a closed subset of $\mathrm{Spec}^*D\ \mathrm{Filt}_\sigma L$.

(a) We first show that the inverse image of a closed set of $\mathrm{Spec}^*D\ \mathrm{Filt}_\sigma L$ under the mapping

$$K: \psi^*L \to \mathrm{Spec}^*D\ \mathrm{Filt}_\sigma L$$

is closed in ψ^*L, i.e., the trace on ψ^*L of a Γ^*-closed subset of L. Since a closed subset of $\mathrm{Spec}^*\mathrm{D}\ \mathrm{Filt}_\sigma L$ can be (uniquely) represented in the form

$$\{\underline{G} \in \mathrm{Spec}^*\mathrm{D}\ \mathrm{Filt}_\sigma L \mid \underline{F} \subseteq \underline{G}\}$$

for some $\underline{F} \in \mathrm{D}\ \mathrm{Filt}_\sigma L$, it suffices to establish the following:

$$\{x \in \psi^*L \mid \underline{F} \subseteq K(x)\} = \bigcap\{\Gamma^*(y) \mid y \in \Delta_{\underline{F}}\} \cap \Psi^*L$$

where

$$\Delta_{\underline{F}} = \{y \in L \mid \text{there is some } G \in \underline{F} \text{ with } G \subseteq \uparrow y\}$$

i.e., $\Delta_{\underline{F}}$ consists of those $y \in L$ which are the lower bounds of some member of $\underline{F}$.

Suppose first that $\underline{F} \subseteq K(x)$ for some $\underline{F} \in \mathrm{D}\ \mathrm{Filt}_\sigma L$ and some $x \in \psi^*L$. Let $y \in L$ be a lower bound of some member G of $\underline{F}$. Assume that

$$\inf\{u_1, \ldots, u_n\} \leq y$$

for some $u_1, \ldots, u_n \in L$ and some $n \in \mathbb{N}$, $n \geq 0$.

Since $\underline{F} \subseteq K(x)$, x is an inner point of G with regard to the Scott topology of L; hence $z \ll x$ for some $z \in G$. It results that $\inf\{u_1, \ldots, u_n\} \leq y \leq z \ll x$. Hence $u_k \leq x$ for some $k \in \{1, \ldots, n\}$, since x is quasi-meet-prime in L. In all, this says that $y \vdash x$ for every $y \in \Delta_{\underline{F}}$, whence

$$\{x \in \psi^*L \mid \underline{F} \subseteq K(x)\} \subseteq \bigcap\{\Gamma^*(y) \mid y \in \Delta_{\underline{F}}\} \cap \Psi^*L$$

In order to prove the inverse inclusion, suppose $\underline{F} \in \mathrm{D}\ \mathrm{Filt}_\sigma L$ and $y \vdash x$ for every $y \in \Delta_{\underline{F}}$. Assume that $G \in \underline{F}$. We have to show that x is an inner point of G with regard to the Scott topology of L in order to prove that $\underline{F} \subseteq K(x)$. Since $\underline{F}$ is a Scott-open subset of $\mathrm{Filt}_\sigma L$, there is some $F \in \underline{F}$ with $F \ll G$, where $\ll$ denotes the way below relation in $\mathrm{Filt}_\sigma L$; hence

$$F \subseteq \uparrow z \subseteq G$$

for some $z \in L$, by Lemma 2.2. We infer from $F \subseteq \uparrow z$ that $z \in \Delta_{\underline{F}}$; hence $z \vdash x$ by hypothesis. Since $z \in G$, there are $u_1, \ldots, u_n \in \mathrm{int}_\sigma G$ ($n \in \mathbb{N} \cup \{0\}$) with

$$\inf\{u_1, \ldots, u_n\} \leq z$$

by the very definition of $\mathrm{Filt}_\sigma L$. Since $z \vdash x$, we can infer that $u_i \leq x$ for some $i \in \{1, \ldots, n\}$; hence $x \in \mathrm{int}_\sigma G$, as claimed.

(b) It remains to show that every basic closed subset of ψ^*L, i.e., the trace on ψ^*L of a basic Γ^*-closed set of L, is an inverse image of some closed subset of $\text{Spec}^*D\ \text{Filt}_\sigma L$ under $K: \psi^*L \to \text{Spec}^*D\ \text{Filt}_\sigma L$.

A basic closed subset of the Γ^*-topology of L is of the form

$$\Gamma^*(x) = \{y \in L \mid x \vdash y\}$$

for some $x \in L$. Using the member

$$H(x) = \{G \in \text{Filt}_\sigma L \mid x \in G\}$$

of $D\ \text{Filt}_\sigma L$ (cf. Lemma 3.5), we shall establish that

$$\{y \in \psi^*L \mid H(x) \subseteq K(y)\} = \psi^*L \cap \Gamma^*(x)$$

First assume that $x \vdash y$ for some $x \in L$ and some $y \in \psi^*L$. Suppose that $G \in H(x)$. We want to show that $G \in K(y)$, i.e., y is an inner point of G with regard to the Scott topology of L. Since $x \in G$, there are $u_1, \ldots, u_n \in \text{int}_\sigma G$ ($n \in \mathbb{N} \cup \{0\}$) with

$$x \geq \inf\{u_1, \ldots, u_n\}$$

by the very definition of $\text{Filt}_\sigma L$ (cf. 2.0). Since $x \vdash y$, we can infer that $u_i \leq y$ for some $i \in \{1, \ldots, n\}$; hence $y \in \text{int}_\sigma G$. This proves

$$\psi^*L \cap \Gamma^*(x) \subseteq \{y \in \psi^*L \mid H(x) \subseteq K(y)\}$$

To prove the inverse inclusion, we assume that $H(x) \subseteq K(y)$ for some $x \in L$ and some $y \in \psi^*L$. Let us assume, to the contrary, that

$$x \text{ not } \vdash y$$

Then there are $u_1, \ldots, u_n \in L$ with $n \in \mathbb{N} \cup \{0\}$ such that

$$\inf\{u_1, \ldots, u_n\} \leq x$$

but $u_k \not\leq y$ for every $k \in \{1, \ldots, n\}$. Since L is a continuous lattice, we have

$$u = \sup\{v \in L \mid v \ll u\}$$

for every $u \in L$; hence there are $v_1, \ldots, v_n \in L$ with

$$v_k \ll u_k \quad \text{and} \quad v_k \not\leq y$$

for every $k \in \{1, \ldots, n\}$. We consider the filter G generated by the Scott-open set

$$\Uparrow v_1 \cup \cdots \cup \Uparrow v_n$$

i.e., $G = \varphi(\Uparrow v_1 \cup \cdots \cup \Uparrow v_n)$. Since

$$\inf\{u_1, \ldots, u_n\} \le x$$

we have $x \in G$; hence, by hypothesis, $y \in \mathrm{int}_\sigma G$. Thus there is some $z \in G$ with $z \ll y$. As a consequence, there are $z_1, \ldots, z_m \in L$ with

$$z \ge \inf\{z_1, \ldots, z_m\}$$

such that for every $j \in \{1, \ldots, m\}$ there is some $k(j) \in \{1, \ldots, n\}$ with $v_{k(j)} \ll z_j$, by the very definition of G.

Since y is quasi-meet-prime, we can infer that $z_j \le y$ for some $j \in \{1, \ldots, m\}$; hence $v_{k(j)} \ll z_j \le y$, contradicting the assumption that $v_k \not\ll y$ for every $k = 1, \ldots, n$. Thus we have $x \vdash y$, as we want. ■

Since for a distributive continuous lattice L, D $\mathrm{Filt}_\sigma L$ is a distributive continuous lattice (by 2.12), we can infer from the result of Hofmann and Lawson [1978] that

$$\underline{O}\ \mathrm{Spec}^* D\ \mathrm{Filt}_\sigma L \cong D\ \mathrm{Filt}_\sigma L$$

Thus we have

4.10 COROLLARY For a distributive continuous lattice L,

$$\underline{O}(\psi^* L) \cong D\ \mathrm{Filt}_\sigma L \cong D\mathbb{D}(L)$$

Here, as before, $P \hookrightarrow I(P)$ denotes a representation of the injective hull of a continuous poset P.

4.11 PROPOSITION For a T_0-space X whose lattice $\underline{O}(X)$ of open subsets is a continuous lattice, $\psi^* X$ is a strongly sober, locally quasi-compact space.

Proof: (1) Since $\psi^* \underline{O}(X)$ is homeomorphic to $\mathrm{Spec}^* D\ \mathrm{Filt}_\sigma \underline{O}(X)$ (by Theorem 4.9), and since $\mathrm{Spec}^* K$ is sober for every complete lattice K (cf. 1.1), $\psi^* \underline{O}(X)$ is sober.

(2) By Corollary 4.10,

$$\underline{O}(\psi^* X) \cong D\ \mathrm{Filt}_\sigma \underline{O}(X)$$

Hence $\underline{O}(\psi^* X)$ is a continuous lattice in which the greatest element is compact and the way below relation is multiplicative. Now, 1.13 "(1) iff (3)" applies. ■

4.12 COROLLARY For a T_0-space X whose lattice $\underline{O}(X)$ of open subsets is a continuous lattice, the canonical embedding

$$\psi_{\psi X}: \psi X \hookrightarrow \psi(\psi X)$$

is a homeomorphism.

Proof: By 1.13(5), $\psi_Y: Y \hookrightarrow \psi Y$ is bijective for every strongly sober locally quasi-compact space. By Proposition 4.11 this applies to $Y := \psi X$. ∎

The above result says that the ψ-extension is "idempotent (up to an isomorphism)" for spaces X with $\underline{O}(X)$ continuous.

4.13 REMARK From the proof of Proposition 4.11 we extract that, in $\underline{O}(\psi^*L)$ (for a distributive continuous lattice L), the pseudo-meet-prime elements are exactly the meet-prime elements:

$$\psi^*\underline{O}(\psi^*L) = \mathrm{Spec}^*\underline{O}(\psi^*L)$$

4.14 REMARK We shall need in the following section some more information about the relation between the homeomorphisms

$$K: \psi^*L \to \mathrm{Spec}^*D\ \mathrm{Filt}_\sigma L$$

and

$$\psi^*_{\psi^*L}: \psi^*X \to \psi^*(\psi^*X) = \psi^*(\underline{O}\psi^*L)$$

for a space X with $L := \underline{O}(X)$ continuous.

Clearly, there is an induced homeomorphism $K': \psi^*(\underline{O}(\psi^*L)) \to \mathrm{Spec}^*D\ \mathrm{Filt}_\sigma L$ such that

$$\begin{array}{ccc} \psi^*L & \xrightarrow{\psi^*_{\psi^*L}} & \psi^*(\underline{O}\psi^*L) \\ & \searrow^{K} & \downarrow{K'} \\ & & \mathrm{Spec}^*D\mathrm{Filt}_\sigma(L) \end{array}$$

commutes.

Since $\psi^*(\underline{O}\psi^*L) = \mathrm{Spec}^*(\underline{O}\psi^*L)$, K' is induced by an isomorphism

$$K'': \underline{O}\psi^*L \to D\ \mathrm{Filt}_\sigma L$$

(which assigns to $V \in \underline{O}\psi^*L$

$$\inf\{K'(p) \mid V \subseteq p \in \mathrm{Spec}^*(\underline{O}\psi^*L)\}$$

where the infimum is taken in D $\mathrm{Filt}_\sigma L)$, such that

$$\begin{array}{ccc} \psi^*(\underline{O}\psi^*L) & \hookrightarrow & \underline{O}\psi^*L \\ K' \big\downarrow & & \big\downarrow K'' \\ \mathrm{Spec}^*\mathrm{DFilt}_\sigma(L) & \hookrightarrow & \mathrm{DFilt}_\sigma(L) \end{array}$$

commutes, where the horizontal maps are the inclusions.

5. THE RELATIONSHIP BETWEEN THE FELL COMPACTIFICATION $\underline{H}(\psi X)$ OF ψX AND $\underline{H}(X)$ WHEN $\underline{O}(X)$ IS A CONTINUOUS LATTICE

For a space X whose lattice $\underline{O}(X)$ of open subsets is a continuous lattice, we have seen in Corollary 4.12 that

$$\psi_{\psi X}: \psi X \to \psi(\psi X)$$

is a bijection, hence a homeomorphism. Since, for a locally quasi-compact space X, ψX has the same points as the Fell compactification $\underline{H}(X)$ of X and $\psi(\psi X)$ has the same points as the Fell compactification of $\underline{H}(\psi X)$ of ψX, it is a natural question whether the given bijection

$$\underline{H}(X) \to \underline{H}(\psi X)$$

is a homeomorphism.

It is convenient for the proofs to use the representation ψ^*X of ψX by open subsets of X, i.e., we shall show that the bijection

$$\psi^*_{(\psi^*X)}: \psi^*X \hookrightarrow \psi^*(\psi^*X)$$

gives a homeomorphism

$$\underline{H}^*(X) \to \underline{H}^*(\psi^*X)$$

where $\underline{H}^*(Y)$ denotes the space of pseudo-meet-prime elements of $\underline{O}(Y)$ with the topology inherited from the Lawson topology of $\underline{O}(Y)$, under the proviso that $\underline{O}(Y)$ is a continuous lattice. [Note that $\underline{H}^*(Y)$ has the same points as ψ^*Y and that, by the remarks in the introduction to this paper, passing to complements relative to Y gives a homeomorphism $\underline{H}^*(Y) \to \underline{H}(Y)$.]

5.1 Recall that the inclusion

$$e: D(L) \hookrightarrow \mathrm{Filt}_\sigma L$$

for a continuous lattice L preserves 1,$\wedge$, and suprema of nonempty up-directed subsets and, as noted in Remark 2.13, e induces a map

$$D(e)\colon D\,\mathrm{Filt}_\sigma L \to DD(L)$$

which takes $\Phi \in D\,\mathrm{Filt}_\sigma L$ into $\Phi \cap D(L)$. This map D(e) also preserves 1,$\wedge$, and suprema of nonempty up-directed subsets; hence it is continuous with regard to the respective Scott topologies.

5.2 Lawson [1979, 9.7] establishes a necessary and sufficient criterion in order that, for continuous posets S and T, a map $f\colon S \to T$ with the property that an inverse image of a Scott-open filter of T is a Scott-open filter of S be right adjoint[†], viz., that the induced map $D(f)\colon D(T) \to D(S)$ preserves the way below relation $\ll$. (Note that a 1,$\wedge$-preserving Scott-continuous map $f\colon S \to T$ between continuous 1,$\wedge$-semilattices S and T automatically has the property that inverse images of Scott-open filters are Scott-open filters.)

5.3 We have already observed in Theorem 2.5 that, for a continuous lattice L, $e\colon D(L) \hookrightarrow \mathrm{Filt}_\sigma L$ preserves the way below relation. Thus we can infer by 5.2, from the commutativity of

$$\begin{array}{ccc}
D(L) & \xrightarrow{\quad e \quad} & \mathrm{Filt}_\sigma(L) \\
{\scriptstyle \varepsilon_{DL}}\downarrow & & \downarrow{\scriptstyle \varepsilon_{\mathrm{Filt}_\sigma(L)}} \\
D^3(L) & \xrightarrow{\quad D^2(e) \quad} & D^2\mathrm{Filt}_\sigma(L)
\end{array}$$

(where the vertical arrows are isomorphisms - cf. 0.6), that

$$D(e)\colon D\,\mathrm{Filt}_\sigma L \to DD(L)$$

is right adjoint; hence so is the composite

$$g := \mu_L \circ D(e)\colon D\,\mathrm{Filt}_\sigma L \to DD(L) \to L$$

where $\mu_L\colon DD(L) \to L$ denotes the isomorphism inverse to $\varepsilon_L\colon L \to DD(L)$.

Since a right adjoint preserves arbitrary infima, it results that $g\colon D\,\mathrm{Filt}_\sigma L \to L$ is continuous with regard to the respective Lawson topologies (the Compendium [III-1.8]).

[†]The terminology of Lawson [1979] is in conflict with the one used above which is generally accepted among categorists: An isotone map $f\colon S \to T$ is right adjoint to $g\colon T \to S$ iff $g(y) \le x$ is equivalent to $y \le f(x)$ for $x \in S$, $y \in T$ (cf. MacLane [1971, I.2, p. 11]).

5.4 LEMMA For a distributive continuous lattice L, the mapping $g: D\ \mathrm{Filt}_\sigma L \to L$ defines, by restriction and corestriction, a bijection

$$d: \mathrm{Spec}^* D\ \mathrm{Filt}_\sigma L \to \psi^* L$$

inverse to K(•) (defined in Lemma 3.4).

Proof: Every meet-prime element F of $D\ \mathrm{Filt}_\sigma L$ is of the form

$$K(x) := \{F \in \mathrm{Filt}_\sigma L \mid x \in \mathrm{int}\ F\}$$

for a unique quasi-meet-prime element x of L, by Proposition 3.9. Evidently,

$$D(e)(K(x)) = \{F \in DL \mid x \in F\} = \epsilon_L(x)$$

where $\epsilon_L: L \to DD(L)$ denotes the canonical isomorphism (cf. 0.6). Thus

$$d(K(x)) = g(K(x)) = (\mu_L \circ D(e))(K(x)) = \mu_L \epsilon_L(x) = x$$

Since, by Proposition 3.9, K(•) is a bijective map

$$\psi^* L \to \mathrm{Spec}^* D\ \mathrm{Filt}_\sigma L$$

we can infer that d is inverse to K. ■

5.5 PROPOSITION For a distributive continuous lattice L, the mapping $K: \psi^* L \to \mathrm{Spec}^* D\ \mathrm{Filt}_\sigma L$ with

$$K(x) = \{F \in \mathrm{Filt}_\sigma L \mid x \in \mathrm{int}_\sigma F\}$$

is a homeomorphism with regard to the topologies inherited from the Lawson topologies of L and $D\ \mathrm{Filt}_\sigma L$, respectively.

Proof: We observe first that

$$d: \mathrm{Spec}^* D\ \mathrm{Filt}_\sigma L \to \psi^* L$$

is continuous with regard to the traces of the Lawson topologies of $D\ \mathrm{Filt}_\sigma L$ and L, respectively (since it restricts and corestricts the Lawson-continuous map $g: D\ \mathrm{Filt}_\sigma L \to L$). By 1.9, both the domain and the codomain of this mapping are compact Hausdorff spaces. (Recall from Remark 2.15 that $\mathrm{Spec}^* D\ \mathrm{Filt}_\sigma L = \psi^* D\ \mathrm{Filt}_\sigma L$.) Thus the inverse K of d is also continuous and, in fact, a homeomorphism. ■

5.6 (i) It has been observed in Remark 4.14 that, for a distributive continuous lattice L, the bijection K′ making

$$\begin{array}{ccc} \psi^*L & \xrightarrow{\psi^*_{\psi^*L}} & \psi^*\underline{O}(\psi^*L) \\ & {\scriptstyle K}\searrow & \downarrow {\scriptstyle K'} \\ & & \mathrm{Spec}^*\mathrm{DFilt}_\sigma(L) \end{array}$$

commutative is induced, by restriction and corestriction, from an isomorphism $K'': \underline{O}(\psi^*L) \to \mathrm{D\ Filt}_\sigma L$. This isomorphism K'' is, of course, a homeomorphism for the respective Lawson topologies; hence K' is a homeomorphism with regard to the traces of these topologies.

Combining this observation with Proposition 5.5 we obtain that, for a distributive continuous lattice L,

$$\psi^*_{\psi^*L}: \psi^*L \to \psi^*\underline{O}(\psi^*L)$$

is a homeomorphism with regard to the traces of the Lawson topologies of L and $\underline{O}(\psi^*L)$, respectively, i.e. (by the remarks in 5.0), a homeomorphism

$$\underline{H}^*(X) \to \underline{H}^*(\psi^*L)$$

provided that X is a space with $L = \underline{O}(X)$.

(ii) Since, by Corollary 4.12, the mapping

$$\psi_{\psi X}: \psi X \to \psi(\psi X)$$

for a space X with $\underline{O}(X)$ a continuous lattice is a homeomorphism (with regard to the genuine topologies of these spaces), it is an order isomorphism with regard to the respective specialization orders. These are the partial orders induced from the lattices $\underline{A}(X)$ and $\underline{A}(\psi X)$, respectively, i.e., the (restricted) inclusion relations.

In all, this gives

5.7 THEOREM For a space X whose lattice $\underline{O}(X)$ of open subsets (ordered by inclusion) is a continuous lattice, the canonical embedding

$$\psi X \hookrightarrow \psi(\psi X)$$

"is" (i.e., determines) a homeomorphism and an order isomorphism

$$\underline{H}(X) \to \underline{H}(\psi X)$$

i.e., an isomorphism in the category of compact ordered spaces.

5.8 REMARK The mapping $K(\cdot): \psi^*L \to \mathrm{Spec}^*\mathrm{D\ Filt}_\sigma L$ extends (with the same definition as given in Lemma 3.4) to a map

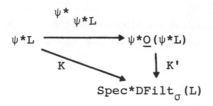

$L \to D\ \mathrm{Filt}_\sigma L$

which is easily shown to be Scott continuous. I do not know whether it is Lawson continuous. (This would give an alternative proof of Theorem 5.7.)

6. THE RELATIONSHIP BETWEEN $\underline{H}(X)$ AND ψX FOR A SPACE X WHOSE LATTICE $\underline{O}(X)$ OF OPEN SUBSETS IS CONTINUOUS. FUNCTORIALITY OF $\underline{H}(\cdot)$ AND $\psi(\cdot)$

For a space X whose lattice $\underline{O}(X)$ of open subsets is a continuous lattice we want to show that the open sets of ψX are precisely the open upper sets of $\underline{H}(X)$.

Owing to the homeomorphisms

$\psi X \to \psi\psi X \qquad \underline{H}(X) \to \underline{H}(\psi X)$

established in Corollary 4.12 and Theorem 5.7, respectively, this question can be reduced to the study of those spaces X for which $\underline{O}(X)$ is a continuous lattice and the canonical mapping $X \to \psi X$ is a homeomorphism, i.e., by 1.13, the strongly sober, locally quasi-compact spaces X.

Thus the result, we claim, is a consequence of an isomorphism between the category whose objects are the strongly sober, locally quasi-compact spaces and whose morphisms are the continuous perfect maps (i.e., those continuous maps that enjoy the property that the inverse image of every saturated quasi-compact subset of the codomain is quasi-compact) and the category of compact ordered spaces and continuous isotone maps.

This isomorphism is implicit in the construction of an isomorphism between the category of compact ordered spaces and isotone continuous maps and the category of distributive continuous lattices with 1 compact and $\ll$ multiplicative and those mappings preserving $\ll$, finite infima, and arbitrary suprema, described in the Compendium [VII-3] (cf. in particular, [VII-3.7(iii)]). (The result is, on the object level, due to Gierz and Keimel [1977].)

We reformulate the key results of the Compendium [VII-3] in order to make visible the ingredients of the desired isomorphism.

For the subset $|\psi^*L|$ of a (distributive) continuous lattice L, let λ' denote the trace of the Lawson topology λ_L of L.

6.1 LEMMA Let L be a distributive continuous lattice in which Spec^*L is closed with respect to the Lawson topology λ_L of L. A subset U of Soec^*L is an open lower set of (ψ^*L, λ') iff

$U = \mathrm{Spec}^*L - \uparrow a$

for some $a \in L$, i.e., iff U is open in the space Spec^*L.

Proof: If Spec^*L is closed in (L,λ_L), then so is $\{1\} \cup \mathrm{Spec}^*L$. Now VII-3.1 of the Compendium applies. ■

6.2 PROPOSITION Let X be a compact p(artially) o(rdered) space. The system $\Omega(X)$ of all open lower sets of X is a topology on $|X|$ that is a (distributive) continuous lattice with the property that $\mathrm{Spec}^*\Omega(X)$ is closed with regard to the Lawson topology on $\Omega(X)$. The specialization order of $(|X|,\Omega(X))$ is the inverse of the order of the pospace X. The mapping

$x \mapsto X - \uparrow x \quad x \in X$

is a bijection $|X| \to |\mathrm{Spec}^*\Omega(X)|$, which is a homeomorphism

$X \to (\psi^*\Omega(X),\lambda')$

where λ' is the trace on $\mathrm{Spec}^*\Omega(X) = \psi^*\Omega(X)$ of the Lawson topology of $\Omega(X)$.

Proof: This is an obvious modification of the Compendium [VII-3.3] (cf. also [VII-3.7 of the Compendium). ■

6.3 The above results 6.1 and 6.2 establish (in view of 1.13) a one-to-one correspondence between compact partially ordered spaces and strongly sober, locally quasi-compact spaces.

It is not difficult to prove, along the lines of the Compendium [pp. 323-324], that a continuous isotone map $f: X \to Y$ between compact partially ordered spaces X and Y induces a continuous map

$f: (X,\Omega(X)) \to (Y,\Omega(Y))$

with the property that the induced map

$\Omega(f): \Omega(Y) \to \Omega(X)$

preserves the way below relation.

Also, the arguments given in the Compendium [p. 324] suffice to show that every continuous perfect map between strongly sober, locally quasi-compact spaces is induced by a (unique) continuous map between the associated compact ordered spaces.

In all, this gives

6.4 THEOREM There is an isomorphism J^* between the category of compact partially ordered spaces and continuous isotone maps and the category of locally quasi-compact, strongly sober spaces and continuous perfect mappings. The functor J^* assigns to a compact ordered space X the space with the same points as X whose open sets are the open lower sets of X, and leaves the morphisms unchanged.

6.5 REMARK Reversing the order defines an automorphism (of order 2) of the category of compact pospaces and isotone continuous maps. Thus, by composition with J^* (of Theorem 6.4), we obtain an isomorphism J from the category of compact pospaces and isotone continuous map to the category of locally quasi-compact, strongly sober spaces and continuous perfect maps. The functor J assigns to a compact ordered space X the space with the same points as X whose open sets are the open upper sets of X, and leaves the morphisms unchanged.

In view of Corollary 4.12 and Theorem 5.7, Lemma 6.1 gives

6.6 THEOREM Let X be a T_0-space with $\underline{O}(X)$ a continuous lattice: (a) The open sets of ψ^*X (in its genuine topology) are precisely the open lower sets of the compact ordered space $\underline{H}^*(X)$.

(b) The open sets of ψX (in its genuine topology) are precisely the open upper sets of the Fell compactification $\underline{H}(X)$.

6.7 For locally compact (noncompact) Hausdorff spaces X and Y, a map $u\colon X \to Y$ extends (uniquely) to a continuous map $u^+\colon X^+ \to Y^+$ of the Alexandrov one-point quasi-compactifications X^+ and Y^+ of X and Y, respectively, such that

$$\begin{array}{ccc} X & \xrightarrow{\;u\;} & Y \\ \downarrow & & \downarrow \\ X^+ & \xrightarrow[\;u^+\;]{} & Y^+ \end{array}$$

commutes and $u^+(\infty_X) = \infty_Y$ (where ∞ denotes the adjoined point) if and only if $u\colon X \to Y$ is continuous and perfect (i.e., the inverse image of a compact subset of Y is compact in X).

Note that u^+ is perfect iff it is continuous, and that a nonconstant map $v\colon \underline{H}(X) \to \underline{H}(Y)$ is isotone iff $v(\infty_X) = \infty_Y$. (Recall the definition of the partial order of X^+ given in the introduction.)

The following result partially extends these facts to spaces X and Y with $\underline{O}(X)$, $\underline{O}(Y)$ a continuous lattice.

Recall that for spaces X,Y with $\underline{O}(X)$, $\underline{O}(Y)$ a continuous lattice, a continuous map $u\colon X \to Y$ is called perfect iff the inverse image map $\underline{O}(u)\colon \underline{O}(Y) \to \underline{O}(X)$ preserves the way below relation (cf. Remark 1.16).

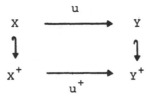

6.8 THEOREM For (T_0-)spaces X and Y for which $\underline{O}(X)$ and $\underline{O}(Y)$ are continuous lattices, a continuous perfect map u: X → Y uniquely extends to a continuous perfect map $\psi(u)$: $\psi X \to \psi Y$ such that

$$\begin{array}{ccc} X & \xrightarrow{\;u\;} & Y \\ {\scriptstyle \psi_X}\downarrow & & \downarrow{\scriptstyle \psi Y} \\ \psi X & \xrightarrow[\psi(u)]{} & \psi Y \end{array}$$

commutes.

Both $\psi(\cdot)$ and $\underline{H}(\cdot)$ extend to functors defined on the category of those (T_0-)spaces whose lattice of open sets is a (distributive) continuous lattice and the continuous perfect mappings. The codomain of ψ is the category of locally quasi-compact, strongly sober spaces and continuous perfect mappings, whereas the codomain of $\underline{H}$ is the category of compact ordered spaces and continuous isotone maps. The functors ψ and $\underline{H}$ are related by the isomorphism J of Remark 6.5:

$$J \circ \underline{H} = \psi$$

Both ψ and $\underline{H}$ are retractions.

Proof: Uniqueness of ψ(u): If there exists a continuous perfect map $\hat{u}$: $\psi X \to \psi Y$ rendering

$$\begin{array}{ccc} X & \xrightarrow{\;u\;} & Y \\ {\scriptstyle \psi_X}\downarrow & & \downarrow{\scriptstyle \psi_Y} \\ \psi X & \xrightarrow{\;\hat{u}\;} & \psi Y \end{array}$$

commutative, then, by Remark 6.5 and Theorem 6.6(b), $\hat{u}$: $\underline{H}(X) \to \underline{H}(Y)$ is a continuous isotone map. Since $\underline{H}(Y)$ is (compact) Hausdorff and $\psi_X[X]$ is dense in $\underline{H}(X)$, $\hat{u}$ is uniquely determined by u, i.e., there is at most one such morphism $\hat{u}$.

By a standard argument, it results from the uniqueness of $\hat{u}$ that ψ and $\underline{H}$ are functors provided that the induced morphism always exists.

When the functoriality of $\psi(\cdot)$ is established, we may infer from the idempotency of ψX for those spaces X with $\underline{O}(X)$ a continuous lattice (Corollary 4.12) that ψ is a retraction. Consequently, $\underline{H} = J^{-1} \cdot \psi$ [by Theorem 6.6(b)] is, under the proviso that it is functorial, also a retraction.

The proof of existence of $\psi(u)\colon \psi X \to \psi Y$ is more subtle. We reduce the problem to a lattice-theoretic question and transfer the solution back.

(1) First note that, since Sob is a full reflective subcategory of $\underline{T}_0$ (and Top, the category of all topological spaces and continuous maps), there is a unique map ${}^s u$ rendering

$$\begin{array}{ccc} X & \xrightarrow{\ u\ } & Y \\ {\scriptstyle \tilde{s}_X}\downarrow & & \downarrow{\scriptstyle \tilde{s}_Y} \\ {}^sX & \xrightarrow{\ {}^su\ } & {}^sY \end{array}$$

commutative. Furthermore, by the same argument, there is a splitting

$$\begin{array}{ccccc} & & {}^sX & & \\ & {\scriptstyle \tilde{s}_X}\nearrow & & \searrow{\scriptstyle \psi'_X} & \\ X & & \xrightarrow{\ \ \psi_X\ \ } & & \psi X \end{array}$$

(and an analogous splitting for $\psi_Y\colon Y \hookrightarrow \psi Y$), since ψX is sober (by Proposition 4.11).

Since $\underline{O}(\tilde{s}_X)\colon \underline{O}({}^sX) \to \underline{O}(X)$ (the inverse image map) is an isomorphism, it results [from the functoriality of $\underline{O}(\cdot)$] that $\psi'_X\colon {}^sX \hookrightarrow \psi X$ is equivalent to $\psi_{{}^sX}\colon {}^sX \to \psi({}^sX)$, since the definition of ψX depends only on the lattice $\underline{O}(X)$.

In all, this implies that we may assume without loss of generality that both X and Y are sober. Now the isomorphism $\underline{O}(\cdot)$ between the category of locally quasi-compact sober spaces and continuous maps and the category of distributive continuous lattices and those maps preserving $1, \wedge$, and arbitrary suprema (Hofmann and Lawson [1978]; the Compendium [V-5.16]) "reduces" the problem to the question of whether there is a map $\underline{O}(\psi Y) \to \underline{O}(\psi X)$ preserving $1, \wedge, \ll$, and arbitrary suprema rendering

$$\begin{array}{ccc} \underline{O}(\psi Y) & \dashrightarrow & \underline{O}(\psi X) \\ {\scriptstyle \underline{O}(\psi_Y)}\downarrow & & \downarrow{\scriptstyle \underline{O}(\psi_X)} \\ \underline{O}(Y) & \xrightarrow[\ \underline{O}(u)\ ]{} & \underline{O}(X) \end{array}$$

commutative.

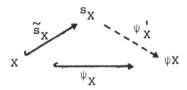

(2) It is convenient, in the following, to substitute

$$L := \underline{O}(Y) \qquad M := \underline{O}(X) \qquad f := \underline{O}(u)$$

i.e., L and M are distributive continuous lattices and f: L → M is a map preserving 1,∧, arbitrary suprema, and (since u is perfect) the way below relation ≪.

Since

$$\underline{O}(\psi Y) \cong \underline{O}(\psi^{*}L) \cong D\ \mathrm{Filt}_{\sigma}L$$

(the last isomorphism noted in Corollary 4.10) and, analogously,

$$\underline{O}(\psi X) \cong D\ \mathrm{Filt}_{\sigma}M$$

what we actually need is a map

$$h'\colon D\ \mathrm{Filt}_{\sigma}L \to D\ \mathrm{Filt}_{\sigma}M$$

preserving 1,∧, ≪, and arbitrary sumprema, and rendering a certain diagram [to be specified in (6) and (7) below] commutative. Such a map is the image under D(·) of a map

$$h\colon \mathrm{Filt}_{\sigma}M \to \mathrm{Filt}_{\sigma}L$$

preserving 1,∧, and suprema of nonempty, up-directed subsets, since h′ has these properties (cf. Lawson [1979, sec. 7]. By Lawson [1979, (9.7)], h′ = D(h) preserves ≪ iff h has a left adjoint, i.e. (since the domain of h is a complete lattice), iff h preserves arbitrary infima.

(3) For distributive continuous lattices L, M and a map f: L → M preserving ≪, 1,∧, and arbitrary suprema, we define a mapping

$$h\colon \mathrm{Filt}_{\sigma}M \to \mathrm{Filt}_{\sigma}L$$

by

$$h(G) := \varphi(f^{-1}[\mathrm{int}_{\sigma}G])$$

for every $G \in \mathrm{Filt}_{\sigma}M$, where $\mathrm{int}_{\sigma}G$ (or int G) denotes the interior of G with regard to the Scott topology of M and φ(·) assigns to a subset of L the smallest filter containing it.

Recall that, for a family $(G_i)_{i\in I}$ of members of $\mathrm{Filt}_{\sigma}M$, we have

$$\bigwedge\{G_i \mid i \in I\} = \varphi(\mathrm{int}_{\sigma}(\bigcap\{G_i \mid i \in I\}))$$

and, if $(G_i)_{i\in I}$ is a nonempty and up-directed family,

$$\bigvee\{G_i \mid i \in I\} = \bigcup\{G_i \mid i \in I\}$$

(3a) For a nonempty and up-directed family $(G_i)_{i\in I}$ of members of $\mathrm{Filt}_\sigma M$, we thus have

$$\begin{aligned} h(\bigvee\{G_i \mid i \in I\}) &= h(\bigcup\{G_i \mid i \in I\}) \\ &= \varphi(f^{-1}[\mathrm{int}_\sigma(\bigcup\{G_i \mid i \in I\})]) \\ &= \varphi(f^{-1}[\bigcup\{\mathrm{int}_\sigma G_i \mid i \in I\}]) \\ &= \varphi(\bigcup\{f^{-1}[\mathrm{int}_\sigma G_i] \mid i \in I\}) \\ &= \bigvee\{\varphi(f^{-1}[\mathrm{int}_\sigma G_i]) \mid i \in I\} \\ &= \bigvee\{h(G_i) \mid i \in I\} \end{aligned}$$

(the third equality, since M is a continuous lattice).

(3b) We shall prove that $h: \mathrm{Filt}_\sigma M \to \mathrm{Filt}_\sigma L$ preserves arbitrary infima.

Note first that for a family $(G_i)_{i\in I}$ of members of $\mathrm{Filt}_\sigma M$

$$\begin{aligned} h(\bigwedge\{G_i \mid i \in I\}) &= h(\varphi(\mathrm{int}(\bigcap\{G_i \mid i \in I\}))) \\ &= \varphi(f^{-1}[\mathrm{int}(\varphi(\mathrm{int}(\bigcap\{G_i \mid i \in I\})))]) \\ &= \varphi(f^{-1}[\mathrm{int}(\bigcap\{G_i \mid i \in I\})]) \end{aligned}$$

(the last equality, since $\mathrm{int}\,\varphi\,\mathrm{int}\,F = \mathrm{int}\,F$ for every filter F of M), and

$$\begin{aligned} \bigwedge\{h(G_i) \mid i \in I\} &= \bigwedge\{\varphi(f^{-1}[\mathrm{int}\,G_i]) \mid i \in I\} \\ &= \varphi(\mathrm{int}(\bigcap\{\varphi(f^{-1}[\mathrm{int}\,G_i]) \mid i \in I\})) \end{aligned}$$

The nontrivial implication is that the latter set is contained in the first. Suppose

$$x \in \varphi(\mathrm{int}(\bigcap\{\varphi(f^{-1}[\mathrm{int}\,G_i]) \mid i \in I\}))$$

Then

$$x \geq x_1 \wedge \cdots \wedge x_n$$

for some $x_1, \ldots, x_n \in L$ with $n \in \mathbb{N} \cup \{0\}$, such that there are $y_1, \ldots, y_n \in L$ with $y_k \ll x_k$ and

$$y_k \in \bigcap\{\varphi(f^{-1}[\text{int } G_i]) \mid i \in I\}$$

for every $k \in \{1, \ldots, n\}$. Since

$$\varphi(f^{-1}[\text{int } G_i]) \subseteq f^{-1}[\varphi(\text{int } G_i)] = f^{-1}[G_i]$$

we infer that

$$f(y_k),\ f(x_k) \in \bigcap\{G_i \mid i \in I\}$$

for every $k = 1, \ldots, n$. Since f preserves the way below relation, we have $f(y_k) \ll f(x_k)$. Hence

$$f(x_k) \in \text{int}(\bigcap\{G_i \mid i \in I\})$$

for every $k = 1, \ldots, n$. Consequently,

$$x_k \in f^{-1}[\text{int}(\bigcap\{G_i \mid i \in I\})]$$

It results that

$$x \in \varphi(f^{-1}[\text{int}(\bigcap\{G_i \mid i \in I\})]$$

since $x \geq x_1 \wedge \cdots \wedge x_n$.

(4) Since

$$h\colon \text{Filt}_\sigma M \to \text{Filt}_\sigma L$$

preserves nonempty up-directed suprema and finite infima, there is an induced map

$$Dh\colon D\,\text{Filt}_\sigma L \to D\,\text{Filt}_\sigma M$$

assigning to every Scott-open filter $\underline{F}$ of $\text{Filt}_\sigma L$ the set of those members of $\text{Filt}_\sigma M$ which are mapped by h into $\underline{F}$. Clearly, Dh preserves nonempty, up-directed suprema and finite infima. Since $h\colon \text{Filt}_\sigma M \to \text{Filt}_\sigma L$ preserves arbitrary infima, h has a left adjoint; hence, by Lawson's criterion [1979, 9.7], Dh preserves the way below relation.

We want to show that Dh preserves arbitrary suprema. We observe first that the smallest element of $D\,\text{Filt}_\sigma L$, viz., $\{L\}$ (since L is a compact element of $\text{Filt}_\sigma L$), is mapped by Dh into $\{M\}$. [If, for some $F \in \text{Filt}_\sigma M$, $L = h(F) = \varphi f^{-1}\text{int } F$, then $0 = x_1 \wedge \cdots \wedge x_n$ and $f(x_i) \in \text{int } F$ for some $x_i \in L$, $i = 1, \ldots, n$. Thus $0 = f(0) = f(x_1) \wedge \cdots \wedge f(x_n) \in \varphi \text{ int } F = F$; hence $F = M$.]

Since Dh preserves suprema of nonempty, up-directed subsets, it suffices now to consider <u>binary suprema</u> in D $\mathrm{Filt}_\sigma L$.

(4a) Let $\underline{F},\underline{G} \in$ D $\mathrm{Filt}_\sigma L$. We show first that

$$\underline{F} \vee \underline{G} = \{F \cap G \mid F \in \underline{F} \text{ and } G \in \underline{G}\}$$

is the supremum of $\underline{F}$ and $\underline{G}$ in D $\mathrm{Filt}_\sigma L$.

Since L is distributive, $\{F \cap G \mid F \in \underline{F} \text{ and } G \in \underline{G}\} \subseteq \mathrm{Filt}_\sigma L$ by Lemma 2.7. If $F \in \underline{F}$, $G \in \underline{F}$, and $V \in \mathrm{Filt}_\sigma L$ such that $F \cap G \subseteq V$, then

$$V = V \dot{\vee} (F \cap G) = (V \dot{\vee} F) \cap (V \dot{\vee} G)$$

where $\dot{\vee}$ denotes the binary supremum in $\mathrm{Filt}_\sigma L$), since $\mathrm{Filt}_\sigma L$ is a distributive lattice (by 2.10). Since $V \dot{\vee} F \in \underline{F}$ and $V \vee G \in \underline{G}$, this shows that

$$\{F \cap G \mid F \in \underline{F} \text{ and } G \in \underline{G}\}$$

is a filter of $\mathrm{Filt}_\sigma L$, since it is also stable under finite intersections.

It remains to show that this set is Scott open in $\mathrm{Filt}_\sigma L$. Since $\underline{F}$ and $\underline{G}$ are Scott open in $\mathrm{Filt}_\sigma L$, there are $F' \in \underline{F}$ and $G' \in \underline{G}$ with

$$F' \in F \quad \text{and} \quad G' \ll G$$

in $\mathrm{Filt}_\sigma L$, i.e. (by 2.2),

$$F' \subseteq {\uparrow} x \subseteq F \quad \text{and} \quad G' \subseteq {\uparrow} y \subseteq G$$

for some $x,y \in L$. It results that

$$F' \cap G' \subseteq {\uparrow} x \cap {\uparrow} y = {\uparrow}(x \vee y) \subseteq F \cap G$$

i.e., $F' \cap G' \ll F \cap G$ in $\mathrm{Filt}_\sigma L$. This shows that

$$\{F \cap G \mid F \in \underline{F} \text{ and } G \in \underline{G}\}$$

is Scott open in $\mathrm{Filt}_\sigma L$; hence it is a member of D $\mathrm{Filt}_\sigma L$. Clearly, it is the smallest member of D $\mathrm{Filt}_\sigma L$ containing both $\underline{F}$ and $\underline{G}$.

(4b) To show that Dh: D $\mathrm{Filt}_\sigma L \to$ D $\mathrm{Filt}_\sigma M$ preserves binary suprema, it suffices to verify the inclusion

$$Dh(\underline{F} \vee \underline{G}) \subseteq Dh(\underline{F}) \vee Dh(\underline{G})$$

Suppose $W \in Dh(\underline{F} \vee \underline{G})$, i.e., $h(W) \in \underline{F} \vee \underline{G}$. Then $h(W) = F \cap G$ for some $F \in \underline{F}$ and some $G \in \underline{G}$, by (4a). We shall consider

$$S = {\uparrow} f[F] = \{z \in M \mid f(x) \le z \text{ for some } x \in F\}$$
$$T = {\uparrow} f[G] = \{z \in M \mid f(x) \le z \text{ for some } x \in G\}$$

We observe first that $S, T \in \mathrm{Filt}_\sigma M$, since f preserves both $\wedge$ and $\ll$. Since

$$h(S) = \varphi \ \mathrm{int}\ f^{-1} \uparrow f[F] \supseteq \varphi \ \mathrm{int}\ F = F$$

we conclude that $h(S) \in \underline{F}$; hence $S \in \mathrm{Dh}(\underline{F})$. Analogously, we obtain $T \in \mathrm{Dh}(\underline{G})$.

Suppose now $p \in S \cap T$. Then we have

$$f(u) \leq p \quad \text{and} \quad f(v) \leq p$$

for some $u \in F$ and some $v \in G$. Thus we have

$$f(u \vee v) \leq p$$

since f preserves (finite) suprema.

Since

$$u \vee v \in F \cap G = h(W) = \varphi \ \mathrm{int}\ f^{-1}[W] \subseteq f^{-1}[W]$$

we can infer $f(u \vee v) \in W$; hence $p \in W$.

In all, this says that $S \cap T \subseteq W$; hence

$$W \in \mathrm{Dh}(\underline{F}) \vee \mathrm{Dh}(\underline{G})$$

(5) We show that, to the mapping $f: L \to M$ which, by hypothesis, preserves $\ll$, 1, $\wedge$, and $\bigvee$, there corresponds a <u>continuous perfect map</u>: Since f preserves $\bigvee$, and L and M are complete lattices, f has a right adjoint $M \to L$ taking $x \in M$ into

$$\sup f^{-1}[\downarrow x]$$

This right adjoint takes meet-prime elements of M into meet-prime elements of L. Thus it defines a map

$$\mathrm{Spec}^* f : \mathrm{Spec}^* M \to \mathrm{Spec}^* L$$

which is continuous with regard to the standard topologies (cf., e.g., the <u>Compendium</u> [IV-1.26]). Indeed,

$$\begin{array}{ccc} \underline{O}\mathrm{Spec}^*(M) & \xrightarrow{\underline{O}\mathrm{Spec}^* f} & \underline{O}\mathrm{Spec}^*(L) \\ \cong\downarrow & & \downarrow\cong \\ M & \xrightarrow{\quad f \quad} & L \end{array}$$

commutes; hence $\mathrm{Spec}^* f$ is a perfect map, since f preserves $\ll$.

Likewise, $\mathrm{Dh}: D\,\mathrm{Filt}_\sigma L \to D\,\mathrm{Filt}_\sigma M$ induces a continuous perfect map

$$\mathrm{Spec}^*\mathrm{Dh}: \ \mathrm{Spec}^* D\,\mathrm{Filt}_\sigma M \to \mathrm{Spec}^* D\,\mathrm{Filt}_\sigma L$$

which assigns to a meet-prime element $\underline{F}$ of $D\ \mathrm{Filt}_\sigma M$ the meet-prime element

$$\sup(Dh)^{-1}[\downarrow\underline{F}]$$

of $D\ \mathrm{Filt}_\sigma L$, where sup (the supremum) is taken in the complete lattice $D\ \mathrm{Filt}_\sigma L$.

(6) We want to show that

$$\begin{array}{ccc} \mathrm{Spec}^*\mathrm{DFilt}_\sigma(M) & \xrightarrow{\mathrm{Spec}^*Dh} & \mathrm{Spec}^*\mathrm{DFilt}_\sigma(L) \\ {\scriptstyle k_M}\uparrow & & \uparrow{\scriptstyle k_L} \\ \mathrm{Spec}^*(M) & \xrightarrow[\mathrm{Spec}^*f]{} & \mathrm{Spec}^*(L) \end{array}$$

<u>commutes</u>, where

$$k_M(x) = \{F \in \mathrm{Filt}_\sigma M \mid x \in \mathrm{int}\ F\}$$

for $x \in \mathrm{Spec}^*M$, and, analogously,

$$k_L(y) = \{G \in \mathrm{Filt}_\sigma L \mid y \in \mathrm{int}\ G\}$$

for $x \in \mathrm{Spec}^*L$. Thus we have to show that, for every $x \in \mathrm{Spec}^*M$,

$$\begin{aligned}(\mathrm{Spec}^*D(h))(k_M(x)) &= \sup(Dh)^{-1}[\downarrow k_M(x)] \\ &= \sup\{\underline{G} \in D\ \mathrm{Filt}_\sigma L \mid Dh(\underline{G}) \subseteq k_M(x)\}\end{aligned}$$

coincides with

$$k_L(\sup f^{-1}[\downarrow x]) = \{F \in \mathrm{Filt}_\sigma L \mid \sup f^{-1}[\downarrow x] \in \mathrm{int}\ F\}$$

Indeed, we shall establish that

$$k_L(\sup f^{-1}[\downarrow x])$$

is the greatest element of

$$\mathfrak{G} := \{\underline{G} \in D\ \mathrm{Filt}_\sigma L \mid Dh(\underline{G}) \subseteq k_M(x)\}$$

Suppose first that

$$F \in Dh(k_L(\sup f^{-1}[\downarrow x])$$

i.e.,

$$\sup f^{-1}[\downarrow x] \in \mathrm{int}\ h(F)$$

Since

$$\text{int } h(F) = \text{int } \varphi \text{ int } f^{-1}[F] = \text{int } f^{-1}[F]$$

this implies

$$x \geq f \sup f^{-1}[\downarrow x] \in f \text{ int } f^{-1}[F] \subseteq \text{int } F$$

since f preserves suprema and preserves $\ll$. Thus we have $x \in \text{int } F$, or, equivalently, $F \in k_M(x)$. In all, this proves that

$$k_L(\sup f^{-1}[\downarrow x]) \in \Theta$$

Now suppose that

$$\underline{G} \in \Theta \quad \text{and} \quad G \in \underline{G}$$

Let

$$S := \uparrow f[G] = \{z \in M \mid f(v) \leq z \text{ for some } v \in G\}$$

We observe first that $S \in \text{Filt}_\sigma M$. Evidently, we have

$$G \subseteq \varphi \text{ int } f^{-1}[S]$$

Hence

$$h(S) = \varphi \text{ int } f^{-1}[S] \in \underline{G}$$

hence $S \in Dh(\underline{G})$. Since $\underline{G} \in \Theta$ (by hypothesis), we infer that $x \in \text{int } S \subseteq S$. By the very definition of S, we may infer that $f(a) \leq x$ for some $a \in G$. Since G is an upper set and $a \in f^{-1}[\downarrow x]$, it results that

$$\sup f^{-1}[\downarrow x] \in G$$

This implies that

$$\sup f^{-1}[\downarrow x] \in \text{int } G$$

or, equivalently,

$$G \in k_L(\sup f^{-1}[\downarrow x])$$

since $\sup f^{-1}[\downarrow x]$ is meet prime. (A member G of $\text{Filt}_\sigma L$ containing a meet-prime element p of L must contain p in its interior; cf. Remark 3.6.)

In all, this proves

$$\underline{G} \subseteq k_L(\sup f^{-1}[\downarrow x])$$

whenever $\underline{G} \in \Theta$, as claimed.

(7) Since, by (6) (and Theorem 4.9)

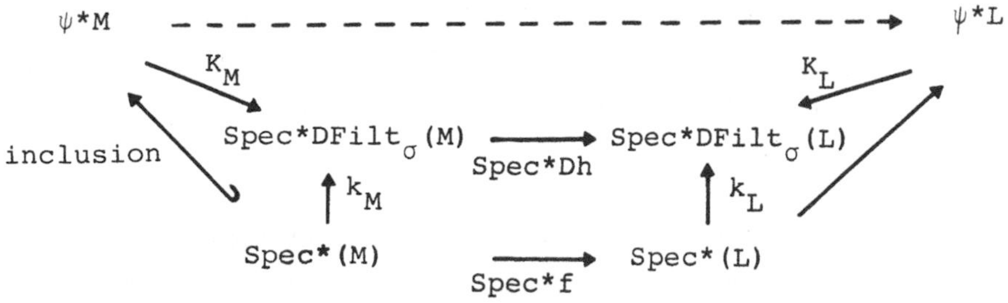

commutes, the <u>dotted arrow</u> $\psi^*(u)\colon \psi^*M \to \psi^*L$ resulting from the fact that K_M and K_L are homeomorphisms (by Theorem 4.9) is, in view of (1) and (2) above, the <u>desired morphism</u> $\psi^*(u)\colon \psi^*X \hookrightarrow \psi^*Y$.

This completes the proof. ∎

6.9 REMARK Possibly there is an alternative (shorter) proof for the existence in Theorem 6.8 relying on the explicit description of the Fell topology of $\underline{H}(X)$ and $\underline{H}(Y)$. When I tried to work it out, I ran into a difficulty which, possibly, cannot easily be circumvented. Also, I hope that the techniques employed in the given proof will turn out to be useful for further research.

6.10 REMARK It may be noted that the functor ψ is not a reflector. Indeed, the full subcategory $\underline{A}$ of locally quasi-compact, strongly sober spaces of the category $\underline{B}$ of locally quasi-compact sober spaces and continuous perfect maps is not reflective in $\underline{B}$:

The one-element space 1 is the terminal object of $\underline{A}$, but it is not preserved by the embedding $\underline{A} \hookrightarrow \underline{B}$, since there is no $\underline{B}$-morphism from a non-quasi-compact object of $\underline{B}$ to 1.

On the other hand, Fell [1962, p. 476] has observed a certain universal property of $\underline{H}(X)$. However, the hypothesis employed there seems to be closer to the conclusion of Theorem 6.8 rather than to its hypothesis.

7. CONCLUDING REMARKS

7.1 Suppose P is a continuous poset. Let

$$e\colon (P,\sigma_P) \hookrightarrow (L,\sigma_L)$$

denote any representation of the injective hull of (P,σ_P), where L is a continuous lattice, by virtue of (the correct half of) the analysis given in Hoffmann [1981a, 3.14]. The closure of e[P] in L with regard to the

Lawson topology (= CL-topology) λ_L of L, (without any topology, but) endowed with the partial order inherited from L, is denoted by C. By corestriction, we obtain an order extension $P \hookrightarrow C$, called the CL- compactification in Hoffmann [1982b].[†] It is shown there that

$$(C, \lambda_L|C) = \underline{H}(X)$$

the Fell compactification of the locally quasi-compact (sober) space $X := (P,\sigma_P)$, where, as noted in Hoffmann [1982b], = (instead of $\cong$) is correct if we choose $(L,\sigma_L) := \gamma(P,\sigma_P)$ (as we shall do here and in the following). We thus have

PROPOSITION For a continuous poset P, the CL-compactification of P endowed with the following topologies:

$$(P,\sigma_P) \hookrightarrow (C,\sigma_L|C)$$

is (equivalent to) the ψ-extension of (P,σ_P).

The Lawson topology λ_L of L induces on P the intrinsic Lawson topology λ_P. Thus the CL-compactification $P \hookrightarrow C$ of a continuous poset P is a bijection iff the Lawson topology λ_P of P is compact (Hausdorff). In view of 1.13 "(3) iff (5)", we can infer

COROLLARY The Scott topology σ_P of a continuous poset P is strongly sober iff the Lawson topology λ_P of P is compact (Hausdorff).

7.2 For a distributive continuous lattice L, we have seen in Theorem 2.12 that

$$D\ \mathrm{Filt}_\sigma L \cong DID(L)$$

is a complete lattice, i.e.,

$$DID(L) \cong IDID(L)$$

It is not unlikely that this is true for arbitrary continuous lattices L or, equivalently, that

$$DIDI(P) \cong IDIDI(P)$$

for arbitrary continuous posets P.[‡]

If this were true, then the number of nonisomorphic continuous posets which can be built up form a given continuous poset P by applying D(•) and

---

[†] Recently, Hofmann and Mislove [1982] have shown by example that C need not be a continuous poset. (See also footnote on p.60.)

[‡] Very recently, I have disproved this by means of a counterexample.

and I(·) would be finite. Indeed, there seems to be some evidence from examples that for every continuous poset P the following sharper formula is valid:

$$DIDI(P) \cong IDID(P)$$

The following example shows at least that this formula would be optimal. Let

$$P := \{a,b\} \cup (0,1]$$

where (0,1], the real numbers x with $0 < x \le 1$, receives the natural order from the order $\le$ of P and

$$a < x \quad \text{and} \quad b < x$$

for $x \in (0,1]$ are the only occurrences of $<$ involving a or b.

We then have the following figures (where $<$ is realized as "strictly below", o indicates a missing point, whereas ● designates an existing point):

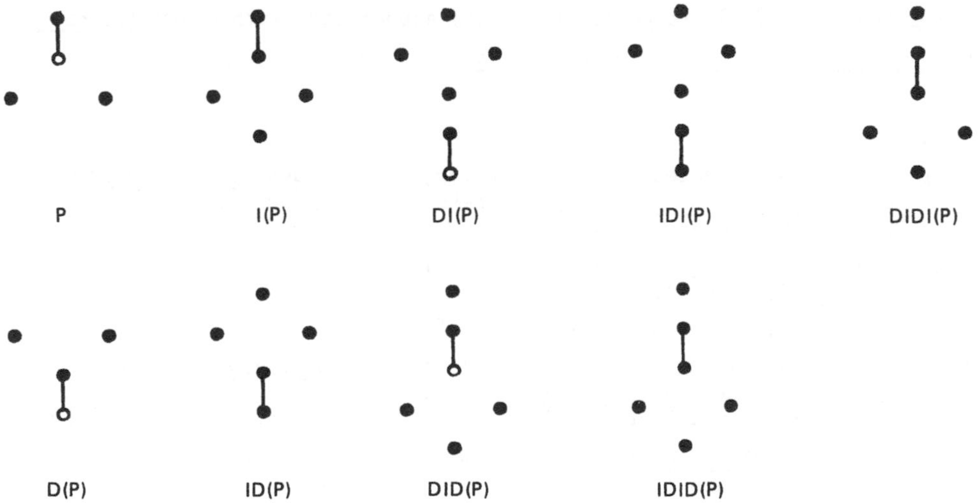

The construction of I(Q) relies upon the observation in Hoffmann [1982c, sec. 1] that the convergence sets of a continuous poset Q are the Scott closures of the Frink ideals of Q. The Frink ideals of Q are easily computed, and so are their Scott closures.

7.3 Is ψX, for arbitrary (not necessarily locally quasi-compact) spaces X, an idempotent construction?

7.4 Certainly desired is an external characterization of the ψ-extension $X \hookrightarrow \psi X$ (as well as an intrinsic characterization).

7.5 Fell [1961, 2.2] has given an interpretation of H(X) in functional analytic terms in the special case that X is the dual space of a C^*-algebra A. It seems to be a natural question whether for every C^*-algebra A there exists an "associated" C^*-algebra A′ such that ψX is the dual space of A′ (and whether there exists a natural morphism between A and A′ inducing, in a sense to be made precise, the extension $X \hookrightarrow \psi X$).

7.6 There is a subtle analogy between the ψ-extension $X \hookrightarrow \psi X$ of a T_0-space X with O(X) continuous and ordinary Hausdorff compactifications, i.e., dense embeddings of a (completely regular Hausdorff) space into a compact Hausdorff space. Partly, this is implicit in Hoffmann [1981b, sec. 8], but this can be pursued further (details will appear in a forthcoming memo or paper).

7.7 For a space X with O(X) an algebraic lattice, the space ψX is a spectral space, i.e., $O(\psi X) \cong DID(O(X))$ is an arithmetic lattice with (quasi-) compact unit (cf. the Compendium [1980, I-4.8]), since both D(•) and I(•) preserve "algebraicity" by Lawson [1979, sec. 8] and Hoffmann [1979d, 6.3], and Hoffmann [1982b], respectively. This association with every distributive algebraic lattice a spectral space.

REFERENCES:

Artin, M., A. Grothendieck, and J. Verdier, Théorie des topos et cohomologie étale des schémas. Springer-Verlag, Lecture Notes in Math. 269: Berlin-Heidelberg-New York, 1972.

Banaschewski, B., Essential extensions of T_0-spaces. General Topology Appl. 7, 233-246 (1977).

Banaschewski, B., Coherent frames. In: [BH], pp. 1-11.

Banaschewski, B., and R.-E. Hoffmann (editors): Continuous Lattices. Proceedings, Bremen 1979. Springer-Verlag, Lecture Notes in Math. 871: Berlin-Heidelberg-New York, 1981. (To be referred to as [BH].)

Birkhoff, G., Lattice Theory. Amer. Math. Soc. Colloquium Publications, 3rd ed., Providence, R.I., 1967.

Bruns, G., Darstellungen und Erweiterungen geordneter Mengen. I. J. reine angew. Math. 209, 167-200 (1962); II., ibid. 210, 1-23 (1962) (Habilitationsschrift Mainz, 1960).

Bruns, G., and J. Schmidt, Zur Äquivalenz von Moore-Smith-Folgen und Filtern. Math. Nachr. 13, 169-186 (1955).

Büchi, J. R., Representation of complete lattices by sets. Portugaliae Math. 11, 151-167 (1952).

Gierz, G., K. H. Hofmann, K. Keimel, J. D. Lawson, M. Mislove, and D. S. Scott, A Compendium of Continuous Lattices. Springer-Verlag, Berlin-Heidelberg-New York, 1980.

Fell, J. M. G., The structure of algebras of operator fields. Acta Math. 106, 233-280 (1961).

Fell, J. M. G., A Hausdorff topology for the closed subsets of a locally compact non-Hausdorff space. Proc. Amer. Math. Soc. 13, 472-476 (1962).

Flachsmeyer, J. Zur Spektralentwicklung topologischer Räume. Math. Ann. 144, 253-274 (1961).

Flachsmeyer, J., Verschiedene Topologisierungen im Raum der abgeschlossenen Mengen. Math. Machr. 26, 321-337 (1964).

Gierz, G. and K. Keimel, A lemma on primes appearing in algebra and analysis. Houston J. Math. 3, 207-224 (1977).

Hochster, M., Prime ideal structure in commutative rings. Trans. Amer. Math. Soc. 142, 43-60 (1969).

Hoffmann, R.-E., Irreducible filters and sober spaces. Manuscripta Math. 22, 365-380 (1977).

Hoffmann, R.-E., On weak Hausdorff spaces. Archiv d. Math. (Basel) 32, 487-504 (1979a).

Hoffmann, R.-E., Essentially complete T_0-spaces (I). Manuscripta Math. 27, 401-432 (1979b).

Hoffmann, R.-E., Projective sober spaces. In [BH], pp. 125-158 (1981a). Corrections in: Zbl. f. Math. 476 (1982), 06004 (Autorreferat).

Hoffmann, R.-E., Continuous posets and adjoint sequences. Semigroup Forum 18, 173-188 (1979c).

Hoffmann, R.-E., Topological spaces admitting a "dual". In Categorical Topology, pp. 157-166. Springer-Verlag, Lecture Notes in Math. 719. Berlin-Heidelberg-New York, 1979d.

Hoffmann, R.-E., Continuous posets, prime spectra of completely distributive complete lattices, and Hausdorff compactifications. In [BH], pp. 159-208 (1981b). Corrections in: Zbl. f. Math. 476 (1982), 06005 (Autorreferat).

Hoffmann, R.-E., Essentially complete T_0-spaces II. A lattice-theoretic approach. Math. Z. 179, 73-90 (1982a).

Hoffmann, R.-E., The injective hull and the CL-compactification of a continuous poset. Preliminary version. In Seminarbericht Mathematik der Fernuniversität Hagen 16, 31-92 (1982b).

Hoffmann, R.-E., Continuous posets: injective hull and MacNeille completion. Manuscript, University of Bremen, 1982c.

Hofmann, K. H., and J. D. Lawson, Irreducibility and generation in continuous lattices. Semigroup Forum 13, 307-353 (1976-7).

Hofmann, K. H., and J. D. Lawson, The spectral theory of distributive continuous lattices. Trans. Amer. Math. Soc. 246, 285-310 (1978).

Hofmann, K. H., and M. W. Mislove, Local compactness and continuous lattices. In [BH], pp. 209-248 (1981).

Hofmann, K. H., and M. W. Mislove, A continuous poset whose compactification is not a continuous poset. The square is the injective hull of a discontinuous CL-compact poset. Seminar on Continuity in Semilattices (SCS), memo, May 28, 1982 (revised July 2, 1982).

Isbell, J. R., Meet-continuous lattices. Symposia Mathematica 16 (Convegno sulla Topologica Insiemsistica e Generale, INDAM, Roma, Marzo 1973), pp. 41-54. Academic Press: London 1974.

Johnstone, P. T., The Gleason cover of a topos, II. J. Pure Appl. Algebra 22, 229-247 (1981).

Keimel, K., and M. Mislove, Several remarks: ... When do the prime elements of a distributive lattice form a closed subset... Seminar on Continuity in Semilattices (SCS), memo, Sept. 30, 1976.

Lawson, J. D., Continuous semilattices and duality. Seminar on Continuity in Semilattices (SCS), memo, Jan. 4, 1977.

Lawson, J. D., The duality of continuous posets. Houston J. Math. 5, 357-394 (1979).

Markowsky, G., A motivation and generalization of Scott's notion of a continuous lattice. Preprint (1977). Revised version in [BH], pp. 298-307 (1981).

MacLane, S., Categories for the Working Mathematician. Springer, Berlin-Heidelberg-New York, 1971.

Mrówka, S., On the convergence of nets of sets. Fund. Math. 45, 237-246 (1958).

Nachbin, L., Topology and Order. Van Nostrand, Princeton, 1965.

Papert, S., Which distributive lattices are lattices of closed sets? Proc. Cambridge Phil. Soc. 55, 172-176 (1959).

Scott, D., Outline of a mathematical theory of computation. Proc. 4th Ann. Princeton Conf. on Information Science and Systems (1970), pp. 169-176.

Scott, D., Continuous lattices. In Toposes, Algebraic Geometry and Logic, pp. 97-136. Springer-Verlag, Lecture Notes in Math. 274: Berlin-Heidelberg-New York, 1972.

Schröder, J., Das Wallman-Verfahren und inverse Limites. Quaest. Math. 2, 325-333 (1977).

Schubert, H., Categories. Springer: Berlin-Heidelberg-New York, 1972.

Simmons, H., A couple of triples. Topology and Appl. 13, 201-223 (1982).

Smyth, M. B., Effectively given domains. Theoretical Computer Science 5, 257-274 (1977).

Wilansky, A., Between T_1 and T_2. Amer. Math. Monthly, 74, 261-266 (1967).

Wilson, R. L., Relationships between continuous posets and compact Lawson posets. Notices Amer. Math. Soc. 24, A-628 (1977).

Wyler, O., Dedekind complete posets and Scott topologies (SCS memo 1977). In [BH], pp. 384-389 (1981).

5

The Trace of the Weak Topology and of the Γ-Topology of L^{op} Coincide on the Pseudo-Meet-Prime Elements of a Continuous Lattice L

RUDOLF-E. HOFFMANN

Universität Bremen
Bremen, Federal Republic of Germany

This is a partial response to a private communication in which Karl H. Hofmann comments on my paper [4], delineating a somewhat different approach to some of the results obtained there.†

Recall that, in a 1,∧-semilattice L,

$$a \vdash b$$

(a is <u>relatively meet prime</u> below b) for a,b ∈ L iff whenever $\inf\{x_1, \ldots, x_n\} \le a$ for $x_1, \ldots, x_n \in L$ ($n \in \mathbb{N}$, the set of natural numbers including 0), then $x_i \le b$ for some $i \in \mathbb{N}$, $0 \le i \le n$. The sets

$$\Gamma^*(x) := \{y \in L \mid x \vdash y\}$$

with x ranging through L, form a (sub-)basis (cf. [2], 1.3(ii)) of the closed sets of the Γ^*-topology (the Γ-<u>topology</u> of L^{op} (cf. [1], sec. 3, [2]). The sets

$$\uparrow x := \{y \in L \mid x \le y\} \qquad x \in L$$

form a subbasis of the closed sets of the lower topology ω_L of L (= the <u>weak topology</u> of L^{op}; cf. [1], sec. 2).

An element p of a complete lattice L is called

(1) <u>meet prime</u> iff $p \vdash p$

(2) <u>pseudo-meet-prime</u> iff p = sup P for a prime ideal P of L

(3) a γ^*-element (i.e., a γ-element of L^{op}) iff $\uparrow p$ is closed in (L, Γ^*) iff $p = \sup\{x \in L \mid x \vdash p\}$ ([2], 1.5, 2.7)

†The first draft of this paper was circulated as a memo in the Seminar on Continuity in Semilattices (SCS), Jan. 9, 1983.

Every meet-prime element is pseudo-meet-prime. Every pseudo-meet-prime element is a γ^*-element ([2]3.4). In a distributive complete lattice, every γ^*-element is a supremum of pseudo-meet-prime elements ([2]3.6).

Endowing the set ψ^*L of pseudo-meet-prime elements of a distributive continuous lattice L with the trace τ of ω_L, Hofmann sketches a proof for

$$\underline{DO}(\psi^*L,\tau) \cong \mathrm{Filt}_\sigma L$$

which closely parallels my result in [4]:

$$\underline{DO}(\psi^*L) \cong \mathrm{Filt}_\sigma L$$

where ψ^*L carries the trace of the Γ^*-topology of L (see Corollary 4.10 in [4]). From a comparison, one may be inclined to infer (erroneously[†]) that these topologies coincide on ψ^*L.

Indeed, arguments very similar to those used in the proof of Theorem 4.9 of [4] provide a "direct proof" of this (which "is bound to exist").

PROPOSITION For a continuous lattice L (not necessarily distributive), the weak topology ω_L of L^{op} and the Γ^*-topology of L have the same trace on the set ψ^*L of pseudo-meet-prime elements of L. For every $x \in L$, we have

$$\psi^*L \cap \uparrow x = \psi^*L \cap \bigcap\{\Gamma^*(y) \mid y \in L,\ y \ll x\}$$

Proof: Since the Γ^*-topology of L is always weaker than ω_L ([1]3.3), the above formula suffices to establish the assertion.

(a) Let $z \in \bigcap\{\Gamma^*(y) \mid y \in L,\ y \ll x\}$, i.e., $y \vdash z$ for every $y \in L$ with $y \ll x$. Thus $y \le z$ for every such y. Since L is a continuous lattice, it results that

$$x = \sup\{y \in L \mid y \ll x\} \le z$$

Hence

$$\bigcap\{\Gamma^*(y) \mid y \in L,\ y \ll x\} \subseteq \uparrow x$$

(b) Now let $z \in \psi^*L \cap \uparrow x$, and let $y \in L$ with $y \ll x$. Then $z = \sup P$ for some prime ideal P of L. However, $x \le \sup P$ implies $y \in P$, since $y \ll x$. On the other hand, since P is a prime ideal, $y \in P$ and $z = \sup P$ imply $y \vdash z$,

[†]An order automorphism (with regard to the specialization partial order) of a space need not be a homeomorphism: every permutation of a T_1-space is an order automorphism.

i.e., $z \in \Gamma^*(y)$ for every $y \in L$ with $y \ll x$. Thus

$$\Downarrow^* L \cap \uparrow x \subseteq \bigcap\{\Gamma^*(y) \mid y \in L, y \ll x\}$$

This completes the proof. ∎

REMARKS 1. For <u>every</u> complete lattice L, the Γ^*-topology and ω_L have the same trace on Spec^*L, the set of meet-prime elements of L (cf. [1] 3.7).

2. It is an open question whether the above proposition holds for <u>all</u> complete lattices L. It is also unknown whether it extends to γ^*L, the set of γ^*-elements of L (i.e., γ-elements of L^{op}). The latter extension is known to be true for <u>completely distributive</u> complete lattices L (cf. [3] 5.5) and also for the lattice of (all) open sets of a Hausdorff space (cf. [2] 2.11(b)).

REFERENCES

1. Hoffmann, R.-E., Essentially complete T_0-spaces. Manuscripta Math. <u>27</u>, 401-432 (1979).

2. Hoffmann, R.-E., Essentially complete T_0-spaces, II. A lattice-theoretic approach. Math. Z. <u>179</u>, 73-90 (1982).

3. Hoffmann, R.-E., The injective hull and the <u>CL</u>-compactification of a continuous poset. Preliminary version in: Seminarbericht Mathematik der Fernuniversität Hagen <u>16</u>, 31-92 (1982). (Canad. J. Math., to appear)

4. Hoffmann, R.-E., The Fell compactification revisited. (This volume, Chapter 4.) [Preliminary version in: Universität Bremen, Mathematik-Arbeitspapiere <u>27</u>, 68-141 (1982).]

6

Complete Distributivity and the Essential Hull of a T_0-Space

KARL H. HOFMANN*

Tulane University
New Orleans, Louisiana

1. BANASCHEWSKI'S LATTICE THEORETICAL REPRESENTATION OF A T_0-SPACE

If X is an arbitrary T_0-space and L the algebraic lattice Filt O(X) of all filters of open sets of X, then the function $x \mapsto \mathcal{U}(x): X \to L$ which associates with a point its neighborhood filter is a topological embedding if L is given its Scott topology $\sigma(L)$. This was established by Banaschewski [1977] and used for the construction of the essential hull of X. Each neighborhood filter $\mathcal{U}(x)$ is completely prime in L. We recall that an element p of a complete lattice L is called completely prime if and only if for any subset T of L the relation inf $T \leq p$ implies that $t \leq p$ for some $t \in T$. We shall denote the set of all completely prime elements of a complete lattice L by $\theta(L)$. It is an almost direct consequence of the definition that an element p of L is contained in $\theta(L)$ iff there is a compact element p^* in L such that L is the disjoint union of $\downarrow p$ and $\uparrow p^*$. (The element p^* is a completely coprime element, and the function $p \mapsto p^*$ is a bijection from $\theta(L)$ onto the set of completely coprime elements.) Each $p \in \theta(L)$ determines a complete lattice morphism $f: L \to 2$ via $f^{-1}(0) = \downarrow p$, and $p \mapsto f$ is a bijection from $\theta(L)$ onto (INF ∩ SUP)(L,2) in the terminology of A Compendium of Continuous Lattices [1980, p. 171, Definition 1.1].

The set $\theta(L)$ may be empty. The following information gives some idea of the lattices in which completely prime elements abound.

*Current affiliation: Technische Hochschule Darmstadt, Darmstadt, Federal Republic of Germany

1.1 PROPOSITION For a complete lattice L the following conditions are equivalent:

(1) $\theta(L)$ is order generating (inf-dense), i.e., $x = \inf(\uparrow x \cap \theta(L))$ for all $x \in L$.

(2) The complete lattice morphisms $L \to 2$ separate the points of L.

(3) There is an embedding $L \to 2^X$ that preserves all infs and sups.

(4) L is an algebraic completely distributive lattice.

(5) L is an algebraic lattice whose dual K(L) is a $0,\vee$-semilattice in which every element is a finite sup of coprimes.

We recall in this context that a coprime p in a $0,\vee$-semilattice K is an element for which the relation $p \leq \sup F$ for a finite set $F \subseteq K$ implies $p \leq x$ for some $x \in F$. The set of all coprimes will be denoted Cospec K. The grounding functor of the category $\mathcal{J}$ of all $0,\vee$-semilattices (with $0,\vee$-morphisms) into the category $\mathcal{P}$ of posets (and monotone maps) has a left adjoint S which associates with a poset P the $\cup$-semilattice $S(P) \subseteq 2^P$ of all $\downarrow F$, $F \subseteq P$ finite, and with the front adjunction $x \mapsto \downarrow x\colon P \to S(P)$. If $\mathcal{J}_p$ denotes the full subcategory of all $0,\vee$-semilattices in which every element is the sup of finitely many coprimes, the corestriction $S\colon \mathcal{P} \to \mathcal{J}_p$ is well defined and yields an equivalence of categories. (Cf. Hofmann, Mislove, and Stralka [1974, pp. 10 ff. and pp. 75 ff.].) We say that S(P) is the free semilattice generated by the poset P. We can thus complement Proposition 1.1 as follows:

1.2 PROPOSITION The equivalent conditions (1)-(5) of Proposition 1.1 are also equivalent to the following:

(6) L is an algebraic lattice whose dual K(L) is a free semilattice over a poset [namely, Cospec K(L) in the induced order].

In any complete lattice L we have $\theta(L) \subseteq \operatorname{Irr} L$, where Irr L is the set of completely irreducible elements (see the Compendium [p. 92, Definition 4.19]). In any algebraic lattice L the set Irr L is the unique smallest order-generating subset (see the Compendium [p. 93, Theorem 4.23]). In view of this remark and condition (1) of Proposition 1.1, we can therefore add to our list:

1.3 PROPOSITION The equivalent conditions (1)-(6) of Propositions 1.1 and 1.2 are also equivalent to the following:

(7) $\theta(L) = \operatorname{Irr} L$.

We now return to Banaschewski's representation $x \mapsto \mathcal{U}(x): X \to L$ of a T_0-space in the arithmetic lattice $L = \text{Filt } O(X)$ equipped with the Scott topology $\sigma(L)$. We showed in Hofmann [1984] how the elements of $\theta(L)$ are characterized in this case:

1.4 LEMMA If $L = \text{Filt } O(X)$ for a T_0-space X, then $\mathcal{U}(x) \in \theta(L)$ for all $x \in X$, and the following conditions are equivalent:

(a) $\mathcal{U} \in \theta(L)$.

(b) There is a unique closed irreducible set $C \subseteq X$ with $\mathcal{U} = \{U \in O(X): U \cap C \neq \emptyset\}$.

If X is sober, then (a) and (b) are in fact equivalent to

(c) $\mathcal{U} = \mathcal{U}(x)$ for some $x \in X$.

Conversely, if (c) is satisfied, then X is sober.

In particular, every sober space X may be viewed as the subspace $\theta(L)$ of some arithmetic lattice L equipped with the Scott topology $\sigma(L)$.

We now pose and answer the question: For which T_0-spaces X is every completely meet-irreducible filter of open sets the neighborhood filter of a point? First, we observe the following:

1.5 PROPOSITION For a T_0-space X the following conditions are equivalent, with $L = \text{Filt } O(X)$:

(1) If $\mathcal{U}$ is a completely meet-irreducible filter of open sets, then there is a unique closed irreducible set C in X with $\mathcal{U} = \{U \in O(X): U \in F = \emptyset\}$.

(2) $\text{Irr } L = \theta(L)$.

(3) Every closed subset of X is a finite union of irreducible closed sets.

(4) L is completely distributive.

Proof: The equivalence of (1) and (2) is clear by Lemma 1.4. The equivalence of (2) and (4) follows from Proposition 1.3. We recall that in the filter lattice $L = \text{Filt } O(X)$ we have $K(L) = O(X)^{op}$ (cf. Hofmann, Mislove, and Stralka [1974]). Therefore, Proposition 1.1 (5) is precisely (3) above; thus Proposition 1.1 shows that (3) and (5) are equivalent. ∎

1.6 THEOREM For a T_0-space X the following conditions are equivalent, where $L = \text{Filt } O(X)$:

(1) Every completely meet-irreducible filter of open sets is the neighborhood filter of a point.

(2) X is sober and every closed set is a finite union of point closures.

(3) The poset $(X,\leq)$ with the specialization order satisfies the following condition:

(FF) For any family $\{F_j\colon j \in J\}$ of finite subsets of X there is a finite subset F such that $\downarrow F = \bigcap\{\downarrow F_j\colon j \in J\}$.

Moreover, $O(X)$ is the upper topology $\omega(X,\leq)$.

REMARK We recall that the upper topology $\omega(P)$ of a poset P is generated by the subbasic sets $P\setminus\downarrow x$, $x \in P$. (Cf. the Compendium, p. 142, Definition 1.1.)

Proof: The equivalence of (1) and (2) above follows from the equivalence of (1) and (3) in Proposition 1.5 in view of Lemma 1.4. Condition (2) is equivalent to the following:

(2′) Every closed subset of X is a finite union of point closures.

This in turn is equivalent to

(2″) The closed subsets of X are precisely the sets $\downarrow F$ relative to the specialization order of X, where F ranges through all finite subsets of X.

In view of the fact that the collection of closed subsets of a topological space is closed under finite unions and arbitrary intersections, condition (2″) is indeed equivalent to condition (3). ■

We shall say that a poset X is finitely fitting if condition (FF) in Theorem 1.6(3) is satisfied. Thus Theorem 1.6 establishes a bijection between those T_0-spaces whose completely irreducible filters of open sets are neighborhood filters of points and finitely fitting posets. Indeed, the lattice of closed sets of such a T_0-space is a free semilattice over a finitely fitting poset.

2. ESSENTIAL AND INJECTIVE HULLS OF T_0-SPACES

Every T_0-space can be embedded via the Banaschewski representation $x \mapsto \mathcal{U}(x)$ into the subspace $\theta(L)$ of an arithmetic lattice L with its Scott topology $\sigma(L)$; namely, $L = \mathrm{Filt}\, O(X)$ will do, as we saw in Section 1. If we denote the smallest subset of L that contains all $\mathcal{U}(x)$ and is closed under the formation of arbitrary sups by λX, then the embedding $X \xrightarrow{e} \lambda X$ is essential (i.e., every continuous map f: $X \to Y$ into a T_0-space for which fe is an embedding, is itself an embedding), and e is the largest possible such embedding (up to equivalence). Therefore, according to Banaschewski, e : $X \to \lambda X$ will be called the <u>essential hull</u> of X. In many important categories, essential hulls are injective, as Banaschewski showed; however, in the case of T_0-spaces this is not always the case. In order to give a succinct characterization of those spaces that have an <u>injective hull</u>, i.e., λX is an injective space in the sense of Scott (cf. Scott [1972] and the <u>Compendium</u> [1980, pp. 121 ff.]), we recall the Scott order $\prec$ on a T_0-space, which is given by $x \prec y$ iff $\uparrow x$ is a neighborhood of y, where $\uparrow x = \{z \in X: \mathcal{U}(x) \subseteq \mathcal{U}(z)\}$ is the upper set of x relative to the specialization order of X. We will use the following convention: If $x \in X$ and $Y \subseteq X$ we shall write $x = \sup_{\lambda X} Y$ whenever $\mathcal{U}(x) = \sup_L\{\mathcal{U}(y): y \in Y\} = \sup_{\lambda X}\{\mathcal{U}(y): y \in Y\}$. With this notation we have (Hofmann [1984]):

2.1 PROPOSITION For a T_0-space X the following conditions are equivalent:

(1) X has an injective hull.

(2) For all $x \in X$ we have $\mathcal{U}(x) = \sup\{\mathcal{U}(y): y \prec x\}$.

(3) $x = \sup_{\lambda X} \downarrow x$ for all $x \in X$.

We now discuss the essential and injective hull of those spaces characterized in Section 1. Recall that we considered spaces that may be viewed as subspaces Irr L of an arithmetic lattice L with the Scott topology $\sigma(L)$. The essential hull of such a space will be precisely the subspace

$$A = \{x \in L: x = \sup(\downarrow x \cap \text{Irr } L)\}$$

This motivates the following observation:

2.2 LEMMA Let L be a complete lattice and suppose that L is inf-generated by $G \subseteq L$, i.e., $x = \inf(\uparrow x \cap G)$ for all $x \in L$. Set

$$A = \{x \in L: x = \sup(\downarrow x \cap G)$$

Then the complete lattice A is inf-generated by G within A, i.e., $x = \inf_A(\downarrow x \cap G)$ for all $x \in A$.

In particular, G is inf- as well as sup-dense in A, whence A is the MacNeille completion of G.

Proof: Let $k: L \to L$ be the kernel operator associated with A. Let $g: L \to A$ be the corestriction of k. Then g is the upper adjoint of the inclusion $d: A \to L$, hence preserves all infs. Now let $x \in A$. By hypothesis, $x = \inf_L(\uparrow x \cap G)$, whence $x = k(x) = g(x) = \inf_A g(\uparrow x \cap G) = \inf_A(\uparrow x \cap G)$. ■

The spaces of Section 1 now allow a particularly simple characterization that describes which ones among them have injective hulls:

2.3 THEOREM Let X be a T_0-space in which every completely irreducible filter of open sets is a neighborhood filter of a point. Then the following conditions are equivalent:

(1) X has an injective hull.

(2) For each point $x \in X$ and each point y with $x \not\leq y$ there is a $z \not\leq y$ with $z \ll x$.

(3) For each $x \in X$ we have $x = \sup \downarrow x$ in X.

Proof: By 1.5 and 1.6 we may consider X as the subspace Irr L of a completely distributive algebraic lattice L. The essential hull λX of X may then be identified with $A = \{x \in L: x = \sup (\uparrow x \cap \text{Irr } L)$ with the topology $\sigma(L)|A\}$. Now (1) is equivalent with

(4) A is a continuous lattice with $\sigma(A) = \sigma(L)|A$.

(Cf. Hofmann [1984].) By Hofmann [1984, Theorem 2.3], condition (4) is equivalent under the present circumstances to the following:

(5) $(\forall x,y \in X)\ x \not\leq y \Rightarrow [\exists z \in X,\ c \in K(L)]\ z \not\leq y$ and $z \leq c \leq x$

But the relation $z \leq c \leq x$ for some $c \in K(L)$ means exactly $z \ll x$. Thus (5) is indeed equivalent to (2). By Proposition 2.1, condition (1) is equivalent to

(3′) For each $x \in X$ we have $x = \sup_A \downarrow x = \sup_L \downarrow x$.

It is clear that (3′) implies (3) (cf. Hofmann [1984, 3.8]). We claim that, conversely, (3) ⇒ (3′): By Lemma 2.2, X = Irr L is inf- and sup-dense in A. Then Hoffmann's Lemma 0.3 [1981-2] applies and shows that the embedding $X \to A$ preserves all existing sups. Thus $\sup_X \downarrow x = \sup_A \downarrow x$. Hence (3) implies (3′). ■

2.4 COMPLEMENT TO 2.3 Under the conditions of Theorem 2.3, X is inf- and sup-dense in its injective hull λX, which is therefore the MacNeille completion of X.

Proof: This is an immediate consequence of Lemma 2.2. ∎

The machinery of essential and injective hulls for the spaces satisfying the general hypothesis of Theorem 2.3 is thus particularly simple. It should be remarked that in view of Theorem 1.6 the spaces occurring in algebraic geometry always fall into this class. Indeed, if a T_0-space X is noetherien, i.e., satisfies the ascending chain condition for closed sets, then every closed set is a finite union of irreducible closed sets. The converse is not true in general. Thus every noetherien sober space satisfies the general hypotheses of Theorem 2.4 and its Complement 2.3.

REFERENCES

Banaschewski, B., Essential extensions of T_0-spaces, Gen. Top. Appl. 7 (1977), 233-246.

Banaschewski, B., and R.-E. Hoffmann, editors, Continuous Lattices, Lecture Notes in Mathematics 871 (1981), Springer-Verlag, Berlin-Heidelberg-New York.

Gierz, G., K. H. Hofmann, K. Keimel, J. D. Lawson, M. Mislove, D. S. Scott, A Compendium of Continuous Lattices, Springer-Verlag, Berlin-Heidelberg-New York, 1980.

Hoffmann, R.-E., Essentially complete T_0-spaces, Manuscr. Math. 22 (1977), 401-452.

Hoffmann, R.-E., Essentially complete T_0-spaces. A lattice theoretical approach, Math. Z. 179 (1982), 73-90.

Hoffmann, R.-E., Projective sober spaces, in Banaschewski and Hoffmann above, 125-158.

Hoffmann, R.-E., Continuous posets, prime spectra of completely distributive spaces and Hausdorff compactifications, in Banaschewski and Hoffmann, above, 159-208.

Hoffmann, R.-E., The CL-compactification and the injective hull of a continuous poset, Preprint 1981-82, to appear.

Hoffmann, R.-E., The Fell compactification revisited, Preprint, 1982.

Hofmann, K. H., Completely distributive algebraic lattices, SCS Memo 11-27-79.

Hofmann, K. H., Order aspects of the essential hull of a topological T_0-space, Annals of Discrete Mathematics 23 (1984), 193-205.

Hofmann, K. H., M. Mislove, and A. Stralka, The Pontryagin Duality of Compact O-Dimensional Semilattices and Its Applications, Lecture Notes in Mathematics 396 (1974), Springer-Verlag, Berlin-Heidelberg-New York.

Hofmann, K. H., and M. Mislove, Free objects in the category of completely distributive lattices, these Proceedings. (See also SCS Memo 11-24-81.)

Hofmann, K. H., and M. Mislove, A continuous poset whose compactification is not a continuous poset, SCS Memo 6-8-82.

Scott, D. S., Continuous Lattices, Lecture Notes in Math. 274 (1972), 97-136, Springer-Verlag, Berlin-Heidelberg-New York.

7

Free Objects in the Category of Completely Distributive Lattices

KARL H. HOFMANN* and MICHAEL MISLOVE
Tulane University
New Orleans, Louisiana

INTRODUCTION

Completely distributive lattices (which we always assume are complete) have attracted attention since the early history of lattice theory, notably through the work of Raney (see, e.g., Raney [1952]). This presumably stems from their rich representation theory and from the inherent symmetry which make them an obvious generalization of the concept of a Boolean algebra. In fact, completely distributive lattices are defined by assuming the most restrictive distributive law possible, which involves infinitary infimum and supremum operations [see condition (cd) in Definition 1.1 below]. Thus they form a variety. Even though recent attention focused on the less symmetric generalization in the form of continuous lattices (see A Compendium of Continuous Lattices [1980]), interest in completely distributive lattices has remained alive (see Dwinger [1981], Hoffmann [1981], Lawson [1979], Liber [1977], and Markowsky [1979]), and was in fact revitalized by the discovery, due independently to Lawson [1979] and to Hoffmann [1981], that there is an equivalence in a certain sense between completely distributive lattices and the so-called continuous posets. (The latter generalize continuous lattices, and have attracted the interest of workers in computer science applications of lattice theory and ordered sets.) Indeed, the equivalence operates via the spectral theory for completely distributive lattices and the duality theory of continuous posets.

The fact that free completely distributive lattices exist (over sets) is a relatively recent discovery, made independently by Markowsky in his

*Current affiliation: Technische Hochschule Darmstadt, Darmstadt, Federal Republic of Germany

1973 dissertation (see Markowsky [1979]), by Dwinger [1981], and implicitly by Liber [1977] in a little-known paper. In fact, in the latter, Liber actually shows that free completely distributive lattices can be constructed over compact Lawson semilattices and over compact spaces. (We thank Klaus Keimel for drawing our attention to Liber's paper.)

The objective of our work here is a systematic presentation of adjoint functor theorems involving the category of completely distributive lattices with the obvious morphisms, those preserving all infima and all suprema. In the process, we develop various adjoint functor theorems, based primarily on one lattice theoretical argument, some of which express universal properties that are somewhat weaker than those expressed in a true adjoint situation. The framework we provide in this fashion allows us to deduce all of the theorems on free constructions by Dwinger, Liber, and Markowsky as special cases. Moreover, we also show the connection of the work of Hoffmann on continuous posets with this subject.

Our paper is organized as follows: The first section is devoted to a careful analysis of the category CD of completely distributive lattices and their morphisms within the framework of Galois adjunctions. We collect the necessary background material on the relationships between continuous lattice, continuous posets, and their morphisms; for a general discussion of free functors with varying grounding categories, it is very important to keep detailed records of the morphisms one uses in each instance.

Next, we actually provide the basic tools of the free construction and discuss their properties in Section 2. We also provide a sufficient measure of generality to derive the existence of free completely distributive lattices as established in the theorems by Dwinger and Markowsky. We then take note of the fact that we encounter in our situation a weaker form of adjoint situations, which we call subadjunctions, and we briefly summarize their category theoretical properties (Section 3).

Next, we introduce Scott continuity into our discussion, a concept which deeply permeates all the theory of continuous lattices and continuous posets. We obtain the relevant adjunction and subadjunction theorems, which now permit the derivation of a form of Liber's theorem, expressed in lattice-theoretical terms, rather than in those of topological algebra. (The equivalence of the two approaches is now well understood (see the Compendium [1980].) We use the cardinality invariant of the weight of a continuous lattice and of a space (which was developed for continuous lattices in the Compendium [Section III-4]) as a measure of the size of a free completely distributive lattice; this appears for many purposes to be more convenient and more suitable than that of cardinality.

Finally, we conclude our discussion by introducing a functor from continuous posets to continuous lattices with the aid of the free constructions introduced earlier, and we show that this functor is the injective hull of a continuous poset. This topic was recently studied extensively by Hoffmann [1983], who thereby pursues a theme initiated by Banaschewski [1977] in his study of essential hulls of T_0-spaces. Our standard reference is the Compendium.

1. THE CATEGORY CD

We begin with a definition of the category of completely distributive lattices; when we fix the objects, the selection of the morphisms follows automatically.

1.1 DEFINITION The objects of the category CD of completely distributive lattices are those complete lattices which satisfy the following identity for all families $\{a_{ij} : i \in I, j \in J\}$:

$$\text{(cd)} \quad \inf_{i\in I} \sup_{j\in J} a_{ij} = \sup_{f\in K} \inf_{i\in I} a_{i,f(i)}$$

where $K = I^J$. The morphisms of CD are those maps between CD objects that preserve all infs and all sups; these are sometimes called complete lattice maps.

REMARKS The identity (cd) is called complete distributivity, and it always holds simultaneously with its opposite identity. Completely distributive lattices therefore share with many other classical objects of lattice theory the feature of being preserved under the passage $L \to L^{op}$ to the opposite lattice, which is so strikingly absent in the more general category of continuous lattices.

The choice of morphisms is indeed dictated by (cd): The operations (infinitary to be sure) entering into the defining equation must be preserved. This choice of morphisms makes the category CD a full subcategory of Inf ∩ Sup, i.e., INF ∩ SUP in the notation of the Compendium (IV-1.1).

The category CD is obviously a variety which is closed under the formation of arbitrary products, subalgebras, and quotients. Cartesian products are the categorical products; equalizers are formed as in the category of sets. The category is also cocomplete, but one must resist the temptation to believe that the coproduct of, say, two CD-objects is the cartesian product with any of the obvious injections as coprojections. It looks suspiciously as though this were the case, but a closer inspection reveals that neither of these injections preserves both 0 and 1, while every CD map must respect them. Coproducts are more complicated; more about that later.

That free objects exist in CD is perhaps not obvious, considering that the category of complete lattices has no free objects. However, the existence of free completely distributive lattices over sets is by now well known through the work of Dwinger [1981], Markowsky [1973,1979], and Liber [1977]. Their structure and cardinality were also determined in these works. However, our emphasis here is slightly different. Completely distributive lattices are special cases of continuous lattices (see the Compendium, p. 59), and it was shown in the Compendium (p. 85) that continuous lattices are free over the category of compact Hausdorff spaces. In fact, Wyler [1981] showed that the category CL of continuous lattices and maps preserving all infs and directed sups is algebraic over the category of compact spaces. The Stone-Čech compactification provides the free compact Hausdorff space over a set; as a consequence, free continuous lattices over sets exist. Indeed, that the category of continuous lattices is algebraic over sets was first observed by Day (see [1975]; for further details see the Compendium, pp. 90 and 285). Our objective now is to show that the category CD is free over the category CL of continuous lattices. As a first step, this requires a better understanding of the basic features of the category CD, which we provide in the remainder of this section. For the language of Galois connections (or adjunctions) used in the following, we refer to the Compendium, p. 18 ff.

1.2 PROPOSITION Let $f\colon L \to M$ be a monotone map between CD-objects.

(i) The following statements are equivalent:

(1) f is a CD-morphism.

(2) f has an upper adjoint $g : M \to L$ and a lower adjoint $d\colon M \to L$.

(ii) If (1) and (2) are satisfied, then the following hold:

(a) g preserves Spec M, and $g|\text{Spec } M : (\text{Spec } M, \geq) \to (\text{Spec } L, \geq)$ preserves directed sups (i.e., is Scott continuous).

(b) d preserves Cospec M, and $d|\text{Cospec } M : \text{Cospec } M \to \text{Cospec } L$ preserves directed sups (i.e., is Scott continuous), where Cospec carries the induced order.

(iii) The following conditions are equivalent:

(3) f preserves Spec.

(4) d preserves finite infs; i.e., d is a cHa-map.

Under these conditions, $f|\text{Spec } L : (\text{Spec } L,\geq) \to (\text{Spec } M,\geq)$ is the upper (!) adjoint of $g|\text{Spec } M$.

(iv) The following conditions are equivalent:

(5) f preserves Cospec

(6) g preserves finite sups.

One of the crucial new facts about completely distributive lattices which has come to light recently is the bijective correspondence between completely distributive lattices and continuous posets; these insights we owe to Lawson [1979] and Hoffmann [1980]. The bijection was established on the object level via spectral theory; more recently it has begun to crystallize which morphisms play a role in this correspondence (see, e.g., Niño [1981]).

In Proposition 1.2(a) and (b) we see that the adjoints of CD-maps produce Scott continuous maps on the spectra (and cospectra). But not every Scott continuous map between continuous posets arises in this fashion. Indeed, let $g : T \to S$ be a Scott continuous map between continuous posets T and S. If we let $L = \sigma(S)$ and $M = \sigma(T)$ be the respective Scott topologies on S and T, then L and M are CD-objects, and the map $f : L \to M$ given by $f(U) = g^{-1}(U)$ is surely a cHa-map. By Proposition 1.2(i), this f is a CD-morphism iff it has a lower adjoint.

Let us pause for a moment and consider the situation more generally on the level of general topology. If $g : Y \to X$ is a continuous map of topological spaces, then we generate a cHa-map $f : O(X) \to O(Y)$ via $f(U) = g^{-1}(u)$. The map f has a lower adjoint $d : O(Y) \to O(X)$ iff every open set V of Y determines an open set $d(V)$ in X so that $d(V) \subset U$ for an open set $U \subset X$ iff $V \subset f(U) = g^{-1}(U)$ iff $g(V) \subset U$. Thus, $d(V) = \bigcap \{U \in O(X) : g(V) \subset U\} = \text{sat } g(V)$, where the <u>saturation</u> sat A of a set A in X is the intersection of all open subsets of X which contain A. We have in fact shown the following:

1.3 LEMMA The cHa-map $O(g) : O(X) \to O(Y)$ induced by the continuous function $g : Y \to X$ via $(O(g))(U) = g^{-1}(U)$ has a lower adjoint if and only if the saturation sat $g(V)$ is open in X for every open subset V of Y.

REMARK If X is T_1, this occurs precisely when g is open (indeed, sat A = A for any subset A in a T_1-space).

For easy reference, we choose the following nomenclature:

1.4 DEFINITION A function $g : Y \to X$ between topological spaces is called quasi-open iff sat g(V) is open in X for each open subset V of Y.

Since the Scott-continuous and quasi-open maps between continuous posets are precisely those arising from CD-morphisms by restricting adjoints to spectra, we declare:

1.5 DEFINITION A map between up-complete posets $c : T \to S$ is called a comorphism iff it is Scott continuous and Scott quasi-open; i.e., iff it preserves directed sups and $\uparrow c(V)$ is Scott open in S for every Scott-open set V in T.

We then have the following remark:

1.6 PROPOSITION The category CD of completely distributive lattices is dually equivalent to the category of continuous posets and comorphisms between them.

We recall some familiar facts from the Compendium (p. 180): If $F : S \to T$ is the upper adjoint of $c : T \to S$, then F is Scott continuous iff $F^{-1}(V)$ is Scott open for each Scott-open subset of V; but $F^{-1}(V) = \uparrow c(V)$, and so F is Scott continuous iff c is Scott quasi-open. In particular, if F is an Inf-map between complete lattices, then F is in fact an $\mathrm{Inf}^{\uparrow}$-map iff its upper adjoint is a comorphism. This leads us to the following definition:

1.7 DEFINITION A map $F : S \to T$ between up-complete posets is called a morphism iff it has a lower adjoint $c : T \to S$ which is a comorphism, iff it has a lower adjoint and is Scott-continuous.

From the Compendium (pp. 182-183), we know that a map $F : S \to T$ between continuous lattices is a morphism in the sense of Definition 1.7 iff it is a CL-morphism, i.e., iff it is in $\mathrm{Inf}^{\uparrow}$. We may think of a morphism between up-complete posets as a pair (F,c) of adjoint maps. The category of continuous posets and morphisms between them contains the category CL as a full subcategory and is a bit smaller than the dual category of continuous posets and comorphisms between them. To be more precise:

1.8 PROPOSITION The subcategory $\underline{CD}_{Spec}$ of $\underline{CD}$ with the same objects as $\underline{CD}$ and all $\underline{CD}$-morphisms between them that preserve spectra is equivalent to the category of all continuous posets and morphisms between them.

The functor that implements the equivalence simply associates with a $\underline{CD}$-object its spectrum and with a Spec-preserving morphism its restriction to spectra. Recall that a $\underline{CD}$-morphism $f : L \to M$ is a $\underline{CD}_{Spec}$-morphism iff the lower adjoint d of f is a cHa-map [see Proposition 1.2(iii)].

This discussion shows that there is a close link between the theory of completely distributive lattices (the category $\underline{CD}$) and the theory of continuous posets and their morphisms. It illustrates that the case for a general study of completely distributive lattices does not rest on their rich structure and symmetry alone, but also on their hierarchical position within the cycle of classes of ordered structures:

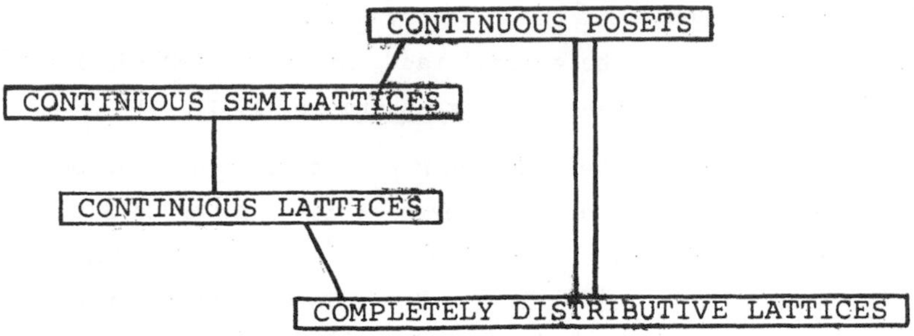

It is not atypical for lattice theory that a strictly increasing chain of classes of ordered structures closes because of an equivalence between a "larger" class with a properly "smaller" class, just as in the example above (cf., e.g., the Compendium, p. 199 ff., and notably pp. 184,185).

2. UNIVERSAL PROPERTIES

Before we prove the essential ingredients necessary for the main freeness theorems for completely distributive lattices, we clarify some functorial preliminaries.

The category $\underline{Set}$ of sets has two exponential self-functors, one contravariant and one covariant. We consider here the covariant functor $X \mapsto 2^X : \underline{Set} \to \underline{Set}$ for which a function $f : X \to Y$ between sets is transformed into the function $2^f : 2^X \to 2^Y$ by $2^f(A) = f(A)$ for any set A in 2^X. Now, every 2^X is, among other things, a complete lattice relative to set inclusion as order, and the maps 2^f preserve arbitrary suprema (unions). Hence the functor in question corestricts to a functor $X \mapsto 2^X$: Set → Sup (in the terminology of the Compendium, 1.1, p. 179).

2.1 DEFINITION Let X be a set and M a complete lattice. For a function $g : X \to M$, we define the map $g^* : 2^X \to M$ by $g^*(A) = \sup g(A)$ for every $A \subset X$.

2.2 PROPOSITION Let $g : X \to M$ be a function from a set into a complete lattice. Then the function $g^* : 2^X \to M$ has an upper adjoint $u : M \to 2^X$ given by $u(m) = g^{-1}(\downarrow m)$.

Proof: Let $A \subset X$ and let $m \in M$. Then $g^*(A) \leq m$ iff $\sup g(A) \leq m$ iff for all $a \in A$ we have $g(a) \leq m$, iff for all $a \in A$ the relation $a \in g^{-1}$ $(\downarrow m)$ holds, iff $A \subset g^{-1}(\downarrow m) = u(m)$. ■

2.3 COROLLARY The function $g^* : 2^X \to M$ of Definition 2.1 preserves arbitrary sups (i.e., is a morphism of the category Sup).

Proof: This follows from the general fact that all lower adjoints automatically preserve suprema (see the Compendium, p. 19). ■

At this point we consider the category Mon of posets and monotone maps between them. As an analogue to the exponential functor on Set, we consider the lower set functor L : Mon → Mon given by $L(X) = \{A \subset X : A = \downarrow A\}$ (cf. the Compendium, 1.3, p. 2); if $f : X \to Y$ is a monotone function, we set $L(f)(A) = \downarrow f(A)$. Since $\downarrow g \circ f(A) = \downarrow g(\downarrow f(A))$ (cf. the Compendium, 1.11, p. 5), indeed L is a covariant functor. If $g : X \to M$ is a Mon-morphism for which M is a complete lattice, then $g^*(A) = \sup g(A)$ gives a well-defined map $g^* : L(X) \to M$, which is the restriction of the map of Definition 2.1. The corestriction of the map u of Proposition 2.2 defines a lower adjoint $u : M \to L(X)$ via $u(m) = g^{-1}(\downarrow m)$.

2.4 THEOREM The functor L : Mon → Sup from the category of posets and monotone maps into the category of complete lattices and sup-preserving maps is left adjoint to the grounding functor Sup → Mon. The front adjunction is $\eta_X : X \to L(X)$ by $\eta_X(x) = \downarrow x$.

Proof: For each $g : X \to |M|$ in Mon, for a complete lattice M, the function $g^* : L(X) \to M$ is in Sup by Corollary 2.3. Since g is monotone, we have $g^*\eta_X(x) = g^*(\downarrow x) = \sup g(\downarrow x) = g(x)$. For any $A \in L(X)$, $A = \bigcup \{\downarrow x : x \in A\}$, so that im η_X Sup-generates $L(x)$. Hence, g^* is the unique Sup-morphism satisfying $g^* \circ \eta_X = g$. ■

To bring the category CD into the picture, we refine the category Mon in the spirit of Galois connections. Namely, the category Mon has a subcategory with the same objects, but fewer morphisms: We define the category Gal of Galois connections to have all posets as objects, and those monotone maps $g : X \to Y$ as morphisms that have a lower adjoint $d : Y \to X$ (see the *Compendium*, p. 18ff). A partial motivation for this category can be found in Proposition 2.2 above; the function $g^* : L(X) \to M$ is in Gal. The next result is our core proposition in this discussion.

2.5 PROPOSITION Let $g : X \to M$ be a morphism in Gal such that M is in fact a completely distributive lattice. Then $g^* : L(X) \to M$ preserves arbitrary infima, as well as arbitrary suprema.

Proof: Let $\{A_j : j \in J\}$ be a family in L(X). Since g^* is monotone, the relation

(i) $$g^*(\bigcap_{j\in J} A_j) \le \inf_{j\in J} g^*(A_j) = \inf_{j\in J} \sup g(A_j)$$

is automatic, and we shall prove the reverse inequality. By complete distributivity of M, we have [using Definition 1.1(cd)] the relation

(ii) $$\inf_{j\in J} \sup g(A_j) = \sup \{\inf_{j\in J} g(a_j) : (a_k)_{k\in J} \in \prod_{k\in J} A_k\}$$

Now, set $m = \inf_{j\in J} g(a_j)$, and note that $m \le g(a_j)$ for all $j \in J$. If $d : M \to X$ is the lower adjoint of g (which is guaranteed by hypothesis), then $m \le g(a_j)$ for all $j \in J$ is equivalent to $d(m) \le a_j$ for each $j \in J$. This means that $d(m) \in A_j$ for each $j \in J$, so that $d(m) \in \bigcap_{j\in J} A_j$. In particular, $m \le g \circ d(m) \in g(\bigcap_{j\in J} A_j)$, whence

(iii) $$\inf_{j\in J} g(a_j) \le m \le \sup g(\bigcap_{j\in J} A_j) = g^*(\bigcap_{j\in J} A_j)$$

for all $(a_k)_{k\in J} \in \prod_{k\in J} A_k$. Hence

(iv) $$\sup \{\inf_{j\in J} g(a_j) : (a_k)_{k\in J} \in \prod_{k\in J} A_k\} \le g^*(\bigcap_{j\in J} A_j)$$

In view of (ii) above, this shows

(v) $$g^*(\bigcap_{j\in J} A_j) = \inf_{j\in J} g^*(A_j) \qquad \blacksquare$$

We recall that L(X) is closed in 2^X under arbitrary unions and intersections, and hence is a complete ring of sets; i.e., L(X) is completely distributive. Thus, we may take 2.3 and 2.5 together and conclude

2.6 THEOREM For each morphism $g : X \to M$ in Gal with M completely distributive, the map $g^* : L(X) \to M$ is a CD-morphism.

As a spin-off of these discussions, we can add a few facets to some known characterizations of completely distributive lattices:

2.7 COROLLARY Let M be a complete lattice. The following conditions are equivalent:

(1) M is completely distributive.

(2) For any family $\{A_j : j \in J\}$ of lower sets from M, we have $\sup \bigcap_{j\in J} A_j = \inf_{j\in J} \sup A_j$.

(3) [Resp., (3′), resp., (3″)] The function $id^* : L(M) \to M$ defined by $id^*(A) = \sup A$ preserves all infima (resp., is a CD-morphism, resp., has a lower adjoint).

(4) For each $m \in M$, we have $m = \sup \{n \in M : n \lll m\}$ (where $n \lll m$ iff for any subset P of M, the relation $m \leq \sup P$ implies $n \leq p$ for some $p \in P$).

(5) For all morphisms $g : X \to M$ in Gal, the map $g^* : L(X) \to M$ is in Inf (and, hence, in Inf $\cap$ Sup).

Proof: That (1) implies (5) is just Theorem 2.6, and (5) implies (3) implies (2) implies (1) are clear. Moreover, the equivalence of (3) with (3′) and (3″) follow from Theorem 2.6 applied to id, and from general properties of Galois adjunctions. Finally, we sketch the equivalence of (3″) and (4):

We define the function $s : M \to L(M)$ by $s(m) = \bigcap \{A \in L(M) : m \leq \sup A\}$. If the function $id^* : L(M) \to M$ has a lower adjoint, then s must be that lower adjoint (see, e.g., the Compendium, 3.4, p. 19). On the other hand, s is the lower adjoint of id^* iff $m = id^*(s(m)) = \sup s(m)$ for all $m \in M$, since $id^* : L(M) \to M$ is surjective (see the Compendium, 3.7, p. 20). But we note that the relation $m = s(m)$ for all $m \in M$ is equivalent to (4). ∎

The equivalence of (1) and (4) belongs to Raney's [1952] repertory of early information on completely distributive lattices.

Our purpose, of course, is the elucidation of universal properties, and for this purpose we define, for a poset X, a natural MON-morphism $\eta_X : X \to L(X)$ by $\eta_X(x) = \downarrow x = \{x\}^-$ (where the closure is with respect to the Scott topology).

2.8 LEMMA For a poset X, the following are equivalent:

(1) X is a complete lattice.

(2) η_X is a Gal-morphism.

Proof: (1) implies (2): The map η_X preserves arbitrary infima, as is readily seen. Hence, by (1) it has a lower adjoint (see the Compendium, 3.4, p. 19).

(2) implies (1): If $d : L(X) \to X$ is a lower adjoint for η_X, then for each lower set $A \subset X$, the element $d(A)$ is the smallest element x in X with $\sup A \leq x$; i.e., $d(A) = \sup A$. Since $\sup A = \sup {\downarrow}A$ for any subset A of X, we see that all subsets of X have suprema, so X is a complete lattice. ∎

We now formulate our first result on the universality of the construction L(X).

2.9 THEOREM For each Gal-morphism $g : X \to M$ into a completely distributive lattice M, there is a unique CD-morphism $g^* : L(X) \to M$ such that $g = g^* \circ \eta_X$.

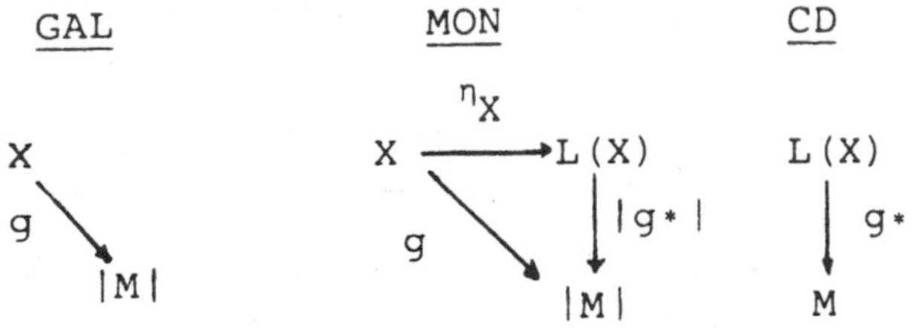

(where $|M|$ denotes the underlying poset of M).

Proof: We define g^* as in 2.1, and then we know from Corollary 2.7 that g^* is a CD-morphism; moreover, $g^*(\eta_X(x)) = \sup g({\downarrow}x) = g(x)$ (since g is monotone). But every lower set $A \in L(X)$ is the union of the principal lower sets which it contains, and so L(X) is generated in CD by the image of η_X, so that the relation $g = g^* \circ \eta_X$ determines g^* uniquely in CD. ∎

The universal property expressed in Theorem 2.9 is strongly reminiscent of an adjoint situation. Indeed, it looks at first sight as though we have shown that L is left adjoint to a grounding functor. However, a closer inspection reveals that there is an obstruction in the form of the candidate η_X for the front adjunction. This map is not in general in Gal; in fact, we have seen in Lemma 2.8 that η_X is in Gal precisely when X is a complete lattice. On the other hand, if g were a Mon-morphism, there would be no reason to believe that g^* is a CD-morphism (cf. Corollary 2.3).

2.10 COROLLARY Under the circumstances of Theorem 2.9, we have a commutative diagram of natural injections:

$$\begin{array}{ccc} \underline{\mathrm{GAL}}(X, |M|) & \xrightarrow{\;g \mapsto g^*\;} & \underline{\mathrm{CD}}(L(X), M) \\ & \searrow^{\text{inclusion}} & \downarrow{\scriptstyle f \mapsto f \circ \eta_X} \\ & & \underline{\mathrm{MON}}(X, |M|) \end{array}$$

The map $(\ )^*$ is a bijection iff η_X is a Gal-morphism, iff X is a complete lattice, in which case its inverse is $f \mapsto f \circ \eta_X$.

As a consequence, we obtain a genuine adjunction (or freeness) theorem as follows:

2.11 THEOREM The grounding functor $|\,|$: CD → Inf from the category CD of completely distributive lattices and complete lattice morphisms into the category Inf of complete lattices and morphisms preserving all infima has a left adjoint L : Inf → CD with injective front adjunction $\eta_X : X \to L(X)$ by $\eta_X(x) = \downarrow x$.

Proof: Indeed, the full subcategory of Gal consisting of complete lattices is precisely Inf (see the Compendium, p. 179). ∎

This theorem allows a proof of the result of Dwinger [1981] and Markowsky [1973,1979] on the existence and structure of free completely distributive lattices over Set. Indeed, we note that the grounding functor Inf → Set has a left adjoint Φ : Set → Inf given by $\Phi(X) = (2^X)^{op}$, i.e., the lattice of all subsets of X with the order $A \le B$ iff $B \subset A$. The front adjunction is the map $x \mapsto \{x\} : X \to \Phi(X)$, and if $f : X \to M$ is a map in Set with M a complete lattice, then the map $f' : \Phi(X) \to M$ by $f'(A) = \inf f(A)$ is the unique Inf-map satisfying $f'(\{x\}) = f(x)$ for all $x \in X$. Thus the functor $L \circ \Phi$: Set → CD is left adjoint to the grounding functor from CD into Set, since left adjoints compose. This yields the results of Dwinger and Markowsky.

2.12 COROLLARY The category CD of completely distributive lattices has a free functor over Set, i.e., for every set X, there is a free completely distributive lattice $L\Phi(X)$ generated by X, and $L \circ \Phi(X) = \{A \subset 2^X : A = A\}$, and the front adjunction is $x \mapsto \{A \subset X : x \in A\}$. The cardinality of $L\Phi(X)$ is $2^{2^{\mathrm{card}\,X}}$.

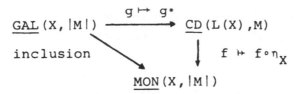

On the other hand, the results of Liber [1977] do not follow from Theorem 2.9, but rather from a variant which we discuss in the following.

3. SUBADJUNCTIONS

The situation we encountered in Theorem 2.9 and Corollary 2.10 will occur in other instances, and so we fix a general framework that fits all of these circumstances. We recall the situation of Theorem 2.9: We have two categories, $\underline{A} = \underline{\text{Gal}}$ and $A' = \underline{\text{Mon}}$ with the same objects, and with $\underline{A} \subset A'$, and we have two functors $U : \underline{B} \to \underline{A}$ and $F : \underline{A} \to \underline{B}$, with $B = \underline{\text{CD}}$, which are nearly adjoints to one another, except that the candidate for the front adjunction, $\eta : 1_{\underline{A}} \to F \circ U$ is in $\underline{A}'$, rather than in $\underline{A}$.

3.1 DEFINITION Let $\underline{A} \subset \underline{A}'$ be two categories with the same objects, and let $U : \underline{B} \to \underline{A}$ be a functor. Suppose that $F : \underline{A} \to \underline{B}$ is a functor and $A \mapsto \eta_A$ a function which associates to each object of $\underline{A}$ an $\underline{A}'$-morphism $\eta_A : A \to UFA$. We say that F is <u>left subadjoint</u> to U iff F satisfies the following universal property:

> For each morphism $f : A \to UB$ in $\underline{A}$, there is a unique $\underline{B}$-morphism $f' : FA \to B$ with $f = (Uf')_{\eta_A}$

3.2 LEMMA If F is left subadjoint to U, then the functions $f \mapsto f'$: $\underline{A}(A,UB) \to \underline{B}(FA,B)$ and $g \mapsto (Ug)_{\eta_A} : \underline{B}(FA,B) \to \underline{A}'(A,UFA)$ are (natural) injections, and their composition is the natural inclusion $\underline{A}(A,UB) \to \underline{A}'(A,UB)$. The image of $1_{FA} \in \underline{B}(FA,FA)$ in $\underline{A}'(A,UB)$ is η_A.

3.3 PROPOSITION Suppose that F is left subadjoint to U. Then the following are equivalent:

(1) $\eta_A : A \to UFA$ is in $\underline{A}$.

(2) $f \mapsto f' : \underline{A}(A,UB) \to \underline{B}(FA,B)$ is a (natural) bijection for all objects B in $\underline{B}$.

Proof: (1) implies (2): If (1) is satisfied, then $g \mapsto (Ug)\eta_A$: $\underline{B}(FA,B) \to \underline{A}(A,UB)$ is well defined and is the inverse of the map in (2) by Lemma 3.2.

(2) implies (1): We know from Lemma 3.2 that the image of $1_{FA} \in \underline{B}(FA,FA)$ in $\underline{A}'(A,UFA)$ is η_A. But, by (2), the function $f \mapsto f'$ maps $\underline{A}(A,UFA)$ surjectively onto $\underline{B}(FA,FA)$. It then follows that η_A is in $\underline{A}(A,UFA)$. ∎

3.4 NOTATION In the set-up of Definition 3.1, we denote by $\underline{A}_0$ the full subcategory of $\underline{A}$ containing all objects A for which η_A is in $\underline{A}$.

3.5 PROPOSITION If F is left subadjoint to U, and if the range of U is contained in $\underline{A}_0$, then the restriction of F to $\underline{A}_0$ is left adjoint to the corestriction $U : \underline{B} \to \underline{A}_0$.

Proof: This is an immediate consequence of Proposition 3.3. ■

4. SCOTT CONTINUOUS FUNCTIONS AND FREE DISTRIBUTIVE LATTICES

We saw in Section 1 that the most important posets in the context of completely distributive lattices (and for many other topological discussions) are the up-complete posets. We therefore let $\underline{U}$ denote the category of all up-complete posets together with all functions preserving directed sups (i.e., functions that are continuous in the Scott topology; see the Compendium, 2.11, p. 15, and 2.18, p. 119). We let $\underline{Up}$ be the subcategory of $\underline{U}$ having the same objects, but whose morphisms are exactly those of Definition 1.7, i.e., those Scott continuous morphisms that also have Scott continuous lower adjoints.

We recall that, for any up-complete poset X, we denote by $\sigma(X)$ the Scott topology on X (see the Compendium, p. 99), and the opposite lattice $\sigma(X)^{op}$ can be concretely realized as the lattice of all Scott closed sets, which we denote by $\Gamma(X)$ (cf. the Compendium, p. 286).

4.1 LEMMA Let X be an up-complete poset. Then the function $\eta_X : X \to \Gamma(X)$ by $\eta_X(x) = \downarrow x$ is Scott continuous, and the following statements are equivalent:

(1) η_X is a $\underline{Up}$-morphism (i.e., η_X has a lower adjoint).
(2) X is a complete lattice.

Proof: Since X is up-complete, η_X preserves directed sups, and hence is Scott continuous (cf. the Compendium, pp. 112 and 119). Condition (1) says precisely that for each $A \in \Gamma(X)$ there is a smallest element $a \in X$ with $A \subset \downarrow a$, i.e., that each $A \in \Gamma(X)$ has a sup. An arbitrary subset of X has a sup iff its Scott closure has a sup, and these two suprema agree. Hence (1) and (2) are equivalent. ■

If X is an up-complete poset, and if $g : X \to M$ is a function from X into the complete lattice M, then we denote the restriction of the map $g^* : 2^X \to M$ to $\Gamma(X)$ again by $g^* : \Gamma(X) \to M$ [where $g^*(A) = \sup g(A)$].

4.2 LEMMA If $g : X \to M$ is a Scott continuous map of up-complete posets, and if M is a complete lattice, then $g^* : \mathcal{T}(X) \to M$ has an upper adjoint $u : M \to \mathcal{T}(X)$ given by $u(m) = g^{-1}(\downarrow m)$. In particular, g^* is in Sup. If, in addition, M is completely distributive and g has a lower adjoint, then g^* is in Inf $\cap$ Sup.

Proof: After Proposition 2.2, it suffices to show that the image of $u : M \to 2^X$ is contained in $\mathcal{T}(X)$. But, since g is Scott continuous and $\downarrow m$ is Scott closed for every $m \in M$, this is immediate from the definition of u. The remainder follows from Proposition 2.5 and the fact that $\mathcal{T}(X)$ is closed in 2^X under the formation of arbitrary intersections. ■

4.3 LEMMA Let $f : S \to T$ be a U-morphism, and define $\mathcal{T}(f) : \mathcal{T}(S) \to \mathcal{T}(T)$ by $\mathcal{T}(f)\ (A) = [\downarrow f(A)]^-$. Then $\mathcal{T}(f)$ has the upper adjoint $B \mapsto f^{-1}(B) : \mathcal{T}(T) \to \mathcal{T}(S)$, whence $\mathcal{T}(f)$ is in Sup. Furthermore, we have $\mathcal{T}(f) = (\eta_T \circ f)^*$ and $\eta_T \circ f = \mathcal{T}(f) \circ \eta_S$. The following also hold:

(i) If T is a complete lattice and if f is in Up, then $\mathcal{T}(f) \in$ Inf $\cap$ Sup.

(ii) If S is a complete lattice, then for every Inf $\cap$ Sup-morphism $F : \mathcal{T}(S) \to \mathcal{T}(T)$, the map $g = F \circ \eta_S : S \to \mathcal{T}(T)$ is an Up-morphism such that $g^* = F$.

The following diagram illustrates the situation:

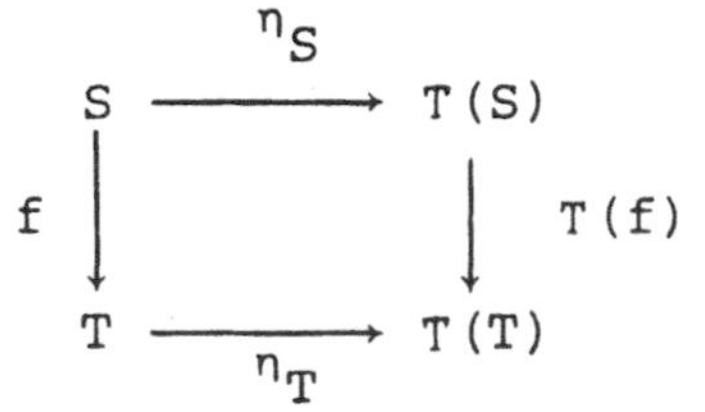

Proof: We have $A \subset f^{-1}(B)$ for $A \in \mathcal{T}(S)$ and $B \in \mathcal{T}(T)$ iff $f(A) \subset B$, iff $[\downarrow f(A)]^- \subset B$, since B is Scott closed. This proves the first claim. Furthermore, we have $\mathcal{T}(f)(A) = [\downarrow f(A)]^- = (\bigcup\{\downarrow f(a) : a \in A\})^- = \sup\{(\eta_T \circ f)(a) : a \in A\} = \sup\,(\eta_T \circ f)(A) = (\eta_T \circ f)^*(A)$. Also, $\eta_T \circ f(s) = \downarrow f(s) = \downarrow[f(\downarrow s)]^-$ as f is monotone, and this set is closed, being a principal ideal. Thus, $\eta_T \circ f(s)$ agrees with $[\downarrow(f(\downarrow s))]^- = \mathcal{T}(f)\ (s)$.

If T is a complete lattice and f is in Up, then η_T is a Up-morphism by Lemma 4.2, and so $f \circ \eta_T$ is a Up-morphism, whence $\mathcal{T}(f) = (f \circ \eta_T)^*$ is in Inf $\cap$ Sup by 4.2. If S is a complete lattice and F is as in (ii), then $g = F \circ \eta_S$ is a Up-morphism since η_S is one by Lemma 4.1. Also, $g^*(A) = \sup g(A) = \sup F \circ \eta_S(A) = F(\sup \eta_S(A)) = F([\bigcup\{\downarrow a : a \in A\}]^-) = F(A^-) = F(A)$.

■

At this stage we can formulate the following theorem:

4.4 THEOREM The functor $\mathcal{T} : \underline{U} \to \underline{Sup}$ is left adjoint to the grounding functor $|\,| : \underline{Sup} \to \underline{U}$ with front adjunction $\eta_X : X \to \mathcal{T}(X)$ by $\eta_X(x) = {\downarrow}x$.

Proof: For each $g : X \to |M|$ in $\underline{U}$ with M a complete lattice, the function $g^* : \mathcal{T}(X) \to M$ is in $\underline{Sup}$ by Lemma 4.2. We have $g^* \circ \eta_X(x) = g^*({\downarrow}x) = \sup g({\downarrow}x) = g(x)$, since g is monotone. Since $\text{im}\, \eta_X = \{{\downarrow}x : x \in X\}$ $\underline{Sup}$-generates $\mathcal{T}(X)$, it follows that g^* is the unique $\underline{Sup}$-morphism satisfying $g^* \circ \eta_X = g$. ∎

For an arbitrary up-complete poset X, the lattice $\mathcal{T}(X)$ is complete and distributive (in fact, its opposite is a complete Heyting algebra). However, from Lawson's [1979] theory, we record the following lemma:

4.5 LEMMA For an up-complete poset X, the following are equivalent:

(1) X is a continuous poset.

(2) $\mathcal{T}(X)$ is a completely distributive lattice. ∎

We denote by $\underline{CP}$ the full subcategory of $\underline{UP}$ containing all continuous posets.

4.6 THEOREM (a) The functor $\mathcal{T} : \underline{CP} \to \underline{CD}$ is left subadjoint to the grounding functor $\underline{CD} \to \underline{CP}$ (with respect to the category $\underline{U}$ containing $\underline{CP}$).

(b) In particular, there is a commutative diagram of natural inclusions

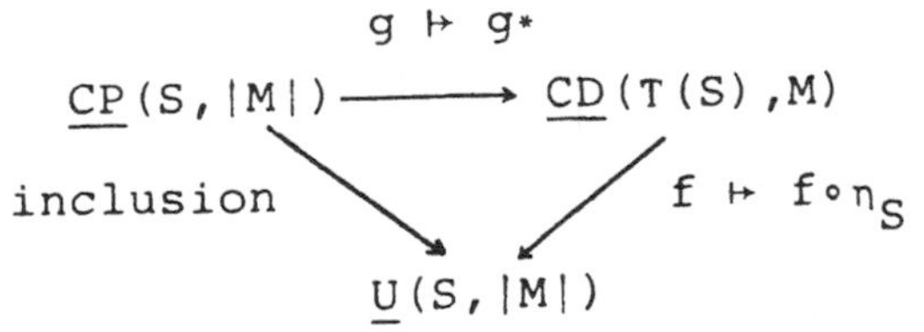

(c) If S is a continuous poset, then the front adjunction η_S identifies S with the cospectrum of the completely distributive lattice $\mathcal{T}(S)$.

Proof: First, for X in $\underline{CP}$, we have $\mathcal{T}(X)$ in $\underline{CD}$ by Lemma 4.5, so, second, if $g : S \to |M|$ is in $\underline{CP}$ with M in $\underline{CD}$, then $g^* : \mathcal{T}(S) \to M$ is in $\underline{CD}$ by Lemma 4.2. Then (a) and (b) are clear from Theorem 4.4 and Section 3. Regarding (c), we know that $A \in \mathcal{T}(S)$, $A \subset S$ is in the cospectrum of $\mathcal{T}(S)$ iff A is an irreducible closed subset of $\Sigma S = (S, \sigma(S))$. Since ΣS is sober (due to the continuity of S; see the Compendium, 1.20, p. 109), then A is a singleton closure, so that $A = {\downarrow}s$ for some $s \in A$. ∎

We recall that $\underline{\text{Inf}}^{\uparrow}$ denotes the category of complete lattices and maps preserving all infima and all directed suprema (see the Compendium, 1.9, p. 182). Thus, from 3.5 and 4.6 we obtain the following freeness theorem, first proved by Liber [1977] in somewhat different terminology:

4.7 THEOREM The functor $\mathcal{T} : \underline{\text{CL}} \to \underline{\text{CD}}$ is left adjoint to the grounding functor $\underline{\text{CD}} \to \underline{\text{CL}}$. ∎

To estimate the size of the free completely distributive lattice, we recall that the most relevant cardinal invariant of a continuous lattice is its weight (see the Compendium, p. 168ff). If L is an infinite distributive continuous lattice, then $w(L) = w(\Lambda L) = w(\sigma L)$, where ΛL is the space L with the Lawson topology (see the Compendium, 4.7, p. 170). If the lattice M is completely distributive, then $\lambda M = \lambda M^{op}$ (see the Compendium, 2.9, p. 318). Thus, $w(M) = w(M^{op})$ for any completely distributive lattice M, and $w(\sigma(L)) = w(\sigma(L)^{op}) = w(\mathcal{T}(L))$ for any continuous lattice L. It follows that generating a free completely distributive lattice by a continuous lattice does not increase the weights:

4.8 COROLLARY If M is a completely distributive lattice that is free over an infinite continuous lattice L, then L and M have equal weight. ∎

Liber [1977] noticed that, from Theorem 4.7, one may deduce that CD has a free functor over compact spaces, since the category CL is free over the category Comp of compact Hausdorff spaces and continuous maps. Indeed, if X is a compact Hausdorff space and if $\Gamma(X)^{op}$ denotes the lattice of compact subsets of X with reverse containment as the order, then $\Gamma(X)^{op}$ is the free CL-object over X (see the Compendium, p. 285). Since $\Gamma(X)^{op}$ is isomorphic to the lattice $O(X)$ of open subsets of X, we know that the weight of the continuous lattice $\Gamma(X)^{op}$ is equal to the weight of the space X (see the Compendium, 4.5, p. 170).

4.9 COROLLARY For each compact Hausdorff space X there is a completely distributive lattice M with $w(M) = w(X)$ (in case X is infinite), and M is freely generated by X. Namely, $M = \mathcal{T}(\Gamma(X)^{op}) \cong \mathcal{T}(O(X))$. ∎

Since the Stone-Čech compactification $\beta : \underline{\text{Set}} \to \underline{\text{Comp}}$ is left adjoint to the grounding functor, one may recover the result of Dwinger and Markowsky too. If X is a set, then $\mathcal{T}(\Gamma(\beta X)^{op}) \cong \mathcal{T}(\text{Id } 2^X)$ is the free completely distributive lattice over the set X, and its weight, for X infinite, is $w(\mathcal{T}(\text{Id } 2^X) = w(\text{Id } 2^X)$ [since $w(\mathcal{T}(L)) = w(\sigma(L)) = w(L)$ for any infinite continuous lattice L; see the Compendium, 4.7, p. 170]. But, $w(\text{Id } 2^X) =$

card 2^X, since $w(L) = \text{card } K(L)$ for an algebraic lattice L (see the Compendium, pp. 169-170), and since card $2^X = 2^{\text{card } X}$, we conclude that $w(T(\Gamma(\beta X)^{op})) = 2^{\text{card } X}$. ∎

4.10 REMARK For an infinite set X, the weight of the free completely distributive lattice over X is $2^{\text{card } X}$.

One should contrast the weight of a free CD-object with its cardinality (see Definition 2.1).

5. COMPLEMENTARY RESULTS

The universal property expressed in Theorem 4.6 concerns a completely distributive lattice M and its cospectrum S = Cospec M. From Theorem 4.7 we know that M is free (over CL) iff S is a lattice, in which case M is freely generated by S. The Lawson and Scott topologies of the continuous poset S are the ones induced from the respective topologies of M (see, e.g., Hoffmann [1982]).

5.1 NOTATION Let M be a complete lattice and S any subset of M. We write

(i) $G(S) = S \cup S^2 \cup S^3 \cup \cdots$ where $S^n = \{s_1 \wedge s_2 \wedge \cdots \wedge s_n : s_i \in S\}$.

(ii) $D(S) = \{\inf X : X \subset S$ is filtered and nonempty$\}$.

(iii) $U(S) = \{\sup X : X \subset S$ is directed and nonempty$\}$.

(iv) $I(S) = \{\inf X : X \subset S\}$.

(v) $\gamma(S) = DI(S)$.

(The letter D stands for "down," and U for "up.")

5.2 LEMMA Let M be a complete lattice and let S be a subset of M.

(a) $\gamma(S) = UDG(S) \cup \{1\}$.

(b) The smallest lim inf closed subsemilattice of M containing S is UDG(S) (see the Compendium, p. 158ff).

(c) If M is continuous, then the lim inf topology and the Lawson topology agree.

Proof: (a) is clear since $DG(S) \cup \{1\} = I(S)$ (see, e.g., the Compendium, 1.5, p. 3).

(b): Let T be the smallest lim inf closed subsemilattice of M containing S. The relation $G(S) \subset T$ is trivial, and $UDG(S) \subset T$ follows from the Compendium, 3.4, p. 153. In order to see the reverse inclusion, we first observe that $T = G(S)^-$. Thus, $t \in T$ means that we have $t = \underline{\lim}\, \mathfrak{F}$ for some ultrafilter $\mathfrak{F}$ on G(S). By the Compendium, p. 159, we conclude that $t \in UDG(S)$.

(c): See the Compendium. 3.9, p. 161.   ∎

5.3 DEFINITION For an up-complete poset S, we define M = T(S) and we set
$\gamma(S)$ = UI(S) ∪ {1} (in M).   ∎

5.4 REMARK If S is an arbitrary continuous poset, we may identify S with
the cospectrum of T(S) via s ↦ ↓s. Under this identification, I(S) is the
poset of all Scott closed subsets of S that are intersections of singleton
closures (principal lower sets), and $\gamma(S)$ is the poset of all $(\cup A_j)^-$,
where A_j is an ascending family of closed lower sets from I(S), and of the
set S itself.

5.5 LEMMA If f : S → T is a CP-morphism, then the map T(f) : T(S) → T(T)
according to 3.3 induces a CL-morphism $\gamma(f)$: $\gamma(S)$ → $\gamma(T)$ by restriction
and corestriction.

    Proof: By Proposition 3.3, T(f) is a CD-morphism, i.e., it preserves
all infs and all sups. In particular, it preserves all infs and all directed
sups, so it is a CL-morphism. Since T(f) (Cospec T(S)) ⊂ Cospec T(T), it
then follows that the CL-subobject T(S) generated by Cospec T(S) is mapped
into the CL-subobject T(T) generated by Cospec T(T).   ∎

    This means that γ is a functor from CP to CL.

5.6 THEOREM The functor γ : CP → CL from the category of continuous posets and their morphisms (in the sense of 1.7) is left subadjoint to the
forgetful functor from CL to CP (with respect to the category U). In particular, if η_S : S → $\gamma(S)$ is given by $\eta_S(s)$ = ↓s, then for each CP-morphism
f : S → L into a continuous lattice L, there is a unique morphism f′ : $\gamma(S)$
→ L in CL so that f′ ∘ η_S = f. The function f ↦ f′ ∘ η_S : CP(S,|L|) →
CL($\gamma(S)$,L) is a natural injection.
    Proof: For a continuous poset S, a continuous lattice L, and a CP-
morphism f : S → L, we obtain the commutative diagram of CP-morphisms:

$$\begin{array}{ccccc}
S & \xrightarrow{\eta_S} & \gamma(S) & \longrightarrow & T(S) \\
f \downarrow & & \gamma(f) \downarrow & & \downarrow T(f) \\
L & \xrightarrow{\eta_L} & \gamma(L) & \longrightarrow & T(L)
\end{array}$$

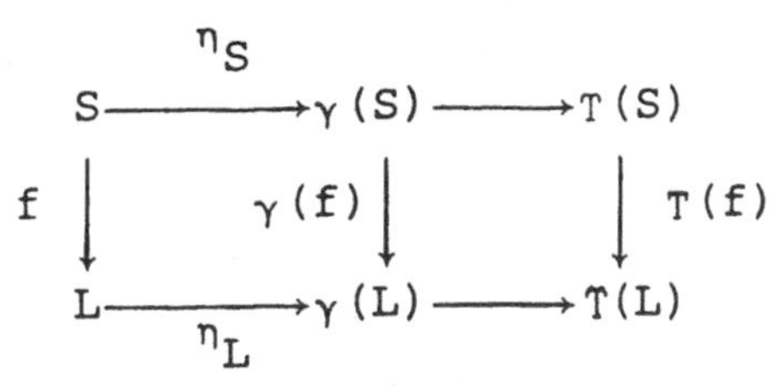

However, since L is a continuous lattice and $\gamma(L)$ is the smallest continuous
lattice subobject of T(L) generated by the image of L, and since a complete
lattice L satisfies the embedding s ↦ ↓s : L → T(L) is an Inf -map, we

conclude that $\eta_L : L \to \gamma(L)$ is an isomorphism. Thus the CL-map $\eta_L^{-1} \circ \gamma(f)$ is the required f', and its uniqueness follows from the fact that $\gamma(S)$ is CL-generated by im η_S. ∎

Theorem 5.6 gives us an example of a left subadjunction that becomes trivial if it is restricted to the largest full subcategory of those S in CP for which η_S is a CP-morphism, for this is exactly the case for continuous lattices by Proposition 3.5.

The functor γ on CP has another significance, as was recently discovered in the work of Hoffmann [1983]. If S is a continuous poset, which we identify via the map η_S with a subposet of $L = \gamma(S)$, then the following conditions are satisfied:

(i) S is join dense in L [i.e., $x = \sup(\downarrow x \cap S)$ for all $x \in L$].

(ii) S is closed in L under the formation of all directed sups.

(iii) If $s \ll_S t$ in S, then $s \ll_L t$ in L.

(iv) L is the smallest CL-subalgebra of L containing S; i.e., $L = \mathrm{UDG}(S) \cup \{1\}$.

Indeed, we may assume that S is Cospec M for a completely distributive lattice M [$\equiv \Gamma(S)$], since S is a continuous poset, and we know that L is the CL-subalgebra of M generated by S by Lemma 5.2 (cf. the Compendium, 1.11, p. 146). This proves (iv). Since Cospec M is join dense in M, it follows that S is join dense in L, i.e., (i) is satisfied. It is well known that the infimum of a filtered set of primes is again a prime, whence S is closed in M under the formation of directed sups, so (ii) follows. Finally, if $s \ll_S t$, then $s \ll_M t$ (indeed, $\Downarrow_M t \cap S$ is directed and has t as its sup (see the Compendium, 1.9, p. 240), hence $s \ll_S t$ implies there is some u with $s \le u \in \Downarrow_M t \cap S$, whence $s \ll_M t$). Then (iii) follows.

Now, by Theorem 1.7 of Hoffmann [1982], we may conclude that L is the injective hull of S in the following sense: For each T_0-space X, Banaschewski [1977] constructed a unique largest essential extension $X \to \lambda X$, which he called the essential hull of X. Now, it is not always the case that the essential hull of a space is injective in the category of T_0-spaces with respect to the class of embeddings (for the concept of injectivity, see the Compendium, p. 121ff., notably Theorem 3.8, p. 124). However, it was shown by Banaschewski [1977] and Hoffmann [1981] that if X is a continuous poset S endowed with the Scott topology, then λX is injective. By a slight abuse of notation, we call the underlying poset obtained by considering λX in the specialization order the injective hull of S. Hence, we have:

5.7 PROPOSITION The map $\eta_S : S \to \gamma(S)$ for a continuous poset S is the injective hull of S. (See also Hoffmann [1983] and Hofmann [1981].)

If S continues to be the cospectrum of a completely distributive lattice M, we point out that there is yet another alternative way of "spanning" $\gamma(S) = L$ within M; this insight we owe to Hoffmann [1982]. We recall that a pseudo-prime element p of a complete lattice L is the supremum of a prime ideal, i.e., the supremum of a prime element in the ideal lattice of L. Dually, a pseudo-coprime element p' is the infimum of a prime filter of L, i.e., the infimum of a prime element in the filter lattice of L (we do not assume that the identity is prime, nor that the zero is coprime in any lattice) (see the Compendium for the concept of pseudo-prime element, and Hoffmann [1982] for the concept of pseudo-coprime). Moreover, it happens that the set of pseudo-coprimes of a completely distributive lattice M is precisely the CL-closure of the set S = Cospec M (see the Compendium, p. 284 and Hoffmann [1982, proposition 3.10]. Notice that Hoffmann improved the conceptual apparatus of pseudo-coprimes somewhat over that in the Compendium in his slightly different treatment of the identity.). From Hoffmann [1982, proposition 3.6], we know that $\gamma(S) = L$ is simply $I(\Psi M)$. Thus we have the following result:

5.8 PROPOSITION Let S be a continuous poset and $\overline{S}$ its CL-compactification [i.e., its CL-closure in $M = \Upsilon(S)$, where we identify S with its image $\eta_S(S)$ in $\Upsilon(S)$]. Then $\gamma(S) = I(\overline{S}) = \{x \in M : x = \inf(\, x \cap \overline{S})\}$.

REMARK A lattice-theoretical characterization of the CL-closure of a subset A of a continuous lattice L is not entirely simple; $\overline{S}$ consists of all lim inf x_j with a net $\{x_j\} \subset S$, or, equivalently, $\overline{S}$ is the set of all lim inf F as F ranges over the ultrafilters on S. (For lim-inf issues, see the Compendium, sections II-1 and III-3).

Proof of Proposition 5.8: A direct proof in the spirit of free constructions is simple: Since $\gamma(S)$ is the smallest CL-closed inf subsemilattice with identity in M that contains S, the inclusion $I(\overline{S}) \subset \gamma(S)$ is clear, as is $\gamma(S) \subset DI(S) \subset I(\overline{S})^-$. The claim then follows from the simple lemma below:

5.9 LEMMA If L is a continuous lattice and $X \subset L$ a CL-closed subset, then $I(X)$ is closed in L.

Proof: We have $x \in I(X)$ iff $x = \inf(\,\downarrow x \cap X)$, so $I(X)$ is the image of the compact space $\Gamma(X)$ under the map $A \mapsto \inf A : \Gamma(X) \to L$ (see the Compendium, 3.9, p. 285). ■

The final version of this paper was completed while both authors enjoyed the hospitality of the Technische Hochschule in Darmstadt, the second as a fellow of the Alexander von Humboldt Stiftung.

REFERENCES

Banaschewski, B., Essential extensions of T_0-space, General Topology and Its Applications 7 (1977), 233-246.

Gierz, G., K. H. Hofmann, K. Keimel, J. D. Lawson, M. Mislove, and D. Scott, A Compendium of Continuous Lattices, Springer-Verlag, Berlin-Heidelberg-New York (1980), 371 pp.

Day, A., Filter monads, continuous lattices, and closure systems, Canadian Journal of Mathematics 27 (1975), 50-59.

Dwinger, P., Structure of completely distributive complete lattices, Proceedings of the Royal Netherlands Academy of Science A 84 (1981), 361-373.

Hoffmann, R.-E., Continuous posets, prime spectra of completely distributive lattices, and Hausdorff compactifications, in Lecture Notes in Mathematics 871 (1981), Springer-Verlag, Berlin-Heidelberg-New York, 159-208.

Hoffmann, R.-E., Essentially complete T_0-spaces II, a lattice theoretical approach, Mathematische Zeitschrift 179 (1982), 73-90.

Hoffmann, R.-E., The CL-compactification and the injective hull of a continuous poset, Preprint 1981/1982, to appear.

Hofmann, K. H., The category CD of completely distributive lattices and their free objects, SCS-Memo 11-24-81.

Lawson, J. D., The duality of continuous posets, Houston Journal of Mathematics 5 (1979), 357-394.

Liber, S. I., On Z-free compact lattices, Issledovaniya Po Algebre 5 (1977), Saratov (in Russian).

Markowsky, G., Free completely distributive lattices, Proceedings of the American Mathematical Society 74 (1979), 227-228.

Niño, J., The category of continuous posets, Tulane University Dissertation (1981), New Orleans.

Raney, G. N., Completely distributive complete lattices, Proceedings of the American Mathematical Society 3 (1952), 667-680.

Wyler, O., Algebraic theories of continuous lattices, in Lecture Notes in Mathematics 871 (1981), Springer-Verlag, Berlin-Heidelberg-New York, 390-413.

8

Discontinuity of Meets and Joins

JOHN ISBELL

State University of New York at Buffalo
Buffalo, New York

1. This paper concerns examples of complete lattices L for which binary meet, or join, from L x L to L, is discontinuous with respect to the Scott topology on L and the product topology on L x L. One example, L_1, is a topology (not, of course, quasi-locally compact) with both meet and join discontinuous. Another, simpler example, L_0, is a sober space in its Scott topology, with join discontinuous (and meet not even separately continuous). I do not know whether L_1 is sober.

The sobriety is of interest because continuity of join is known to imply it [1,II.1.12]; so we are refuting the converse (in answer to a question of Hoffmann).

More interesting, and unfinished, is the following system of implications. Note first that if the lattice L is not algebraically meet continuous (MC), then meet is (trivially) discontinuous. Now if L has discontinuous join, or is MC with discontinuous meet, then the product topology on L x L (from the Scott topology on L) differs from the Scott topology on L x L; and that implies that both join and meet from L^4 to L^2 are discontinuous. [The last, with respect to the Scott topology on L^2 and product topology on $(L^2)^2$.]

2. Scott observed that join V: L x L → L is Scott continuous on any complete lattice L, i.e., continuous with respect to the Scott topologies of L and of L x L, and that the same holds for meet if and only if L is MC. (This is in [3], 2-6 and its proof.) Therefore, if L is MC, discontinuity of either for the product topology implies that that is not the Scott topology.

But the product topology is contained in the Scott topology of $L \times L$. Thus a difference between them requires a set (or net) of points (x_α, y_α) having a limit point (a,b) in the product topology that is not a limit point in the Scott topology. Pause and observe that this applies as well to a product $M \times N$ of two different MC lattices. Now consider the Scott topology on $M \times N$, and the product topology on $(M \times N)^2$. The points $(x_\alpha, 1, 1, y_\alpha)$ have $(a, 1, 1, b)$ as a limit point; but $\wedge$ takes them to (x_α, y_α), and $(a, 1) \wedge (1, b) = (a, b)$ is not a limit point. Exactly the same happens along the set (or net) of points $(x_\alpha, 0, 0, y_\alpha)$ if we apply join.

3. For the (proven) sober example L_0 we need only, by 2 above, complete lattices M_0, N_0 with Scott $\neq$ product topology on $M_0 \times N_0$ which is sober in the Scott topology. The elements of M_0 will be ordered pairs (n, α) with $n < \omega_0$ and $\alpha < \omega_1$, elements (α) with $\alpha < \omega_1$, and a least element 0 and a greatest element 1. [Don't confuse 0 and 1 with (0) and (1).] The successors of (α) are 1 and all (β), $\beta \geq \alpha$. The successors of (n,α) are all (m,α), $m \geq n$, and (α) and its successors. So the successors of a nonzero element are well ordered. A nonempty set then has a least upper bound, viz., the smallest successor of some element that is an upper bound (and 1 is certainly an upper bound). Since the empty set also has a least upper bound 0, M_0 is indeed a complete lattice.

N_0 is "the opposite." More fully, its elements are pairs (α,n) with $(n,\alpha) \in M_0$, elements (n) with $n < \omega_0$, and 0 and 1. The successors of (n) are 1 and all (m) with $m \geq n$; the successors of (α,n) are all (β,n), $\beta \geq \alpha$, and (n) and its successors. Like M_0, N_0 is a complete lattice with the successors of a nonzero element well ordered.

In the product topology on $M_0 \times N_0$, each neighborhood of $(1,1)$ contains some point $((n,\alpha), (\alpha,n))$. For a neighborhood of $1 \in N_0$ contains almost all (n); and since a countable set of countable ordinals is bounded, it contains a "corner" of N_0, $\{(\alpha,n): \alpha \geq \gamma, n \geq c\}$ for some γ, c. Evidently, every neighborhood of 1 in M_0 meets (the "opposite" of) this corner.

In the Scott topology, however, the closure of the set S of all $((n,\alpha), (\alpha,n))$ is the union of $M_0 \times \{0\}$, $\{0\} \times N_0$, and the set of all (p,q) with $p \neq 0$, $q \neq 0$, less than some element of S. It contains that union (which we don't need to know) trivially. Conversely, the union $\mathcal{U}$ is a lower set. If D is a directed subset of $\mathcal{U}$ and is contained in $M_0 \times \{0\}$ or in $\{0\} \times N_0$, so is its join. If not, it has an element (p,q), more fully $((k,\kappa), (\lambda,\ell))$, less than some $((n,\alpha), (\alpha,n))$. But this requires $\alpha = \kappa$ and $n = \ell$. The successors of (p,q) in D are cofinal; to be in $\mathcal{U}$ they must precede $((\ell,\kappa), (\kappa,\ell))$, and so does their join.

As for sobriety, observe first that a Scott-closed set H is the set of predecessors of its set J of maximal elements. We shall see that if H is nonempty, then J has a point j interior to its closure in H; thus if $H \neq \{j\}^-$, then H has two disjoint relatively open subsets, i.e., it is not irreducible.

Note that

(a) In N_0, a set J of incomparable elements has a Scott-closed set of predecessors (nonzero ones precede a unique element of J).

(b) In general, the Scott closure of a set can be built up by alternately taking predecessors and directed joins, and at limit ordinals, unions.

Consider closed $H \subset M_0 \times N_0$ in (Case I) J contains a point (p,q) with p of the form (n,α). Then p has a greatest proper predecessor p_- and a least proper successor p_+. $J - \{(p,q)\}$ partitions into the sets J_1, J_2, J_3 of points with first coordinate m satisfying, respectively, $m \geq p_+$, $m = p$, $m \wedge p \leq p_-$. The second coordinate projection of $H \cap (\{p_+\} \times N_0)$ is a closed set A not containing q, and the predecessors of J_1 are contained in $M_0 \times A$. The predecessors of J_2 are closed by (a) above. For J_3, since $C = \{m \in M_0 : m \wedge p \leq p_-\}$ is not closed under directed join, we have to notice $p!$, the smallest successor of p that is a directed join of elements of C. If $p = (n,\alpha)$, $p!$ is (α) or $(\alpha + 1)$; anyway, $p! \neq p$. The second coordinate projection of $H \cap (\{p!\} \times N_0)$ is a closed set B not containing q, and the closure of J_3 is [via (b) above] contained in $(C \times N_0) \cup (M_0 \times B)$; so (p,q) is not in it, and is interior to its closure in H.

Otherwise (Case II) the first coordinates of elements of J are well ordered. Choose (p,q) in J with the smallest possible p. This is just like Case I except that J_3 is empty.

4. The topology L_1 will be a product lattice $M_1 \times N_1$, the topology of a coproduct. M_1 is an extension of M_0. It is a natural topology on the opposite lattice M_0^* (the order reversal comes because points naturally give <u>closed</u> sets, larger points having larger closures). It seems easier to see what is happening if we leave M_0 alone and define M_1 as the set of upper sets of M_0 which have only finitely many minimal elements, ordered by <u>reverse</u> inclusion.

We must show this is a topology. In view of the reverse inclusion order (= inclusion of complements), we want M_1 closed under finite union (trivial) and arbitrary intersection. The intersections are certainly upper sets. Moreover, a nonzero element of M_1 (the zero is all of M_0) has finite width; every upper set contained in it belongs to M_1, completing the proof. (Remark: M_1 is the principal topology on M_0^*.) Similarly N_1, defined in the same way on N_0, is a topology.

In $M_1 \times N_1$ the principal upper sets [x] give $([(n,\alpha)], [(\alpha,n)])$ meeting every product neighborhood of ([1],[1]) as before; indeed, $x \to [x]$ from M_0 to M_1 (or N_0 to N_1) preserves joins. But, as before, the Scott closure of this set S of principal upper sets consists just of $M_1 \times \{[0]\}$, $\{[0]\} \times N_1$, and the set of all (p,q) with $p \neq [0]$, $q \neq [0]$, which precede some element of S. For such a p has finitely many minimal elements $(k_i,\varkappa_i)$ [besides a possible $(\varkappa_0)$ not affecting the successors of p in S]; similarly, q has finitely many minimal elements (λ_j,ℓ_j), and the successors in S are at most the $([(\ell_j,\varkappa_i)], [(\varkappa_i,\ell_j)])$. Accordingly, a directed subset of our union that contains (p,q) must lie under one of those.

5. The first version of this paper ended with some hopeful remarks on the related problem of constructing an MC lattice which should admit no topology making finite meets and all joins continuous. Now one need only refer to the solution of that problem [5], which depends essentially on a further idea introduced by Erné and Gatzke in a preliminary version of [4].

REFERENCES

1. G. Gierz, K. H. Hofmann, K. Keimel, J. D. Lawson, M. Mislove, and D. S. Scott, A Compendium of Continuous Lattices, Springer-Verlag, New York, 1980.
2. J. Isbell, Function spaces and adjoints, Math. Scand. 36 (1975), 317-339.
3. D. S. Scott, Continuous lattices, in Lecture Notes in Mathematics 274, Springer-Verlag, Berlin, 1972, 97-136.
4. M. Erné and H. Gatzke, Convergence and continuity in partially ordered sets and semilattices, this volume.
5. J. Isbell, A frame with no admissible topology, Math. Proc. Camb. Phil. Soc., to appear.

9
Vietoris Locales and Localic Semilattices

PETER T. JOHNSTONE
University of Cambridge
Cambridge, England

0. INTRODUCTION

It is well known (among aficionados of continuous lattices, at least) that the Vietoris space construction defines a monad on the category of compact Hausdorff spaces, and that the algebras for this monad are exactly the compact Lawson semilattices. Now when one is working constructively (inside a topos, if you wish) it is by now well established [2,4,5] that various constructions and theorems that are classically applied to compact Hausdorff spaces (and that classically involve nonconstructive principles such as the prime ideal theorem) must be considered instead in the context of compact regular locales; that is, we must forget about the points of the spaces and concentrate on their open-set lattices. Accordingly, in seeking to understand how one might define "Lawson semilattices" in a topos, I began by investigating whether the Vietoris-space monad admitted a constructive generalization to the category of compact regular locales. [A different approach was indicated by Porta and Wyler [11], who (implicitly) defined Lawson semilattices in a topos as algebras for the filter monad on the topos itself; but since the resulting category does not admit a forgetful functor to compact regular locales, this approach seems to me unsatisfactory.]

I soon found that there was indeed a constructive theory of "Vietoris locales," reducing in the classical case to the theory of Vietoris spaces; but I was surprised to discover that the Vietoris-locale construction made perfectly good sense (and defined a perfectly good monad) on the whole category of locales, and not just on compact regular locales. The construction thus becomes of interest even in the classical case, since it enables us to define a generalization of Lawson semilattices whose underlying spaces need not be compact or Hausdorff. Although the investigation of these generalized Lawson semilattices is far from complete, it seems worthwhile to present this account of what is known so far. In keeping with my original aim of constructivizing the theory of Lawson semilattices, I shall work for as long as possible within the category of locales, and turn to consideration of points only in the third and final section of the paper; however, I have inserted a number of asides (helpful ones, I hope) in the first two sections, to indicate the spatial constructions that would correspond to the localic ones under discussion.

An earlier account of Vietoris locales (restricted to the compact regular case) was incorporated as section III 4 of my book Stone Spaces [6]; and a preliminary version of this paper, entitled "The Vietoris monad on the category of locales," has appeared in [7]. However, I have tried to make the present version more or less self-contained.

1. VIETORIS LOCALES

Our terminology regarding frames and locales will be that developed in [4] and [6]: we define a frame to be a complete lattice A in which the infinite distributive law

$$a \wedge \bigvee S = \bigvee \{a \wedge s \mid s \in S\}$$

holds for all $a \in A$, $S \subseteq A$. A frame homomorphism is a map preserving finite meets and arbitrary joins; we write Frm for the category of frames. The category Loc of locales is defined simply as the opposite of Frm; we adopt the convention that if $f\colon A \to B$ is a continuous map of locales (= a morphism in Loc), we write $f^*\colon B \to A$ for the corresponding frame homomorphism, and $f_*\colon A \to B$ for its right adjoint.

The category Frm is algebraic (i.e., monadic over Set); in particular, this means that we can define frames by specifying generators and relations, where the relations are written in a language having symbols for finite meets and arbitrary joins (though not for infinite meets). For an explicit account of how to construct frames from generators and relations, see [4] or [6]. Of course, if a frame A has been defined by generators and relations, we may then define frame homomorphisms $A \to B$ for any B by specifying their effect on the generators and checking that they preserve the relations.

We define the Vietoris locale V(A) of a locale A to be the frame generated by abstract symbols t(a) and m(a), $a \in A$, subject to the following relations:

(i) $t(a \wedge b) = t(a) \wedge t(b)$ for all $a,b \in A$, and $t(1_A) = 1_{V(A)}$.

(ii) $t(\bigvee S) = \bigvee\{t(s) \mid s \in S\}$ whenever $S \subseteq A$ is directed.

(iii) $m(\bigvee S) = \bigvee\{m(s) \mid s \in S\}$ for all $S \subseteq A$ (including $S = \emptyset$).

(iv) $m(a \wedge b) \geq t(a) \wedge m(b)$ for all $a,b \in A$.

(v) $t(a \vee b) \leq t(a) \vee m(b)$ for all $a,b \in A$.

Informally, one should think of $V(A)$ as the space of compact subspaces of A (modulo some technical complications, which will be investigated in Section 3), $t(a)$ as the set of compact subspaces contained in the open set a, and $m(a)$ as the set of compact subspaces which meet a.

1.1 PROPOSITION The assignment $A \to V(A)$ defines a functor $\underline{\text{Loc}} \to \underline{\text{Loc}}$; and there are natural transformations $i\colon \text{id}_{\underline{\text{Loc}}} \to V$ and $\mu\colon V^2 \to V$ making (V,i,μ) into a monad on $\underline{\text{Loc}}$.

Proof: Given a locale map $f\colon A \to B$, we define $V(f)\colon V(A) \to V(B)$ by specifying the effect of $V(f)^*$ on the generators of $V(B)$. Specifically, we put

$$V(f)^*(t(b)) = t(f^*(b)) \qquad V(f)^*(m(b)) = m(f^*(b))$$

Since f^* preserves finite meets and all joins, it is clear that $V(f)^*$ preserves the relations between the $t(b)$ and $m(b)$. It is also clear that $V(\text{id}_A) = \text{id}_{V(A)}$ and $V(fg) = V(f)V(g)$ whenever fg is defined; thus V is a functor.

Similarly, we define locale maps $i_A\colon A \to V(A)$ and $\mu_A\colon V(V(A)) \to V(A)$ by

$$i_A^*(t(a)) = i_A^*(m(a)) = a$$

$$\mu_A^*(t(a)) = t(t(a)) \qquad \mu_A^*(m(a)) = m(m(a))$$

It is again easy to verify that these definitions preserve the relations (i) - (v); the fact that they define natural transformations and satisfy the equations for a monad follows from the fact that a locale map h into $V(A)$ is uniquely determined by the effect of h^* on the $t(a)$ and $m(a)$. ∎

1.2 LEMMA There is a natural isomorphism $V(A +_\ell B) \cong V(A) \times_\ell V(B)$, where $+_\ell$ and $\times_\ell$ denote coproduct and product in $\underline{\text{Loc}}$.

Proof: We recall that $A +_\ell B$ is simply the cartesian product of A and B with the product ordering, whereas $A \times_\ell B$ is the frame generated by symbols "$a \times b$," $a \in A$ and $b \in B$ ("open rectangles"), subject to the relations

(vi) "1×1" = 1 and "$a_1 \times b_1$" $\wedge$ "$a_2 \times b_2$" = "$(a_1 \wedge a_2) \times (b_1 \wedge b_2)$".

(vii) $"\bigvee S \times b" = \bigvee\{"a \times b" \mid a \in S\}$.

(viii) $"a \times \bigvee T" = \bigvee\{"a \times b" \mid b \in T\}$.

We define a map $q\colon V(A) \times_\ell V(B) \to V(A +_\ell B)$ by

$$q^*(t(a,b)) = "t(a) \times t(b)" \qquad q^*(m(a,b)) = "m(a) \times 1" \vee "1 \times m(b)"$$

Let us verify that q^* preserves relations (iv) and (v) (the others are easy). We have

$$\begin{aligned}
q^*m((a_1,b_1) \wedge (a_2,b_2)) &= "m(a_1 \wedge a_2) \times 1" \vee "1 \times m(b_1 \wedge b_2)" \\
&\geq ["t(a_1) \times 1" \wedge "m(a_2) \times 1"] \vee ["1 \times t(b_1)" \wedge \\
&\quad "1 \times m(b_2)"] \\
&\geq ["t(a_1) \times t(b_1)" \wedge "m(a_2) \times 1"] \vee ["t(a_1) \times \\
&\quad t(b_1)" \wedge "1 \times m(b_2)"] \\
&= "t(a_1) \times t(b_1)" \wedge ["m(a_2) \times 1" \vee "1 \times m(b_2)"] \\
&= q^*t(a_1,b_1) \wedge q^*m(a_2,b_2) \\
q^*t((a_1,b_1) \vee (a_2,b_2)) &= "t(a_1 \vee a_2) \times t(b_1 \vee b_2)" \\
&\leq "t(a_1) \times t(b_1 \vee b_2)" \vee "m(a_2) \times t(b_1 \vee b_2)" \\
&\leq "t(a_1) \times t(b_1)" \vee "t(a_1) \times m(b_2)" \vee "m(a_2) \times \\
&\quad t(b_1 \vee b_2)" \\
&\leq "t(a_1) \times t(b_1)" \vee "1 \times m(b_2)" \vee "m(a_2) \times 1" \\
&= q^*t(a_1,b_1) \vee q^*m(a_2,b_2).
\end{aligned}$$

In the opposite direction, we define maps $r_1\colon V(A +_\ell B) \to V(A)$ and $r_2\colon V(A +_\ell B) \to V(B)$ by

$$r_1^*t(a) = t(a,1) \qquad r_1^*m(a) = m(a,0)$$

and similarly for r_2^*; once again, it is straightforward to verify that the relations (i) - (v) are preserved. Also,

$$\begin{aligned}
(r_1,r_2)^*q^*t(a,b) &= (r_1,r_2)^*("t(a) \times t(b)") \\
&= r_1^*t(a) \wedge r_2^*t(b) \\
&= t(a,1) \wedge t(1,b) = t(a,b)
\end{aligned}$$

and

$$(r_1,r_2)^*q^*m(a,b) = (r_1,r_2)^*(\text{"}m(a) \times 1\text{"} \vee \text{"}1 \times m(b)\text{"})$$
$$= r_1^*m(a) \vee r_2^*m(b)$$
$$= m(a,0) \vee m(0,b) = m(a,b)$$

so $q \circ (r_1,r_2) = id_{V(A +_\ell B)}$; and similar arguments identify r_1q and r_2q with the product projections from $V(A) \times_\ell V(B)$. Thus q is an isomorphism with inverse (r_1,r_2); its naturality in A and B is obvious from the form of the definition. ■

If we think of A and B as being (the open-set lattices of) spaces X and Y, then the isomorphism of Lemma 1.2 is the obvious bijection between compact subspaces $K \subseteq X + Y$ and pairs $(K \cap X, K \cap Y)$ of compact subspaces of X and Y. Another natural construction, given compact subspaces $K \subseteq X$ and $L \subseteq Y$, is to form the product $K \times L \subseteq X \times Y$; corresponding to this, we have

1.3 LEMMA There is a natural map $d: V(A) \times_\ell V(B) \to V(A \times_\ell B)$ defined by

$$d^*t(c) = \bigvee\{\text{"}t(a) \times t(b)\text{"} \mid \text{"}a \times b\text{"} \le c\}$$

and

$$d^*m(c) = \bigvee\{\text{"}m(a) \times m(b)\text{"} \mid \text{"}a \times b\text{"} \le c\}$$

for c any element of $A \times_\ell B$ (not necessarily an open rectangle).

Proof: We begin with a warning: the definition does not imply that

$$d^*t(\text{"}a \times b\text{"}) = \text{"}t(a) \times t(b)\text{"}$$

since we always have $\text{"}0 \times 1\text{"} = 0 \le \text{"}a \times b\text{"}$ in $A \times_\ell B$, whereas $t(0) \ne 0$, and so $\text{"}t(0) \times t(1)\text{"} \not\le \text{"}t(a) \times t(b)\text{"}$ unless $b = 1$. [However, the corresponding assertion with m in place of t is valid; and later on we shall be dealing with the sublocale of V(A) obtained by imposing the extra relation $t(0) = 0$, for which this warning will not apply.]

To verify relation (i), note that if $\text{"}a_1 \times b_1\text{"} \le c_1$ and $\text{"}a_2 \times b_2\text{"} \le c_2$, then

$$\text{"}t(a_1) \times t(b_1)\text{"} \wedge \text{"}t(a_2) \times t(b_2)\text{"} = \text{"}t(a_1 \wedge a_2) \times t(b_1 \wedge b_2)\text{"}$$
$$\le d^*t(c_1 \wedge c_2)$$

whence by the infinite distributive law we deduce $d^*t(c_1) \wedge d^*t(c_2) \le d^*t(c_1 \wedge c_2)$; the reverse inequality is trivial. Next, we note that every element of $A \times_\ell B$ is the join of the open rectangles it contains; so to verify that $d^*m(\cdot)$ preserves joins [i.e., that d^* preserves relation (iii)] it suffices to consider only those joins appearing in the relations (vii) and (viii). And in these cases we have

$$\begin{aligned} d^*m(\text{"}\bigvee S \times b\text{"}) &= \text{"}m(\bigvee S) \times m(b)\text{"} \\ &= \text{"}\bigvee\{m(a) \mid a \in S\} \times m(b)\text{"} \\ &= \bigvee\{\text{"}m(a) \times m(b)\text{"} \mid a \in S\} \end{aligned}$$

etc. The verification of (ii) is similar to (iii), using the fact that every element of $A \times_{\ell} B$ is a directed join of elements that are finite joins of open rectangles; we omit the details.

The verification of (iv) is similar to (i): if "$a_1 \times b_1$" $\le c_1$ and "$a_2 \times b_2$" $\le c_2$, then "$t(a_1) \times t(b_1)$" $\wedge$ "$m(a_2) \times m(b_2)$" $\le$ "$m(a_1 \wedge a_2) \times m(b_1 \wedge b_2)$" $\le d^*m(c_1 \wedge c_2)$. In verifying (v), we reduce as in (ii) and (iii) to the case of a "basic" join of type (vii) or (viii); then we have

$$\begin{aligned} \text{"}t(a) \times t(b_1 \vee b_2)\text{"} &\le \text{"}t(a) \times t(b_1)\text{"} \vee \text{"}t(a) \times m(b_2)\text{"} \\ &\le \text{"}t(a) \times t(b_1)\text{"} \vee \text{"}t(0) \times m(b_2)\text{"} \vee \text{"}m(a) \times m(b_2)\text{"} \\ &\le \text{"}t(a) \times t(b_1)\text{"} \vee \text{"}t(0) \times 1\text{"} \vee \text{"}m(a) \times m(b_2)\text{"} \\ &\le d^*t(\text{"}a \times b_1\text{"}) \vee d^*m(\text{"}a \times b_2\text{"}) \end{aligned}$$

from which the desired result follows. ■

We now turn our attention to the question of which topological properties of a locale are inherited by its Vietoris locale. For the most important separation properties, the question is easily answered:

1.4 PROPOSITION A locale A is regular (resp., completely regular, zero-dimensional) iff its Vietoris locale V(A) is.

Proof: We give the argument for regularity; the others are exactly similar. The map $i_A \colon A \to V(A)$ is an inclusion, since its inverse image is clearly surjective; so since regularity is inherited by sublocales ([4], lemma 4.2), it is immediate that A inherits regularity from V(A). In the converse direction, suppose that A is regular, i.e., for all $a \in A$ we have

$$a = \bigvee\{b \in A \mid b \mathrel{\bar{<}} a\}$$

where $b \mathrel{\bar{<}} a$ means that there exists c with $b \wedge c = 0$ and $a \vee c = 1$. But if these relations hold, then we have $t(b) \wedge m(c) \le m(0) = 0$ and $t(a) \vee m(c) \ge t(1) = 1$ in V(A), so $t(b) \mathrel{\bar{<}} t(a)$; and similarly, $m(b) \mathrel{\bar{<}} m(a)$. Since the set $\{b \mid b \mathrel{\bar{<}} a\}$ is directed, it now follows from relations (ii) and (iii) that each of the generators of V(A) satisfies the regularity axiom. But in any locale, the subset of elements that satisfy the regularity axiom is closed under finite meets and arbitrary joins; so $V(A)$ is regular. ∎

Note that if a and b are complementary elements of A, then it follows from the proof of Proposition 1.4 that $t(a)$ and $m(b)$ are complementary in $V(A)$; in particular, $t(0)$ and $m(1)$ are complementary. $V(A)$ thus decomposes as the disjoint union (= locale coproduct) of two clopen sublocales which we shall denote $V_0(A)$ and $V^+(A)$, and which are obtained by adjoining the relations $t(0) = 1$ and $t(0) = 0$, respectively, to those in the definition of $V(A)$.

1.5 LEMMA $V_0(A)$ is isomorphic to the terminal object $\underline{\Omega}$ of <u>Loc</u>.

Proof: As we have just observed, the relation $t(0) = 1$ forces $m(1) = 0$; and then from (i) and (iii) we obtain $t(a) = 1$ and $m(a) = 0$ for all a. Thus we have used up all the generators of $V(A)$; but we have also used up all the relations (i)-(v) since they are implied by the relations $t(a) = 1$ and $m(a) = 0$. So we are left with the free frame on no generators, i.e., the terminal object of <u>Loc</u>. ∎

$V^+(A)$ should be thought of as "the space of nonempty compact subspaces of A," which explains why its complement should consist of a single point. Note that in $V^+(A)$ we have $t(a) = t(a) \wedge m(1) \leq m(a)$ for all $a \in A$; note also that V^+ is a functor <u>Loc</u> → <u>Loc</u>, that $i_A \colon A \to V(A)$ factors through $V^+(A)$ (in fact, V^+ is a submonad of V), and that the decomposition $V(A) \cong \Omega +_{\ell} V^+(A)$ is natural in A. From the classical theory of hyperspaces [10], we should expect $V^+(A)$ to be connected whenever A is (more generally, that the only complemented elements of $V(A)$ should be finite Boolean combinations of those that come from complemented elements of A), but I do not at present have a proof of this.

We turn next to local compactness; recall that a locale is said to be <u>locally compact</u> iff it is a continuous lattice (cf. [1]).

1.6 PROPOSITION If A is a locally compact locale, then $V(A)$ is locally compact.

Proof: We have to show that each basic element of $V(A)$ (i.e., each finite meet of generators) is expressible as a join of elements way below itself. (Here it is not sufficient to prove the result merely for the generators, as the way-below relation may not be stable under finite meets.) We shall write $w(a;b_1,\ldots,b_n)$ for the basic element $t(a) \wedge m(b_1) \wedge \cdots \wedge$

$m(b_n)$; note that by relation (iv) we have $w(a;b_1,\dots,b_n) = w(a;(a \wedge b_1), \dots ,(a \wedge b_n))$, so there is no loss of generality in supposing (as we do henceforth) that $b_i \le a$ for all i. We shall show that if $c \ll a$ and $d_i \ll b_i$ for all i, then $w(c;d_1,\dots,d_n) \ll w(a;b_1,\dots,b_n)$; by relations (ii) and (iii), this is clearly sufficient to show that V(A) is locally compact.

Let S be a directed subset of V(A); for convenience, we shall assume that S is a lower set, and therefore an ideal. Consider the set

$$F(S) = \{w(a;b_1,\dots,b_n) \mid (\forall c \ll a)(\forall d_i \ll b_i)(w(c;d_1,\dots,d_n) \in S)\}$$

Clearly, every $w \in F(S)$ satisfies $w \le \bigvee S$; we shall show that F(S) is closed under arbitrary joins in V(A), and so is the set of basic elements in a principal ideal of V(A), whose generator must be the join of S. It is straightforward to verify that F(S) is closed under joins arising from relations (ii) and (iii) in the definition of V(A); the problem arises with relation (v), which in this context can be taken to say that

$$w(a \vee b_1;b_2,\dots,b_n) = w(a;a \wedge b_2,\dots,a \wedge b_n) \vee w(a \vee b_1;b_2,\dots,b_n)$$

where $b_i \le a \vee b_1$ for $i \ge 2$. Suppose the two right-hand terms are in F(S), and we are given $c \ll a \vee b_1$ and $d_i \ll b_i$ $(i \ge 2)$; we must show that $w(c;d_2,\dots,d_n) \in S$. Since A is a continuous lattice, we can find $c_1 \ll a$ and $d_1 \ll b_1$ such that $c \le c_1 \vee d_1$; and since A is distributive, we can actually achieve $c = c_1 \vee d_1$. Similarly, for each $i \ge 2$ we can write $d_i = e_i \vee f_i$ with $e_i \ll a \wedge b_i$, $f_i \ll b_1 \wedge b_i$. Now we have

$$w(c_1;e_2,\dots,e_n) \in S \qquad w(c;d_1,e_2,\dots,e_n) \in S$$

by assumption, whence $w(c;e_2,\dots,e_n) \in S$ since S is an ideal. But we also have

$$w(c;f_2,f_2,e_3,\dots,e_n) = w(c;f_2,e_3,\dots,e_n) \in S$$

since f_2 is way below both b_1 and b_2; on taking the join of this element with $w(c;e_2,e_3,\dots,e_n)$ we obtain

$$w(c;d_2,e_3,\dots,e_n) \in S$$

Continuing inductively, we eventually arrive at $w(c;d_2,\dots,d_n) \in S$, as required.

The result claimed earlier now follows easily; for if $c \ll a$, $d_i \ll b_i$ and $w(a;b_1,\dots,b_n) \le \bigvee S$ for some ideal $S \subseteq V(A)$, we have $w(a;b_1,\dots, b_n) \in F(S)$ and hence $w(c;d_1,\dots,d_n) \in S$. ∎

One can prove in general that (global) compactness is inherited by $V(A)$ from A, but the proof (in the absence of any other hypotheses on A) would appear to require a transfinite induction similar to that used to prove the Tychonoff theorem in [4]. Since such inductions do not fit in well with our notions of constructivity, we shall therefore restrict ourselves to proving the result under the additional hypothesis that A is locally compact, in which case the techniques developed in the last proof enable us (as in section 3 of [4]) to reduce the induction to a single step.

1.7 PROPOSITION If A is compact and locally compact, so is $V(A)$.

Proof: After Proposition 1.6, it remains only to show global compactness of $V(A)$. Let S be an ideal of $V(A)$ with join 1; then in the notation developed in the proof of Proposition 1.6 we have $1 = t(1) \in F(S)$, i.e., $t(c) \in S$ for all $c \ll 1$. But since A is compact, we have $1 \ll 1$ in A, and hence $t(1) \in S$. ∎

We record one further result concerning compactness:

1.8 LEMMA If A is a coherent locale, so is $V(A)$.

Proof: If B is a distributive lattice, let $V_f(B)$ denote the distributive lattice generated by abstract symbols $t(b)$ and $m(b)$ ($b \in B$), subject to the finitary parts of the relations defining the Vietoris locale; that is,

$$t(1) = 1 \qquad t(b \wedge c) = t(b) \wedge t(c)$$
$$m(0) = 0 \qquad m(b \vee c) = m(b) \vee m(c)$$
$$m(b \wedge c) \geq t(b) \wedge m(c)$$
$$t(b \vee c) \leq t(b) \vee m(c)$$

Then it is easy to see that if $A = \mathrm{Idl}(B)$ is the (coherent) locale of ideals of a distributive lattice B, we have $V(A) \cong \mathrm{Idl}(V_f(B))$, and so $V(A)$ is coherent. ∎

The final result of this section concerns the embedding $i: A \to V(A)$. As we remarked in [6], if A is compact and regular, then it follows at once from Proposition 1.4 that i is a closed embedding. But in fact we do not need to assume compactness:

1.9 PROPOSITION If A is regular, then the embedding $i: A \to V(A)$ is closed.

Proof: Let x be an element of $V(A)$. By definition, $i^*(x) = \bigvee S$, where

$$S = \{(a \wedge b_1 \wedge \cdots \wedge b_n) \mid w(a;b_1,\ldots,b_n) \leq x\}$$

From this it follows easily that $i_*(c)$ is the join in $V(A)$ of all those $w(a;b_1,\ldots,b_n)$ such that $a \wedge b_1 \wedge \cdots \wedge b_n \le c$. To show that i is closed, we have to show that $i_*(i^*(x)) = x$ for all $x \ge i_*(0)$.

So suppose $x \ge i_*(0)$. First we note that $s \in S$ iff $w(s;s) \le x$, since $s = a \wedge b_1 \wedge \cdots \wedge b_n$ implies $w(s;s) \le w(a;b_1,\ldots,b_n)$. But we also have $t(s) = t(0) \vee w(s;s)$ and $t(0) \le i_*(0) \le x$, so $s \in S$ iff $t(s) \le x$. Next we show that S is directed; suppose $a = a_1 \vee a_2$ with $a_1 \in S$, $a_2 \in S$. Then

$$t(a) = t(a_1) \vee w(a;a_2) = t(a_1) \vee \bigvee\{w(a;b) \mid b \mathrel{\overline{<}} a_2\}$$

But if c is such that $a_2 \vee c = 1$ and $b \wedge c = 0$, then

$$w(a;b) = w(a_2;b) \vee w(a;c,b) \le t(a_2) \vee w(a;c,b)$$

and $w(a;c,b) \le i_*(0) \le x$. So from $t(a_1) \le x$ and $t(a_2) \le x$ we deduce $t(a) \le x$, as required.

It now follows that $t(\bigvee S) = \bigvee\{t(s) \mid s \in S\} \le x$, so S is actually a principal ideal $\downarrow(s_0)$, say. Next consider an element of the form $w(1;b_1,\ldots,b_n)$ $(n \ge 1)$ with $b_1 \wedge \cdots \wedge b_n \le s_0$; we must show that $w(1;b_1,\ldots,b_n) \le x$. But if $c_i \mathrel{\overline{<}} b_i$ for each i, then $c_1 \wedge \cdots \wedge c_n \mathrel{\overline{<}} s_0$, i.e., we can find d such that $d \vee s_0 = 1$ and $d \wedge c_1 \wedge \cdots \wedge c_n = 0$. Then

$$w(1;c_1,\ldots,c_n) = w(s_0;c_1,\ldots,c_n) \vee w(1;d,c_1,\ldots,c_n)$$

and both terms on the right are below x, so $w(1;c_1,\ldots,c_n) \le x$. But this is true for all $c_i \mathrel{\overline{<}} b_i$, so by regularity we have $w(1;b_1,\ldots,b_n) \le x$. It now follows easily that we have $w(a;b_1,\ldots,b_n) \le x$ for any basic element such that $a \wedge b_1 \wedge \cdots \wedge b_n \le s_0 = i^*(x)$, i.e., $x = i_*(i^*(x))$. ∎

1.10 COROLLARY For a regular locale A, V(A) is compact iff A is.

Proof: One direction follows from Proposition 1.7, since a compact regular locale is locally compact. The converse follows from Proposition 1.9, since a closed sublocale of a compact locale is compact. ∎

I do not know any example of a noncompact locale A such that V(A) is compact. However, there are many examples to show that the embedding i need not be closed in the absence of regularity.

2. LOCALIC SEMILATTICES

In this section, our aim is to extend to our more general context the identification of algebras for the Vietoris monad on (compact Hausdorff spaces) with a certain class of topological semilattices (see [13]). We begin with the free algebras:

2.1 PROPOSITION For any A, V(A) carries a semilattice structure in Loc.

Proof: We have already observed that V(A) has a distinguished point p_0, namely the composite $\Omega \cong V_0(A) \to V(A)$ (cf. Lemma 1.5). For the binary semilattice operation, we take the composite

$$n\colon V(A) \times_\ell V(A) \xrightarrow{q} V(A +_\ell A) \xrightarrow{V(\nabla)} V(A)$$

where q is the isomorphism of Lemma 1.2 and ∇ is the codiagonal map; explicitly,

$$n^*t(a) = \text{"}t(a) \times t(a)\text{"} \qquad n^*m(a) = \text{"}m(a) \times 1\text{"} \vee \text{"}1 \times m(a)\text{"}$$

It is clear from the form of the definition that n is commutative and associative; to show it is idempotent, we have to consider $n \circ \Delta$, where $\Delta\colon V(A) \to V(A) \times_\ell V(A)$ is the diagonal map. We have

$$\Delta^*n^*t(a) = \Delta^*(\text{"}t(a) \times t(a)\text{"}) = t(a) \wedge t(a) = t(a)$$
$$\Delta^*n^*m(a) = \Delta^*(\text{"}m(a) \times 1\text{"} \vee \text{"}1 \times m(a)\text{"}) = m(a) \vee m(a) = m(a)$$

so $n \circ \Delta = id_{V(A)}$, as required. Finally, we must consider the composite

$$n \circ (id \times p_0)\colon V(A) \cong V(A) \times_\ell \Omega \to V(A) \times_\ell V(A) \to V(A)$$

The inverse image of this map sends t(a) to "$t(a) \times p_0^*t(a)$" = "$t(a) \times 1$" $\in V(A) \times_\ell \Omega$, which corresponds to $t(a) \in V(A)$ under the canonical isomorphism; and it sends m(a) to "$m(a) \times 1$" $\vee$ "1×0" = "$m(a) \times 1$", which similarly corresponds to m(a). So the composite above is the identity, i.e., p_0 is a unit for n. ∎

In spatial terms, the operation n corresponds to the binary union operation on compact subspaces (and p_0 is of course the empty subspace). So we shall tend to think of $(V(A),n,p_0)$ as a join semilattice rather than a meet semilattice, even though this goes against the usual custom in topological algebra. Note that the semilattice structure on V(A) is natural in A; that is, we can consider the functor V as taking values in the category of localic semilattices. But, in fact, more than this is true:

2.2 PROPOSITION Any algebra for the monad (V,i,μ) on Loc has a natural semilattice structure; i.e., the forgetful functor from V-algebras to locales factors through the category of localic semilattices.

Proof: Given a V-algebra (A,α), we shall consider the distinguished point

$$x\colon \Omega \xrightarrow{p_0} V(A) \xrightarrow{\alpha} A$$

and the binary operation

$$s\colon A \times_\ell A \xrightarrow{i \times i} V(A) \times_\ell V(A) \xrightarrow{n} V(A) \xrightarrow{\alpha} A$$

Before proving that (A,s,x) is a semilattice, let us observe that any V-algebra homomorphism $f\colon (A,\alpha) \to (B,\beta)$ is a homomorphism for the induced operations on A and B; for the maps p_0, $i \times i$, and n are all known to be natural in A. Next, we observe that the operations induced on a free V-algebra $(V(A),\mu_A)$ are those already defined in Proposition 2.1; that is, the diagrams

$$\begin{array}{ccc} \Omega & \xrightarrow{p_0} & V(V(A)) \\ & {\scriptstyle p_0}\searrow & \downarrow{\scriptstyle \mu} \\ & & V(A) \end{array} \qquad \text{and} \qquad \begin{array}{ccc} V(A) \times_1 V(A) & \xrightarrow{n} & V(A) \\ \downarrow{\scriptstyle i \times i} & & \uparrow{\scriptstyle \mu} \\ V(V(A)) \times_\ell V(V(A)) & \xrightarrow{n} & V(V(A)) \end{array}$$

commute. The first of these is trivial; for the second, we have

$$\begin{aligned} (i \times i)^* n^* \mu^* t(a) &= (i \times i)^* n^* t(t(a)) \\ &= (i \times i)^*(\text{"}t(t(a)) \times t(t(a))\text{"}) \\ &= \text{"}t(a) \times t(a)\text{"} = n^* t(a) \end{aligned}$$

and similarly for m(a).

Now since $\alpha\colon (V(A),\mu_A) \to (A,\alpha)$ is a homomorphism of V-algebras, it is also a homomorphism $(V(A),n,p_0) \to (A,s,x)$. And it is (split) epi in <u>Loc</u>; hence s and x satisfy all the equations that are true of n and p_0, in particular, those of the theory of semilattices. ∎

In the classical case, the forgetful functor from V-algebras in (compact Hausdorff spaces) to compact topological semilattices is full as well as faithful; this is one of the key results in the "algebraic" approach to the study of Lawson semilattices [13]. However, its proof depends heavily on the monotone convergence theorem ([6], VII 1.5), and therefore appears to be rather special to the compact case; I do not know whether the result is true at our level of generality. However, I can prove a weaker, but still remarkable, result: that a V-algebra structure on a locale is uniquely determined by the semilattice structure it induces (so that we can think of V-algebras as a particular class of localic semilattices, rather than localic semilattices with extra structure). The proof of this result will occupy the rest of this section.

First, we observe that if a V-algebra structure α on A induces a given semilattice structure (s,x), then the effect of α on the clopen point p_0 is determined, since we have $\alpha p_0 = x$. So we are reduced to showing that α is uniquely determined on the complementary sublocale $V^+(A)$; and henceforth we shall work largely in terms of V^+ rather than V. Our aim is to show that, in spatial terms, the map necessarily sends a compact subset K to its least upper bound in the semilattice ordering.

To do this, we introduce a sublocale $C^+(A)$ of $A \times_\ell V^+(A)$, namely, the equalizer of

$$A \times_\ell V^+(A) \xrightarrow{i \times id} V^+(A) \times_\ell V^+(A) \xrightarrow{d} V^+(A \times_\ell A) \xrightarrow{V^+(s)} V^+(A)$$
$$\searrow_{\pi_1} \qquad\qquad A \xrightarrow{\qquad\qquad} \nearrow_{i}$$

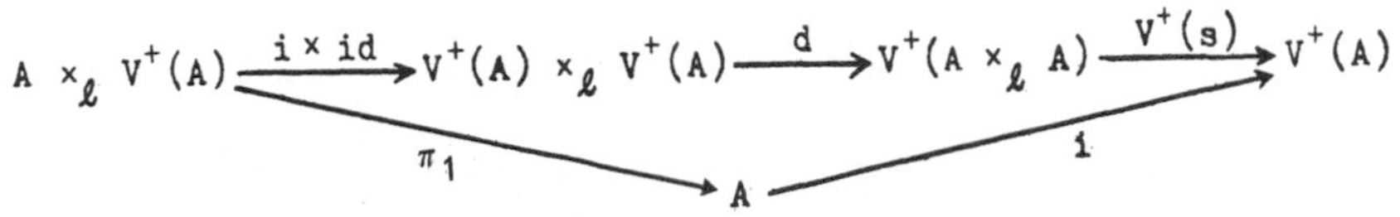

In spatial terms, $C^+(A)$ can be thought of as the space of pairs (x,K) such that $\{x \vee y \mid y \in K\} = \{x\}$, i.e., such that x is an upper bound for K. Clearly, the definition of $C^+(A)$ is functorial on the category of localic semilattices.

In spatial terms, the next two lemmas say, respectively, that $\alpha(K)$ is an upper bound for K, and that it is the least such.

2.3 LEMMA Suppose the semilattice operation on A is induced by a V^+-algebra structure α. Then the map $(\alpha, id)\colon V^+(A) \to A \times_\ell V^+(A)$ factors through $C^+(A)$.

Proof: We have to show that the diagram

$$V^+(A) \xrightarrow{(i\alpha, id)} V^+(A) \times_\ell V^+(A) \xrightarrow{d} V^+(A \times_\ell A) \xrightarrow{V^+(s)} V^+(A)$$
$$\searrow \qquad\qquad A \xrightarrow{\qquad\qquad} \nearrow_{i}$$

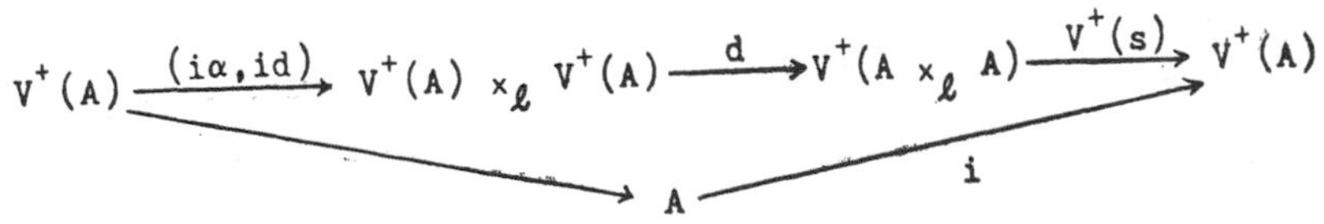

commutes. But it suffices to do this in the case when (A, α) is a free V^+-algebra, since α induces a natural transformation to the above diagram from the corresponding one with $V^+(A)$ in place of A, and $V^+(\alpha)$ is (split) epi. Thus we are reduced to proving that

$$\begin{array}{ccccccc} V^+(V^+(A)) & \xrightarrow{(i\mu,id)} & V^+(V^+(A)) \times_\ell V^+(V^+(A)) & \xrightarrow{d} & V^+(V^+(A) \times_\ell V^+(A)) & \xrightarrow{V^+(n)} & V^+(V^+(A)) \\ & \searrow^{\mu} & & V^+(A) & & \nearrow_{i} & \end{array}$$

commutes.

Since $m(\cdot)$ preserves all joins and $t(\cdot)$ preserves finite meets and directed joins, it is sufficient to compute the effect of the inverse image maps on elements of the form $m(w)$ and $t(z)$, where w is a finite meet of generators of $V^+(A)$ and z is a finite join. Let us first consider $m(w(a;$ $b_1,\ldots,b_n))$. Clearly, the composite $\mu^* i^*$ sends this element to $t(t(a)) \wedge$ $m(m(b_1)) \wedge \cdots \wedge m(m(b_n))$, since μ^* preserves finite meets. On the other hand, $V^+(n)^*$ sends it to

$$m(\text{"}t(a) \times t(a)\text{"} \wedge [\text{"}m(b_1) \times 1\text{"} \vee \text{"}1 \times m(b_1)\text{"}] \wedge \cdots \wedge [\text{"}m(b_n) \times 1\text{"} \vee \text{"}1 \times m(b_n)\text{"}]) = m(\bigvee\{\text{"}w(a;b_{j_1},\ldots,b_{j_k}) \times w(a;b_{j_{k+1}},\ldots,b_{j_n})\text{"} \mid (j,k) \in S\})$$

$$= \bigvee\{m(\text{"}w(a;b_{j_1},\ldots,b_{j_k}) \times w(a;b_{j_{k+1}},\ldots,b_{j_n})\text{"}) \mid (j,k) \in S\}$$

where S is the set of all partitionings of $\{1,2,\ldots,n\}$ into two disjoint sets $\{j_1,\ldots,j_k\}$ and $\{j_{k+1},\ldots,j_n\}$. d^* then sends this element to

$$\bigvee\{\text{"}m(w(a;b_{j_1},\ldots,b_{j_k})) \times m(w(a;b_{j_{k+1}},\ldots,b_{j_n}))\text{"} \mid (j,k) \in S\}$$

and $(i\mu,id)^*$ sends this to

$$\bigvee\{t(t(a)) \wedge m(m(b_{j_1})) \wedge \cdots \wedge m(m(b_{j_k})) \wedge m(w(a;b_{j_{k+1}},\ldots,b_{j_n})) \mid (j,k) \in S\}$$

It is clear that each term in this disjunction is below $\mu^* i^* w(a;b_1,\ldots,b_n)$, since $w(a;b_{j_{k+1}},\ldots,b_{j_n}) \leq m(b_{j_r})$ for each $r \geq k+1$. On the other hand, one of the terms in the disjunction is

$$t(t(a)) \wedge m(m(b_1)) \wedge \cdots \wedge m(m(b_n)) \wedge m(t(a))$$

which is equal to $\mu^* i^* w(a;b_1,\ldots,b_n)$ since we are working with V^+ and thus have $t(t(a)) \leq m(t(a))$. So the inverse image maps commute on elements of the form $m(w)$.

Next consider an element of the form $t(t(a_1) \vee \cdots \vee t(a_n) \vee m(b))$. The map $\mu^* i^*$ sends this to $t(t(a_1)) \vee \cdots \vee t(t(a_n)) \vee m(m(b))$; $V^+(n)^*$ sends it to

$$t(\text{"}t(a_1) \times t(a_1)\text{"} \vee \cdots \vee \text{"}t(a_n) \times t(a_n)\text{"} \vee \text{"}m(b) \times 1\text{"} \vee \text{"}1 \times m(b)\text{"})$$

and d^* sends this to

$$\text{"}t(t(a_1)) \times t(t(a_1))\text{"} \vee \cdots \vee \text{"}t(t(a_n)) \times t(t(a_n))\text{"} \vee \text{"}t(m(b)) \times 1\text{"} \vee$$
$$\text{"}1 \times t(m(b))\text{"}$$

(since we are working with V^+ rather than V). Then $(i\mu, id)^*$ sends this to

$$t(t(a_1)) \vee \cdots \vee t(t(a_n)) \vee m(m(b)) \vee t(m(b))$$

and the last term in this disjunction can be dropped for the same reason as before. Thus the result is established. ∎

2.4 LEMMA Under the same hypotheses as Lemma 2.3, the composites

$$C^+(A) \longrightarrow A \times_\ell V^+(A) \xrightarrow{\ id \times \alpha\ } A \times_\ell A \xrightarrow{\ s\ } A$$
$$\underbrace{\hspace{6cm}}_{\pi_1}$$

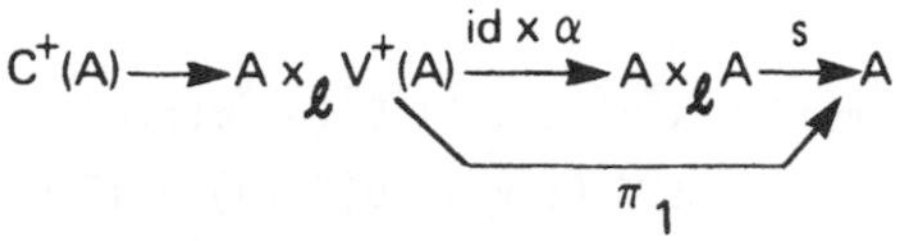

are equal.

Proof: Here we cannot, as in Lemma 2.3, reduce to the case when (A, α) is a free V^+-algebra, since C^+ is a functor on the category of localic semilattices, and so (since α does not in general have a splitting in this category) there is no reason that $C^+(\alpha)$ should be epi. However, it turns out that we can argue directly, by considering the diagram

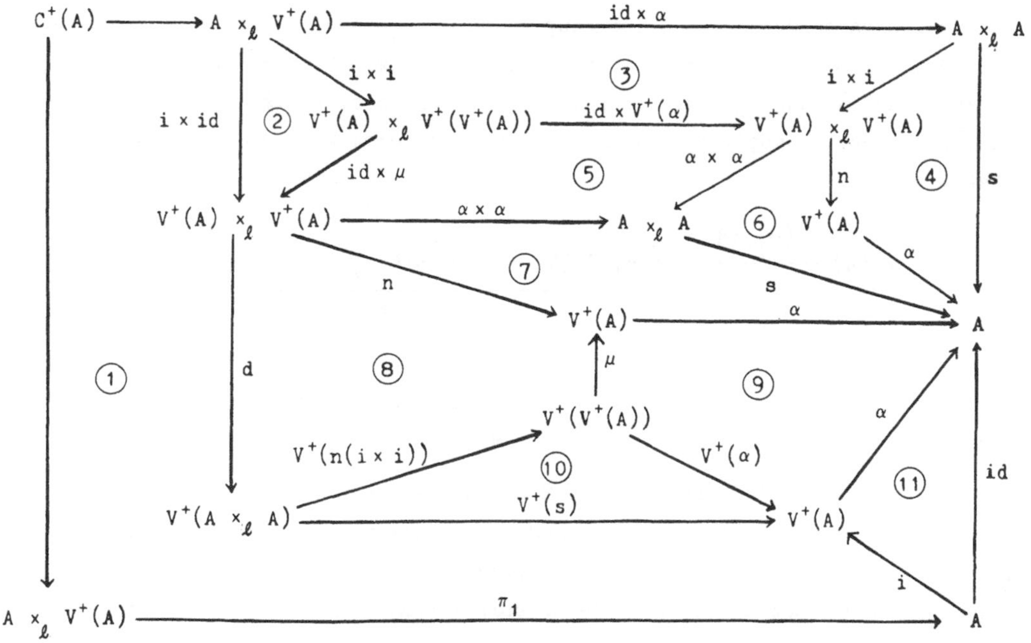

In this diagram, cell 1 commutes by the definition of $C^+(A)$, cell 2 by one of the monad identities, cell 3 by naturality of i, cells 4 and 10 by the definition of s, cells 5 and 9 by the associativity of α, cells 6 and 7 because α is a semilattice homomorphism, and cell 11 by the unit law for α. So it remains to check the commutativity of cell 8, which we do by explicit computation.

We have

$$\begin{aligned} d^*V^+(i \times i)^*V^+(n)^*\mu^*t(a) &= d^*V^+(i \times i)^*V^+(n)^*t(t(a)) \\ &= d^*V^+(i \times i)^*t(\text{"}t(a) \times t(a)\text{"}) \\ &= d^*t(\text{"}a \times a\text{"}) \\ &= \text{"}t(a) \times t(a)\text{"} = n^*t(a) \end{aligned}$$

and

$$\begin{aligned} d^*V^+(i \times i)^*V^+(n)^*\mu^*m(a) &= d^*V^+(i \times i)^*V^+(n)^*m(m(a)) \\ &= d^*V^+(i \times i)^*m(\text{"}m(a) \times 1\text{"} \vee \text{"}1 \times m(a)\text{"}) \\ &= d^*m(\text{"}a \times 1\text{"} \vee \text{"}1 \times a\text{"}) \\ &= \text{"}m(a) \times m(1)\text{"} \vee \text{"}m(1) \times m(a)\text{"} \\ &= \text{"}m(a) \times 1\text{"} \vee \text{"}1 \times m(a)\text{"} = n^*m(a) \end{aligned}$$

where the last step in each case is justified by the fact that we are working with V^+ rather than V. ∎

2.5 THEOREM Let (A,s,x) be a localic semilattice. Then there is at most one V-algebra structure on A inducing s and x.

Proof: Suppose α_1 and α_2 both induce s and x. As previously observed, we know that α_1 and α_2 agree on p_0, so it suffices to prove that they agree on $V^+(A)$. But by applying Lemma 2.3 to α_1 and Lemma 2.4 to α_2, we deduce that

$$V^+(A) \xrightarrow{(\alpha_1, id)} A \times_\ell V^+(A) \xrightarrow{id \times \alpha_2} A \times_\ell A \xrightarrow{s} A, \qquad \pi_1 : V^+(A) \to A$$

commutes, i.e., $s(\alpha_1,\alpha_2) = \alpha_1$; and similarly, $s(\alpha_2,\alpha_1) = \alpha_2$. Since s is commutative, we deduce that $\alpha_1 = \alpha_2$. ∎

We may also deduce from Lemmas 2.3 and 2.4 that if (A,α) and (B,β) are V-algebras and f: A → B is a homomorphism for the induced semilattice structures on A and B, then $f \circ \alpha \geq \beta \circ V(f)$ in the semilattice ordering on maps $V(A) \to B$. But in the absence of some localic substitute for the monotone convergence theorem, it does not seem possible to improve this inequality to an equality.

Another obvious problem that arises from Theorem 2.5 is to find some way of characterizing, without reference to the functor V, the class of localic semilattices that admit V-algebra structures. As yet I have made very little progress on this problem outside the classical (= compact Hausdorff) case, although we shall see in Section 3 below that the class includes all finite localic semilattices.

3. POINTS OF VIETORIS LOCALES

In the two preceding sections, we have frequently hinted that the points of V(A) may be identified with (certain) compact sublocales of A. The time has now come to make this explicit.

3.1 LEMMA Let $f: B \to A$ be a locale morphism whose domain is compact. Then there exists a point $p_f: \Omega \to V(A)$ defined by

$$p_f^*t(a) = 1 \text{ iff } f^*(a) = 1_B \qquad p_f^*m(a) = 1 \text{ iff } f^*(a) \neq 0_B$$

Proof: As usual, we have simply to verify that the relations (i)-(v) are preserved. (i) and (iii) are trivial, and (ii) follows easily from compactness of B. For (iv), suppose $p_f^*t(a) = p_f^*m(b) = 1$; then $f^*(a) = 1$ and $f^*(b) \neq 0$, whence $f^*(a \wedge b) \neq 0$ and so $p_f^*m(a \wedge b) = 1$. Similarly for (v), if $p_f^*t(a \vee b) = 1$, then $f^*(a) \vee f^*(b) = 1$ and so either $f^*(a) = 1$ or $f^*(b) \neq 0$, i.e., $p_f^*t(a) = 1$ or $p_f^*m(b) = 1$. ∎

We remark that, despite the apparent use of the law of excluded middle in the verification of (v) above, the argument can in fact be made constructive if we assume that B is not merely compact but an open locale (i.e., one such that the unique locale map $B \to \Omega$ is open). The proof of this fact (which depends on compactness of B) will appear elsewhere [8]; in the present paper, we shall be content from now on to assume classical logic whenever it is convenient to do so.

Note also that, if $h: B' \to B$ is a locale epimorphism (i.e., h^* is injective), then the points p_f and p_{fh} of V(A) are equal; since an epimorphic image of a compact locale is compact, it folows that we may (by forming the image factorization of our original f) restrict our attention to points of the form p_u where u is the inclusion of a compact sublocale of A. In this case, we may rewrite the definition of p_u^* in terms of the nucleus j that corresponds to this sublocale (see [6]): $p_u^* t(a) = 1$ iff $j(a) = 1$, and $p_u^* m(a) = 1$ iff $a \not\leqslant j(0)$. (In cases like this we shall sometimes write p_j rather than p_u, so that we do not have to give an explicit name to the inclusion $A_j \to A$.) On the other hand, if $g: A \to A'$ is any locale map, then it is easy to see that p_{gf} and $V(g) \circ p_f$ are equal as points of $V(A')$; in conjunction with what we have just said about image factorizations, this tells us that the effect of $V(g)$ on points of the form p_u corresponds to taking the images under g of compact sublocales of A. We leave to the reader the (easy) verification of the following results:

3.2 LEMMA (i) The point p_0 which is the unit of the semilattice structure on $V(A)$ is induced as in Lemma 3.1 by the unique map $1 \to A$, where 1 is the one-element frame (= the initial object of $\underline{\text{Loc}}$).

(ii) If $x: \Omega \to A$ is a point of A, then the composite $i_A \circ x$ is equal to p_x.

(iii) If $f: B \to A$ and $g: C \to A$ are two maps with compact domains, then
$n \circ (p_f, p_g)$ is induced by the map $(f, g): B +_\ell C \to A$.

(iv) If $f: C \to A$ and $g: D \to B$ are maps with compact domains, then $d \circ (p_f, p_g)$ is induced by the map $f \times g: C \times_\ell D \to A \times_\ell B$.

In the converse direction, how do we associate a compact sublocale of A with an arbitrary point of $V(A)$? We first tackle this problem "generically," by constructing a "membership relation" between A and $V(A)$. Let $E(A) \to A \times_\ell V(A)$ be the equalizer of

$$A \times_\ell V(A) \xrightarrow{\ 1 \times \text{id}\ } V(A) \times_\ell V(A) \xrightarrow{\ n\ } V(A)$$
$$\underbrace{\phantom{A \times_\ell V(A) \xrightarrow{\qquad\qquad} V(A) \times_\ell V(A)}}_{\pi_2}$$

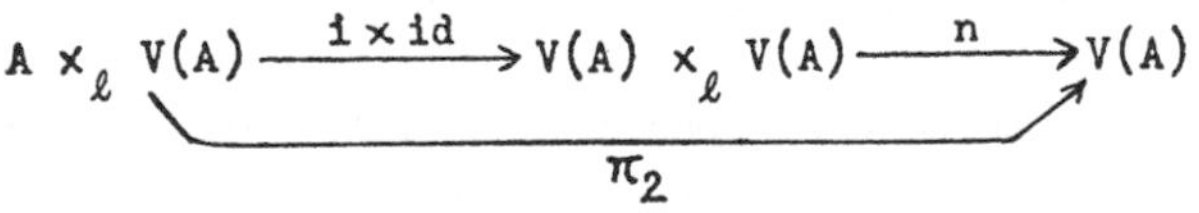

3.3 LEMMA (i) $E(A)$ may be obtained from $A \times_\ell V(A)$ by imposing the additional relations "$1 \times t(a)$" = "$a \times t(a)$" and "$a \times 1$" = "$a \times m(a)$" for all $a \in A$.

(ii) For any $f: A \to B$, there is a commutative square

$$
\begin{array}{ccc}
E(A) & \longrightarrow & A \times_\ell V(A) \\
\downarrow & & \downarrow{\scriptstyle f \times V(f)} \\
E(B) & \longrightarrow & B \times_\ell V(B)
\end{array}
$$

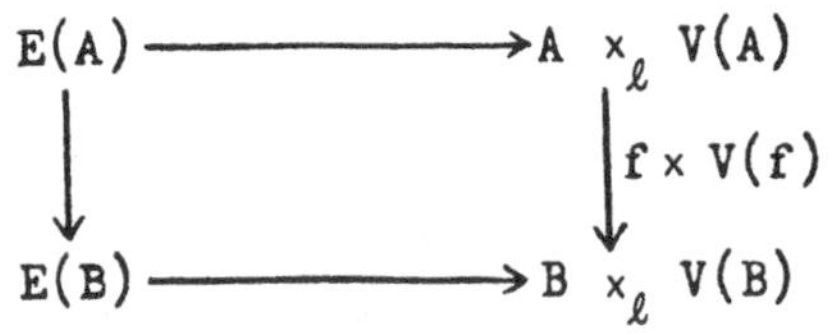

that is a pullback if f is an inclusion.

(iii) There is a pullback square

$$\begin{array}{ccc} A & \xrightarrow{\ \Delta\ } & A \times_{\ell} A \\ \downarrow & & \downarrow{\scriptstyle id \times i} \\ E(A) & \longrightarrow & A \times_{\ell} V(A) \end{array}$$

(iv) The composite $E(A) \to A \times_{\ell} V(A) \xrightarrow{\pi_2} V(A)$ factors through $V^{+}(A)$.

Proof: (i) We have $(i \times id)^{*}n^{*}t(a) =$ "$a \times t(a)$" and $\pi_2{}^{*}t(a) =$ "$1 \times t(a)$"; similarly, $(i \times id)^{*}n^{*}m(a) =$ "$a \times 1$" $\vee$ "$1 \times m(a)$" and $\pi_2{}^{*}m(a) =$ "$1 \times m(a)$". Clearly, the effect of imposing the relations given above is to identify these pairs of elements of $A \times_{\ell} V(A)$, i.e. [since the $t(a)$ and $m(a)$ generate $V(A)$], to form the coequalizer of the given pair of frame homomorphisms.

(ii) Follows directly from the definition; for all the maps involved in it are natural in A, and $V(f)$ is an inclusion [i.e., $V(f)^{*}$ is surjective] if f is.

(iii) We may compute the pushout of $E(A)$ along $(id \times i)^{*}$ in <u>Frm</u> by finding the images under $(id \times i)^{*}$ of the relations established in (i). If we do this, we obtain "$1 \times a$" = "$a \times a$" = "$a \times 1$" for all $a \in A$, which defines the diagonal sublocale of $A \times_{\ell} A$.

(iv) Follows trivially from the relations in (i) and the definition of $V^{+}(A)$. ∎

Now suppose we have a point $y\colon \Omega \to V(A)$. By pulling back $E(A)$ along the map $A \cong A \times_{\ell} \Omega \xrightarrow{id \times y} A \times_{\ell} V(A)$, we obtain a sublocale $A_{j(y)}$ of A; specifically, $A_{j(y)}$ is obtained from A by imposing the relations $a = 1$ whenever $y^{*}t(a) = 1$ and $a = 0$ whenever $y^{*}m(a) = 0$. In other words, $A_{j(y)}$ is the intersection of all open sublocales $A_{u(a)}$ for which $y^{*}t(a) = 1$ and of all closed sublocales $A_{c(b)}$ for which $y^{*}m(b) = 0$; equivalently again, $j(y)$ is the join [in the lattice $N(A)$ of all nuclei on A] of all such nuclei $u(a)$ and $c(b)$.

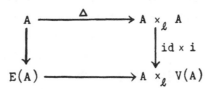

If y happens to be either p_0 or a point of the form ix (where x is a point of A), then it follows from parts (iv) and (iii) of Lemma 3.3 [plus (i) and (ii) of Lemma 3.2] that we have $p_{j(y)} = y$. But in general we do not yet know even whether $A_{j(y)}$ is compact, let alone whether it induces the right point of $V(A)$; the problem is that the relations we have imposed may imply some further relations of the form $a = 1$, which we do not wish to be true. To show that this is not the case, we need a result to the effect that A has "enough compact sublocales"; i.e., that we can (subject to appropriate compatibility conditions) find a compact sublocale A_j of A with specified values for $\{ a \in A \mid j(a) = 1\}$ and $\{b \in A \mid b \leq j(0)\}$. We deal first with the former set, which (following Simmons [12]) we denote $\nabla(j)$ and call the admissible filter of the nucleus j. Although Macnab [9] has given a characterization of the class of filters in a locale that are admissible for some nucleus, the following result appears to be new:

3.4 LEMMA (i) A sublocale A_j of A is compact iff the filter $\nabla(j)$ is open in the Scott topology on A (i.e., inaccessible by directed joins).
(ii) For any Scott-open filter F in a locale A, there exists a nucleus j on A with $\nabla(j) = F$.

Proof: (i) Since j transforms (directed) joins in A to joins in A_j, this is a straightforward translation of the fact that A_j is compact iff the trivial filter $\{1\}$ is open in its Scott topology.

(ii) Let j be the join of the open nuclei u(a), a ∈ F. Then it is clear that $F \subseteq \nabla(j)$; to show the reverse inclusion, we must consider how j may be computed from the u(a). For any b ∈ A, define

$$d(b) = \bigvee\{u(a)(b) \mid a \in F\} = \bigvee\{a \to b \mid a \in F\}$$

Since F is codirected, this join is directed; hence $d(b) \in F$ implies $(a \to b) \in F$ for some $a \in F$, and hence (since F is a filter) $b \in F$. The operation d is not necessarily idempotent, but we do have $b \leq d(b) \leq j(b)$ for any b; also, $b = d(b)$ implies

$$b \in \bigcap \{A_{u(a)} \mid a \in F\} = A_j$$

i.e., $b = j(b)$. So we may compute j(b) by iterating the operation d (transfinitely) until it converges; that is, we define an ordinal sequence of elements $d_\alpha(b)$ by

$$d_0(b) = b \qquad d_{\alpha+1}(b) = d(d_\alpha(b))$$

$$d_\lambda(b) = \bigvee\{d_\alpha(b) \mid \alpha < \lambda\} \qquad \text{if } \lambda \text{ is a limit ordinal}$$

and then $j(b) = d_\gamma(b)$ for the least γ such that $d_\gamma(b) = d_{\gamma+1}(b)$. In particular, if $b \in \nabla(j)$, then there exists α such that $d_\alpha(b) \in F$; we have already seen that the least such α cannot be a successor, but it cannot be a (nonzero) limit ordinal either, since the join in the definition of $d_\lambda(b)$ is directed. So $j(b) = 1$ implies $b = d_0(b) \in F$. ∎

The nucleus constructed in the proof of Lemma 3.4(ii) is clearly the unique smallest j with $\nabla(j) = F$, since for any j we have $a \in \nabla(j)$ iff $u(a) \le j$. Nuclei that are expressible as joins of open nuclei (equivalently, sublocales that are intersections of open sublocales) are called <u>fitted</u> by Isbell [3] and Simmons [12]. However, since we are interested in controlling the element j(0) as well as the filter $\nabla(j)$, we shall consider the larger class of nuclei that are obtainable as the join of a closed and a fitted nucleus; we shall call these <u>semifitted</u>.

3.5 LEMMA Let b be an element of a locale A, and F an admissible filter in A (for example, a Scott-open filter). Then there exists a semifitted nucleus j with $j(0) = b$ and $\nabla(j) = F$ iff

$$(a \vee b) \in F \Rightarrow a \in F \Rightarrow (a \to b) = b$$

for all $a \in A$.

Proof: If such a j exists, then for any $a \in F$ we have $u(a) \le j$, and so

$$(a \to b) = u(a)(b) \le j(b) = b$$

Similarly, we have $c(b) \le j$ and so $(a \vee b) \in F$ implies $j(a) = j(c(b)(a)) = 1$, i.e., $a \in F$. Conversely, suppose the conditions are satisfied. Then by assumption we have a fitted nucleus $k = \bigvee\{u(a) \mid a \in F\}$ with $\nabla(k) = F$. The join of k with the closed nucleus c(b) is simply the composite $j = k \circ c(b)$ ([12], lemma 6); now $j(0) = k(b) = b$ since we have $b \in A_k = \bigcap\{A_{u(a)} \mid a \in F\}$, and $j(a) = 1$ implies $(a \vee b) = c(b)(a) \in \nabla(k) = F$, whence $a \in F$. ∎

3.6 COROLLARY For any point y of V(A), the sublocale $A_{j(y)}$ defined earlier is compact, and $p_{j(y)} = y$.

Proof: Define $F = \{a \in A \mid y^*t(a) = 1\}$ and $b = \bigvee\{a \in A \mid y^*m(a) = 0\}$. Since $y^*t(\cdot)$ preserves finite meets and directed joins, F is a Scott-open filter, and since $y^*m(\cdot)$ preserves all joins, $\{a \in A \mid y^*m(a) = 0\}$ is the principal ideal generated by b. Also, if $(a \vee b) \in F$, then

$$y^*t(a) = y^*(t(a) \vee m(b)) \ge y^*t(a \vee b) = 1$$

by relation (v), and so $a \in F$; and similarly, if $a \in F$, then

$$y^*m(a \to b) = y^*(t(a) \wedge m(a \to b)) \le y^*m(a \wedge (a \to b)) \le y^*m(b) = 0$$

by relation (iv), and hence $(a \to b) = b$. Thus we have verified the conditions of Lemma 3.5, and so there is a semifitted nucleus on A [which is clearly j(y), by the description of it we gave earlier] such that $\nabla(j(y)) =$

F and $j(y)(0) = b$. But this clearly implies that [$A_{j(y)}$ is compact and] $p_{j(y)} = y$. ■

If A is a regular locale, then every sublocale of A is fitted (and every compact sublocale of A is closed); but in general a compact sublocale of an arbitrary locale need not be semifitted; for a simple counterexample, take A to be a four-element totally ordered set $\{0,a,b,1\}$ with $a < b$, and A_j to be the sublocale $\{0,b,1\}$. However, any sublocale has a "semifitted hull," which is the intersection of all the open or closed sublocales that contain it, and which clearly defines the same point of V(A) as the original sublocale; so we may combine Lemma 3.1 and Corollary 3.6 to obtain

3.7 THEOREM For any locale A, the space of points of the Vietoris locale V(A) may be regarded as the set of compact semifitted sublocales of A, equipped with the topology generated by the sets

$$T(a) = \{A_j \mid u(a) \leq j\} \qquad M(a) = \{A_j \mid c(a) \nleq j\}$$

where a ranges over all elements of A.

Proof: The only part that requires further comment is the identification of the topology. But the set of points corresponding to the generator t(a) of V(A) is

$$\{A_j \mid p_j^*(t(a)) = 1\} = \{A_j \mid a \in \nabla(j)\} = \{A_j \mid u(a) \leq j\} = T(a)$$

and m(a) similarly corresponds to the set M(a) defined above. ■

Note that if we were to impose the "Vietoris topology" (i.e., the topology defined in the statement of Theorem 3.7) on the set of all compact sublocales of A, the result would in general not be a T_0-space; for an arbitrary compact sublocale would be contained in exactly the same open sets as its semifitted hull. Thus we can regard the space of points of V(A) as the T_0 reflection (equivalently, the sober reflection) of the space of all compact sublocales of A with the Vietoris topology.

If A is spatial (i.e., isomorphic to the open-set lattice $\Omega(X)$ of a space X, which we may without loss of generality assume to be sober), then it would clearly be useful to have a description of the points of V(A) in terms of subspaces of X rather than sublocales of A. One might have hoped that a compact semifitted sublocale of a spatial locale would always be spatial, but the following example (due to Isbell) shows that it need not be:

3.8 EXAMPLE Let X be a Hausdorff space without isolated points, and let X^* be constructed from X by adjoining a new closed point ∞ whose neighborhoods are all cofinite subsets of X^*. It is not hard to see that X^* is sober; moreover, any sublocale of $\Omega(X^*)$ that contains the point ∞ must be compact. Let B be the intersection in $\Omega(X^*)$ of the open sublocales $\Omega(X^* - \{x\})$, $x \in X$; then B is compact by the last remark, and fitted in $\Omega(X^*)$. B is also dense in $\Omega(X^*)$, being an intersection of dense sublocales; but it has only one point, and so cannot be spatial.

Similarly, it is not always true that the Vietoris locale of a spatial locale is spatial. For if A is the locale of (euclidean) open subsets of the rationals, then it is not hard to see that we have an isomorphism $A \cong A +_\ell A$, and hence a closed embedding

$$A \times_\ell A \xrightarrow{i \times i} V(A) \times_\ell V(A) \xrightarrow{q} V(A +_\ell A) \cong V(A)$$

by Lemma 1.2 and Proposition 1.9. But $A \times_\ell A$ is not spatial ([6], II 2.14); hence V(A) cannot be spatial either.

By Proposition 1.6 (and the theorem [1] that locally compact locales are spatial), the second of these difficulties disappears in the case when A is locally compact. Local compactness alone will not dispel the first difficulty (for X^* is locally compact if X is), but if we strengthen it to stable local compactness (i.e., if we assume that the way-below relation in A is stable under binary meets), then all is well:

3.9 LEMMA Let A be a stably locally compact locale. Then any compact semifitted sublocale of A is spatial.

Proof: Let A_j be such a sublocale, and a,b two elements of A_j with $a \nleq b$. We have to find a point p of A_j with $p^*(a) = 1$ and $p^*(b) = 0$. But since A_j is semifitted, its points correspond to points q of A such that $q^*(c) = 1$ for all $c \in \nabla(j)$ and $q^*(j(0)) = 0$; thus we wish to find such a q^* that additionally sends a to 1 and b to 0, or equivalently to find a completely prime filter in A that contains a and $\nabla(j)$ but not b or $j(0)$. [In fact, since $b \geq j(0)$, the last condition is redundant.]

By local compactness of A plus (countable) dependent choice, there exists a descending sequence of elements

$$a \gg a_1 \gg a_2 \gg \cdots$$

with $a_i \nleq b$ for all i. Now consider the filter F generated by $\nabla(j)$ together with $\{a_i \mid i = 1,2,3,\ldots\}$; we claim first that F is open in the Scott topology. For if $S \subseteq A$ is directed and $\bigvee S \in F$, we have $\bigvee S \geq a_i \wedge c$ for some $c \in \nabla(j)$ and some i; then since $\nabla(j)$ is Scott open (as A_j is compact) we can find $c' \ll c$ in $\nabla(j)$, and by stability of $\ll$ we have $a_{i+1} \wedge c' \ll a_i \wedge c$. Then some member of S is above $a_{i+1} \wedge c'$, and hence in F. It is clear from the construction that F contains $\nabla(j)$ and a; but it does not contain b, since from $a_i \wedge c \leq b$ $[c \in \nabla(j)]$ we could deduce $a_i \leq (c \to b) = b$ because $b \in A_j$. Now we may use a result of Banaschewski ([1], lemma 3) to enlarge F to a completely prime filter that still misses b, as required. ∎

3.10 COROLLARY If X is a stably locally compact space (i.e., a locally compact sober space in which the intersection of two compact subsets is compact, for example, a locally compact Hausdorff space), then the Vietoris locale of (the open-set lattice of) X may be identified with the space of compact semifitted subspaces of X, equipped with the Vietoris topology generated by the sets

$$T(U) = \{K \mid K \subseteq U\} \qquad M(U) = \{K \mid K \cap U \neq \emptyset\}$$

where U ranges over all open subsets of X.

Even Corollary 3.10 is not the end of the story; for there still remains the problem of identifying which (compact) subspaces of an arbitrary space are semifitted. It is not hard to see that the fitted subspaces (i.e., those expressible as intersections of open subspaces) are precisely the "Alexandrov-open" sets, i.e., the upper sets in the specialization ordering on X. Since closed sets are lower sets in the specialization ordering, semifitted subspaces must be convex in this ordering; they are also closed under directed joins, but these two properties do not in general suffice to characterize them (even if the topology on X is the Scott topology for its specialization ordering).

We conclude the paper by looking briefly at finite locales (equivalently, at finite T_0-spaces). Every finite T_0-space is stably locally compact (so Corollary 3.10 applies); every subspace of a finite T_0-space is compact, and the semifitted ones are precisely the convex ones in the specialization ordering. If X is such a space, and $f: X \to Y$ is a continuous map from X to a (T_0) topological semilattice Y, then f has a unique extension to a semilattice homomorphism $g: P(X) \to Y$, where P(X) is the set of all subsets of X, defined by

$$g(K) = \bigvee\{f(x) \mid x \in K\}$$

3.11 LEMMA With the notation developed above, g is continuous for the Vietoris topology on P(X).

Proof: Let V be an open subset of Y, and K a subset of X such that $g(K) \in V$. We may assume $K \neq \emptyset$, since the empty set is isolated in the Vietoris topology. For each $x \in K$, let U_x be the smallest open neighborhood of x, i.e., the upward closure of x in the specialization ordering. Then by continuity of f and the join operation of Y, we have

$$g(W_x) \geq g(\{x\}) = f(x)$$

(in the specialization ordering on Y, which is in general different from the semilattice ordering) for any nonempty subset W_x of U_x. Hence if W is any set contained in the union U of $\{U_x \mid x \in K\}$ and such that $W_x = W \cap U_x$ is nonempty for each $x \in K$ (i.e., if W belongs to the open set

$$T(U) \cap \bigcap\{M(U_x) \mid x \in K\})$$

then $g(W) = \bigvee\{g(W_x) \mid x \in K\} \geq \bigvee\{f(x) \mid x \in K\} = g(K)$ (where once again the inequality sign refers to the specialization ordering, though the join signs refer to the semilattice ordering), and so $g(W) \in V$. Thus we have found a Vietoris-open neighborhood of K mapping into V. ∎

Recalling that P(X) is in general not a T_0-space in the Vietoris topology, and that its T_0 reflection is (the space of points of) the Vietoris locale of X, we deduce that the latter is the free T_0 topological semilattice generated by X. Moreover, it is not hard to see that the monad structure on the functor V, and its natural semilattice structure, agree with those on the free-T_0-semilattice functor for finite T_0-spaces; so we deduce

3.12 COROLLARY Every finite localic semilattice admits a (unique) algebra structure for the Vietoris monad.

In view of Lemma 1.8, it seems reasonable to hope that the result of Corollary 3.12 might be extendable from finite locales to coherent locales. But it cannot be extended much further, since we know that it fails for compact regular locales.

REFERENCES

1. B. Banaschewski, The duality of distributive continuous lattices. Canad. J. Math. 32 (1980), 385-394.

2. B. Banaschewski and C. J. Mulvey, Stone-Čech compactification of locales, I. Houston J. Math. 6 (1980), 301-312.

3. J. R. Isbell, Atomless parts of spaces. Math. Scand. 31 (1972), 5-32.

4. P. T. Johnstone, Tychonoff's theorem without the axiom of choice. Fund. Math. 113 (1981), 21-35.

5. P. T. Johnstone, The point of pointless topology. Bull. Amer. Math. Soc. 8 (1983), 41-53.

6. P. T. Johnstone, Stone Spaces. Cambridge Studies in Advanced Math. 3, Cambridge University Press, 1982.

7. P. T. Johnstone, The Vietoris monad on the category of locales. In Continuous Lattices and Related Topics, Universität Bremen, Mathematik-Arbeitspapiere 27 (1982), 162-179.

8. P. T. Johnstone, Open locales and exponentiation. In Applied Category Theory (Proceedings of Denver Special Session, January, 1983), to appear.

9. D. S. Macnab, Modal operators on Heyting algebras. Algebra Universalis 12 (1981), 5-29.

10. S. B. Nadler, Hyperspaces of Sets. Pure and Applied Math. Monographs 49, Marcel Dekker, Inc., New York, 1978.

11. H. J. Porta and O. Wyler, On compact space objects in topoi. In Categorical Aspects of Topology and Analysis, Lecture Notes in Math. 915, Springer-Verlag, Berlin-Heidelberg-New York (1982), 375-385.

12. H. Simmons, A framework for topology. In Logic Colloquium '77, Studies in Logic Vol. 96, North-Holland Publishing Co., Amsterdam, 1978, 239-251.

13. O. Wyler, Algebraic theories of continuous lattices. In Continuous Lattices, Lecture Notes in Math. 871, Springer-Verlag, Berlin-Heidelberg-New York, 1981, 390-413.

10

On the Exponential Law for Function Spaces Equipped with the Compact-Open Topology

PANOS TH. LAMBRINOS
School of Engineering
Democritos University of Thrace
Xanthi, Greece

1. PRELIMINARIES

No separation axioms are preassumed, unless the contrary is stated explicitly; e.g., compact or regular spaces need not be T_2-spaces, etc. A space is weakly locally compact if each point has a compact neighborhood. A basic locally compact space has a neighborhood basis consisting of compact sets for each of its points. In general, a weakly locally compact space need not be basic locally compact, though these two concepts coincide in T_2- or regular spaces, in which case the term locally compact can be used instead without confusion.

Given arbitrary topological spaces Y and Z, $\mathcal{C}_{co}(Y,Z)$ denotes the set $\mathcal{C}(Y,Z)$ of continuous functions on Y to Z, equipped with the compact-open topology $\mathcal{T}_{co}$. Given an ordered triple of spaces (X,Y,Z), the corresponding exponential function $E_{XYZ} = E: \mathcal{C}(X \times Y,Z) \to \mathcal{C}(X, \mathcal{C}_{co}(Y,Z))$ is a well-defined injection, where for every $f \in \mathcal{C}(X \times Y,Z)$ its adjoint $E(f) = \hat{f}: X \to \mathcal{C}_{co}(Y,Z)$ is defined for each $x \in X$ by $\hat{f}(x)(y) = f(x,y)$ for all $y \in Y$. The exponential law holds for a given ordered triple (X,Y,Z) whenever the corresponding exponential function E_{XYZ} is a bijection. Given spaces Y and Z, $\mathcal{T}_{co}$ is jointly continuous on $\mathcal{C}(Y,Z)$ whenever the evaluation function e: $\mathcal{C}_{co}(Y,Z) \times Y \to Z$ is continuous, where $e(f,y) = f(y)$ for all $(f,y) \in \mathcal{C}(Y,Z) \times Y$; or, equivalently, if for every space X the exponential law holds for the triple (X,Y,Z).

It should be noted that throughout, we are using the standard definition of the compact-open topology (see, e.g., [5], [10], [12], [22], [4]), since our point of view is purely topological rather than categorical (as e.g., in [3]). For the same reason and for the sake of clarity, we choose to call a k-space any (not necessarily T_2) space equipped with the weak topology with respect to the collection of all its compact subsets (following, e.g., Wilansky [22] and Brown [4]; i.e., a space Y is a k-space provided a set A is closed in Y iff $A \cap K$ is closed in K for every compact subset $K \subset Y$. For a discussion of slightly differing notions of k-space (or compactly generated space), especially in connection with certain "weak Hausdorffness" notions, see [16],[9],[24],[8] and the literature cited there.

In view of several applications, one is concerned with conditions sufficient to make the exponential function E_{XYZ} continuous, or an embedding, or a bijection, or even a homeomorphism with respect to the compact-open topology.

In this respect, recall the following standard results (cf. [5],[10], [12]), due mainly to Fox [6], Arens [1], and Jackson [11], obtained under certain separation conditions on the domains of the function spaces involved.

Let (X,Y,Z) be a given ordered triple of arbitrary topological spaces. Then:

(i) The exponential function E_{XYZ} is continuous if X is a T_2-space.

(ii) E_{XYZ} is actually an embedding if both X and Y are T_2-spaces.

(iii) (Fundamental Lemma) The compact-open topology $\mathcal{T}_{co}$ on $\mathcal{C}(Y,Z)$ is always jointly continuous on all basic locally compact subspaces L in Y, i.e., the evaluation e: $\mathcal{C}_{co}(Y,Z) \times L \to Z$ is continuous for every such subspace L in Y ([22], p. 280). Consequently, if Y is a T_2-space or a regular space, then $\mathcal{T}_{co}$ is jointly continuous on all compact subsets K in Y.

(iv) (Exponential Law) E_{XYZ} is a bijection if either (a) Y is a T_2- or regular locally compact space or (b) Y is a T_2- or regular space and $\mathcal{C}_{co}(Y,Z) \times Y$ is a k-space or (c) Y is a T_2- or regular space and $X \times Y$ is a k-space. In particular, if either (a) or (b) above is satisfied, then $\mathcal{T}_{co}$ is jointly continuous on $\mathcal{C}(Y,Z)$ [2].

(v) If both X and Y are T_2-spaces, then E_{XYZ} is a homeomorphism, if, in addition, either Y is locally compact, or $\mathcal{C}_{co}(Y,Z) \times Y$ is a k-space, or $X \times Y$ is a k-space.

2. MOTIVATION AND MAIN RESULT

There are two obvious ways to attack the natural question of how to obtain the standard results (i)-(v) under different or relaxed separation assumptions.

On the one hand, one can drop altogether any separation conditions on the domains and still obtain the desired results, at the reasonable cost of imposing a certain "weak regularity" assumption only on the range space (see the main theorem below). We are motivated in this direction, partly by our success in establishing the exponential law for the bounded open topology ([14], or [15] th. 2.7) using a separation condition (namely, regularity) only on the range space. Indeed, in some applications of the exponential law (e.g., in most formulations of Ascoli-type theorems), it is already customary to assume (complete) regularity of the range (cf. [2], [12],[18],[19], with the notable exception of [21]).

Further motivation in the same direction comes from the fact that the separation of properties of the compact-open topology depend exclusively on the corresponding separation properties of the range space, and so it appears more natural to impose separation conditions only on the range. This is especially so in view of the increasing popularity of using in the domains of function spaces "weaker local compactness" concepts, which do not include any separation as part of their definition; recall, for instance, the recently recognized importance of the class of core-compact (or quasi-locally compact) spaces, which are exactly the exponential objects of the category Top ([8], p. 16; [7], ch. II).

We advocate this line of thought in Section 4, culminating in the following main theorem. Notice that a compact regular (KR) space is a space whose compact subsets are regular subspaces, the intuitive idea for this "weak regularity" being that our basic building blocks (i.e., the compact subsets) ought to be well structured "internally." Clearly, the class of KR spaces properly contains the classes of T_2- and regular spaces.

2.1 MAIN THEOREM Let Z be a KR space. Then, for arbitrary spaces X and Y, the exponential function $E_{XYZ}: \mathcal{C}_{co}(X \times Y, Z) \to \mathcal{C}_{co}(X,\mathcal{C}_{co}(Y,Z))$ is an embedding. Furthermore, E_{XYZ} is a homeomorphism if also either Y is weakly locally compact, or $\mathcal{C}_{co}(Y,Z) \times Y$ is a k-space, or $X \times Y$ is a k-space.

On the other hand, a more traditional approach to obtaining the results (i)-(v), by simply relaxing the separation conditions in the domains, has been followed by Brown ([4], pp. 155-156) who obtained essentially (i)-(v) using the concept of an hl (Hausdorff-like) space, i.e., a space whose compact subsets are basic locally compact subspaces (cf. also Wilansky's hint in [22], p. 280). It is clear that any T_2- or regular (more generally, any KR) space is an hl space, though a basic locally compact space need not be an hl space. Nevertheless, the results (i)-(v) remain valid if we replace "hl" by "basic locally compact." Furthermore, justifying a referee's speculation that these results could be unified using the even broader concept of an hl^* space (i.e., compact subsets are contained in basic locally compact subspaces), one can verify that this is indeed the case.

In fact, the situation is even better. In Section 3 we further strengthen such improved results by using a considerably broader concept than hl^*, that of a space satisfying Wilker's condition (D) [23].

3. PROPERTIES OF THE COMPACT-OPEN TOPOLOGY ON FUNCTION SPACES WHOSE DOMAINS SATISFY CERTAIN WEAK CONDITIONS

A space X satisfies condition (D) if given any compact subset K in X and open subsets A_1, A_2 in X such that $K \subset A_1 \cup A_2$, there are compact sets K_1, K_2 in X with $K_1 \subset A_1$, $K_2 \subset A_2$, $K \subset K_1 \cup K_2$. We call such spaces Wilker spaces. The class of Wilker spaces properly contains the class of hl^* spaces; hence, in particular, all basic locally compact spaces and all hl spaces (see [23] for the importance of Wilker spaces in a categorical setting). Recall from [13], [14], or [15] that the class of k-spaces is contained properly in the class of ℓ-spaces (i.e., boundedly generated spaces).

Using the above terminology, we have the following standard results. Let (X,Y,Z) be any ordered triple of spaces. Then

- (i^*) The exponential function E_{XYZ} is continuous if X is a Wilker space.
- (ii^*) E_{XYZ} is actually an embedding if the product $X \times Y$ is a Wilker space.
- (iii^*) The compact-open topology $\mathcal{T}_{co}$ on $\mathcal{C}(Y,Z)$ is jointly continuous on all compact subsets K in Y provided Y is an hl^* space.
- (iv^*) (Exponential Law) E_{XYZ} is a bijection if either (a) Y is basic locally compact, (b) Y is hl^* and $\mathcal{C}_{co}(Y,Z) \times Y$ is a k-space, (c) Y is hl^* and $X \times Y$ is a k-space, (d) Y is regular and $\mathcal{C}_{co}(Y,Z) \times Y$ is an ℓ-space, or (e) Y is regular and $X \times Y$ is an ℓ-space. In particular, if either (a), (b), or (d) above is satisfied, then $\mathcal{T}_{co}$ is jointly continuous on $\mathcal{C}(Y,Z)$.
- (v^*) If X is an hl^* space, then E_{XYZ} is a homeomorphism if also one of the preceding conditions (a)-(e) is satisfied.

The proof of (i^*) and (ii^*) above is based on the following Jackson-type lemma (cf. [5], p. 264), and on the fact that Wilker's property is preserved by projections. I am indebted to T. Kouphos and to B. Papadopoulos for a proof of Lemma 3.1(b) below, and for examples of spaces that are basic locally compact but not hl, and of Wilker spaces that are not hl^* spaces; also to J. S. Yang for inspiring comments leading essentially to (iv^*,d) above.

3.1 SUBBASIS LEMMA (a) Let X be a Wilker space and let $\mathcal{S}$ be a subbasis of the topology $\mathcal{O}(T)$ of an arbitrary topological space T. Then the collection of all sets $(K,R) = \{f \in \mathcal{C}(X,T)\colon K \subset f^{-1}(R)\}$, where K runs through the compact subsets in X and $R \in \mathcal{S}$, generates the compact-open topology on $\mathcal{C}(X,T)$.

(b) Let $X \times Y$ be a Wilker space and let Z be an arbitrary space. Then the compact-open topology on $\mathcal{C}(X \times Y,Z)$ is generated by the collection of all sets $(K \times L,P)$, where K (resp., L) runs through the compact subsets in X (resp., in Y) and $P \in \mathcal{O}(Z)$.

The assertion (iii^*) is clear. Subsequently, the exponential law (iv^*) follows from (iii^*) using standard techniques; note that in a regular space bounded subsets have compact closures ([13,14]). Finally, the exponential homeomorphism (v^*) follows from the preceding, by observing that the class of hl^* spaces is finitely productive.

4. PROPERTIES OF THE COMPACT-OPEN TOPOLOGY ON FUNCTION SPACES WHOSE RANGE SATISFIES SOME WEAK SEPARATION PROPERTY

First we prove a simple proposition, which appears to have escaped being recorded in the literature in its full generality.

4.1 PROPOSITION Given arbitrary spaces Y and Z, let $\mathcal{K}$ be a compact subset in $\mathcal{C}_{co}(Y,Z)$. Then the set $\mathcal{K}(K) = \cup\{f(K)\colon f \in \mathcal{K}\} = e(\mathcal{K} \times K)$ is compact for every compact subset K in Y.

Proof: Let $\mathcal{A} = \{A_i\colon i \in I)$ be an open covering of $\mathcal{K}(K)$ in Z. Denote by $\mathcal{A}^* = \{A_j^*\colon j \in I_F\}$ the covering consisting of all finite unions of members of $\mathcal{A}$. Observe that the collection of the sets $\{(K,A_j^*)\colon j \in I_F\}$ is an open covering of the $\mathcal{T}_{co}$-compact set $\mathcal{K}$. Indeed, for every $g \in \mathcal{K}$, the compact set g(K) is covered by $\mathcal{A}$, and so it is contained in some member of $\mathcal{A}^*$. Now extract a finite subcovering $\{(K,A_{j(\lambda)}^*)\colon j(\lambda) \in I_F\}$ of $\mathcal{K}$, and observe that $\mathcal{K}(K)$ is covered by the finite union $\cup A_{j(\lambda)}^*$ of finite unions of members of $\mathcal{A}$.

We remark that, in view of the fundamental Lemma 4.7 below, the preceding assertion is immediate if Z is a KR space (similarly, if Y is hl^*; cf. (iii^*) above).

4.2 LEMMA For every space Y, $\mathcal{C}_{co}(Y,Z)$ is KR iff Z is a KR space.

Proof: Necessity follows easily since Z is embedded in $\mathcal{C}_{co}(Y,Z)$ and subspaces of KR spaces are KR spaces themselves. Conversely, let Z be a KR space, let $\mathcal{K}$ be a $\mathcal{T}_{co}$-compact subset in $\mathcal{C}_{co}(Y,Z)$, and let $(K,P) \cap \mathcal{K}$ be a subbasic neighborhood of an arbitrary $f \in \mathcal{K}$. We need determine a $\mathcal{T}_{co}$-neighborhood of f in $\mathcal{K}$ whose relative closure is contained in $(K,P) \cap \mathcal{K}$.

Observe that $\mathcal{K}(K) = \cup\{g(K): g \in \mathcal{K}\}$ is actually a compact regular subspace of Z, containing the open [in $\mathcal{K}(K)$] neighborhood $P \cap \mathcal{K}(K)$ of its compact subset f(K). Therefore, there exists an open set R in Z such that

$$f(K) \subset R^* = R \cap \mathcal{K}(K) \subset \overline{R^*} \cap \mathcal{K}(K) \subset P \cap \mathcal{K}(K)$$

We claim that $\overline{(K,R) \cap \mathcal{K}} \cap \mathcal{K} \subset (K,\overline{R^*}) \cap \mathcal{K}$. To verify this, suppose $g \in \mathcal{K} - (K,\overline{R^*})$. Then there is a point $p \in K$ such that $g \in (p, Z - \overline{R^*}) \in \mathcal{T}_{co}$. Observe that $(p, Z - \overline{R^*}) \cap (K,R) \cap \mathcal{K} = \emptyset$, since otherwise, for some $h \in (K,R) \cap \mathcal{K}$, $h(p) \notin \overline{R^*}$ and $h(p) \in h(K) \subset R \cap \mathcal{K}(K) = R^*$, which is absurd. Consequently, $g \notin \overline{(K,R) \cap \mathcal{K}}$, thus establishing the claim.

It is now clear that the relative closure of the neighborhood $(K,R) \cap \mathcal{K}$ of f in $\mathcal{K}$ is contained in the initial subbasic neighborhood $(K,P) \cap \mathcal{K}$ of f, as required. ∎

4.3 FIRST SUBBASIS LEMMA Let X be an arbitrary space and let $\mathcal{S}$ be a subbasis of the topology $\mathcal{O}(T)$ of a KR space T. Then the collection of all sets (K,R), where K is compact in X and $R \in \mathcal{S}$, generates the compact-open topology on $\mathcal{C}(X,T)$.

Proof: Let $f \in (K,P)$, where K is a compact subset of X and $P \in \mathcal{O}(T)$. We need determine a finite intersection of sets (K_λ, R_λ) containing f and contained in (K,P), where each $R_\lambda \in \mathcal{S}$ and each K_λ is compact in X. The open set P can be expressed as the union of a collection $\{Q_i: i \in I\}$ of basic open sets in $\mathcal{O}(T)$, where each Q_i is the intersection of finitely many subbasic open sets $R_i^j \in \mathcal{S}$, $j \in J_i$. Observe that the compact set f(K) is a regular subspace of T. For each point $x \in K$, there is some Q_{i_x} containing f(x). Choose an open set $A_x \in \mathcal{O}(T)$ such that

$$f(x) \in A_x^* = A_x \cap f(K) \subset \overline{A_x^*} \cap f(K) \subset Q_{i_x} \cap f(K)$$

The collection of open (in K) sets $\{f^{-1}(A_x^*) \cap K: x \in K\}$ covers K. Extract a finite subcovering, say $\{f^{-1}(A_{x(\lambda)}^*) \cap K: \lambda \in \Lambda\}$ of K and set $f^{-1}(\overline{A_{x(\lambda)}^*}) \cap K = K_\lambda$ for each $\lambda \in \Lambda$. Each such set K_λ is compact, and obviously

$$f \in \bigcap_\lambda (K_\lambda, Q_{i_{x(\lambda)}}) = \bigcap_\lambda \bigcap_j (K_\lambda, R^j_{i_{x(\lambda)}}) \subset (K,P)$$

as required. ■

4.4 PROPOSITION Given an ordered triple of spaces (X,Y,Z), the exponential function $E_{XYZ}: \mathcal{C}_{co}(X \times Y,Z) \to \mathcal{C}_{co}(X,\mathcal{C}_{co}(Y,Z))$ is continuous, provided Z is a KR space.

4.5 SECOND SUBBASIS LEMMA Let X,Y be arbitrary spaces and Z a KR space. Then the compact-open topology on $\mathcal{C}(X \times Y,Z)$ is generated by the collection of all sets $(K \times L,P)$, where K (resp., L) runs through the compact subsets of X (resp., Y) and $P \in \mathcal{O}(Z)$.

Proof: Let (C,P) be a subbasic $\mathcal{T}_{co}$-open neighborhood of an arbitrary point $f \in \mathcal{C}(X \times Y,Z)$, where C is a compact subset of $X \times Y$ and $P \in \mathcal{O}(Z)$. We need determine a finite intersection of sets $(K_\lambda \times L_\lambda, P_\lambda)$ containing f and contained in (C,P), where each K_λ (resp., L_λ) is a compact subset of X (resp., Y) and each $P_\lambda \in \mathcal{O}(Z)$. Observe that the projections $\pi_x(C) = A$ and $\pi_y(C) = B$ of C to the factor spaces X and Y are compact sets, and thus $f(A \times B)$ is a compact regular subspace of Z.

For each point $(x,y) \in C$, choose an open set R_{xy} in Z such that

$$f(x,y) \in R^*_{xy} = R_{xy} \cap f(A \times B) \subset \overline{R^*_{xy}} \cap f(A \times B) \subset P \cap f(A \times B)$$

Since each $f^{-1}(R^*_{xy}) \cap (A \times B)$ is an open (in $A \times B$) neighborhood of (x,y), choose an open neighborhood N^*_{xy} (resp., M^*_{xy}) of x (resp., y) in A (resp., B) such that

$$(x,y) \in N^*_{xy} \times M^*_{xy} \subset f^{-1}(R^*_{xy}) \cap (A \times B)$$

Extract a finite subcovering, say $\{N^*_\lambda \times M^*_\lambda: \lambda \in \Lambda\}$ of the open (in $A \times B$) covering $\{N^*_{xy} \times M^*_{xy}: (x,y) \in C\}$ of $C \subset A \times B$. Set, for each $\lambda \in \Lambda$, $K_\lambda = \overline{N^*_\lambda} \cap A$ and $L_\lambda = \overline{M^*_\lambda} \cap B$. Each K_λ and each L_λ is compact, and clearly,

$$f(K_\lambda \times L_\lambda) \subset f(\overline{N^*_\lambda \times M^*_\lambda}) \cap f(A \times B) \subset \overline{R^*_{x_\lambda y_\lambda}} \cap f(A \times B) \subset P$$

Therefore, $f \in \bigcap_\lambda (K_\lambda \times L_\lambda, P) \subset (C,P)$, as required. ■

4.6 PROPOSITION Given an ordered triple of spaces (X,Y,Z), the corresponding exponential function is an embedding, provided Z is a KR space.

4.7 LEMMA Let Y be an arbitrary space and Z a KR space. Then the compact-open topology on $\mathcal{C}(Y,Z)$ is jointly continuous on compacta, i.e., for every compact $K \subset Y$, the restricted evaluation $e: \mathcal{C}_{co}(Y,Z) \times K \to Z$ is continuous.

Proof: Let $(f,y) \in \mathcal{C}(Y,Z) \times K$ and let P be an open (in Z) neighborhood of f(y). Observe that f(K) is a compact regular subspace of Z. Choose an open set R_y in Z such that $f(y) \in R_y^* = R_y \cap f(K) \subset \overline{R_y^*} \cap f(K) \subset P \cap f(K)$. Denote by N_y the open (in K) neighborhood $f^{-1}(R_y^*) \cap K$ of y, and by K_y the compact set $f^{-1}(\overline{R_y^*}) \cap K$. Since $(K_y,P) \times N_y$ is an open [in $\mathcal{C}_{co}(Y,Z) \times K$] neighborhood of (f,y) contained in $e^{-1}(P)$, the proof is complete. ∎

4.8 PROPOSITION Given any space Y and any KR space Z, the compact-open topology is jointly continuous on $\mathcal{C}(Y,Z)$ if either

(i) Y is weakly locally compact, or

(ii) $\mathcal{C}_{co}(Y,Z) \times Y$ is a k-space.

Note that Proposition 4.8.i above is known for Z completely regular (i.e., uniformizable; see [19] th.2.32), or at least regular ([20]; cf. [17] lemma 2.3); and that it may fail for arbitrary Z, unless Y satisfies stronger assumptions. Thus, theorem 2.30 in [19] is false in general.

4.9 EXPONENTIAL LAW Let Z be a KR space and let X,Y be arbitrary spaces. Then the exponential function E_{XYZ} is a bijection if, in addition, either (i) Y is weakly locally compact, or (ii) $\mathcal{C}_{co}(Y,Z) \times Y$ is a k-space, or (iii) $X \times Y$ is a k-space.

Proof: Cases (i) and (ii) are immediate from Proposition 4.8 above. Case (iii) follows from Lemma 4.7, since for any continuous function $\hat{f}$: $X \to \mathcal{C}_{co}(Y,Z)$ the composition $e \circ (f \times 1_Y) = \hat{f}$ has continuous restrictions on the compact subsets of $X \times Y$. ∎

Finally, 4.6 and 4.9 imply the main theorem 2.1. Furthermore, Proposition 4.6 and (iv^* d,e) above imply the following mixed result, involving separation assumptions on both the range and the domain.

4.10 COROLLARY Let Z be a KR space and let Y be regular. Then the exponential function is a homeomorphism provided either $X \times Y$ or $\mathcal{C}_{co}(Y,Z) \times Y$ is an ℓ-space.

The preceding theorems describe a wide range of standard situations in which results related to the exponential law can be applied. We conclude with some examples.

4.11 APPLICATIONS (i) The homotopy classes of $\mathcal{C}(Y,Z)$ coincide with the path components of $\mathcal{C}_{co}(Y,Z)$ provided that either (a) Y is an hl^* k-space or (b) Z is a KR space and Y is a k-space.

(ii) Recall that the usual composition operation $T: \mathcal{C}_{co}(X,Y) \times \mathcal{C}_{co}(Y,Z) \to \mathcal{C}_{co}(X,Z)$, where $T(f,g) = g \circ f$, is continuous if Y is basic locally compact for arbitrary X,Z. Then T is also continuous if one of the following sets of conditions is satisfied (where $\mathcal{C}_{co}(A,B)$ is denoted by B^A):

1. Z is a KR space, $Z^Y \times Y$ is a k-space or Y is weakly locally compact, and also either (a) X is basic locally compact, or (b) X is hl^* and $Y^X \times X$ is a k-space, or (c) X is regular and $Y^X \times X$ is an ℓ-space.
2. Y is hl^*, $Z^Y \times Y$ is a k-space and also either (a) or (b) or (c) above holds.
3. X is weakly locally compact or $Y^X \times X$ is a k-space and also either (a) Y is a KR space and $Z^Y \times Y$ is a k-space, or (b) Y is regular and $Z^Y \times Y$ is an ℓ-space.

Indeed, each set of conditions guarantees the continuity of the evaluation functions $e_{YZ}: Z^Y \times Y \to Z$ and $e_{XY}: Y^X \times X \to Y$; hence the continuity of a suitable function on $Y^X \times Z^Y \times X$ to Z whose continuous adjoint is T.

(iii) Observe that, with respect to the compact-open topology, the natural bijection $h: (Y \times Z)^X \to Y^X \times Z^X$ is always continuous, where for each $f \in (Y \times Z)^X$, $h(f) = (\pi_Y \circ f, \pi_Z \circ f)$, $\pi_Y: Y \times Z \to Y$ and $\pi_Z: Y \times Z \to Z$ are the usual projections, and for each pair $(p,q) \in Y^X \times Z^X$ there is a unique product function $\langle p,q \rangle \in (Y \times Z)^X$ [defined for each $x \in X$ by $\langle p,q\rangle(x) = (p(x),q(x))$] such that $h(\langle p,q \rangle) = (p,q)$. Then the continuous bijection h is actually a homeomorphism provided that either X is a Wilker space, or Y and Z are both KR spaces.

I am grateful to the referee for comments leading to a clearer and unified presentation of Section 3.

REFERENCES

1. Arens, R. F. A topology for spaces of transformations. Ann. of Math. 47 (1946), 480-495.

2. Bagley, R. W., and J. S. Yang. On k-spaces and function spaces. Proc. Amer. Math. Soc. 17 (1966), 703-705.

3. Booth, P., and J. Tillotson. Monoidal closed, cartesian closed and convenient categories of topological spaces. Pacific J. Math. 88 (1980), 35-53.

4. Brown, R. Elements of Modern Topology. McGraw-Hill, London, 1968.

5. Dugundji, J. Topology. Allyn and Bacon, Boston, 1966.

6. Fox, R. On topologies for function spaces. Bull. Amer. Math. Soc. 51 (1945), 429-439.

7. Gierz, G., K. H. Hofmann, K. Keimel, J. D. Lawson, M. Mislove, and D. S. Scott. A Compendium of Continuous Lattices. Springer-Verlag, Berlin-Heidelberg-New York, 1980.

8. Herrlich, H. Categorical Topology 1971-1981. Proc. Fifth Prague Topol. Symp. 1981 (General Topology and Its Relations to Modern Analysis and Algebra V, J. Novak, ed.). Heldermann-Verlag, Berlin, 1982, 279-383.

9. Hoffmann, R.-E. On weak Hausdorff spaces. Arch. Math. 32 (1979), 487-504.

10. Hu, S. T. Elements of General Topology. Holden-Day, San Francisco-London-Amsterdam, 1964.

11. Jackson, J. R. Spaces of mappings on topological products with applications to homotopy theory. Proc. Amer. Math. Soc. 3 (1952), 327-333.

12. Kelley, J. L. General Topology. Springer-Verlag, New York-Heidelberg-Berlin, 1955.

13. Lambrinos, P. Th. Boundedly generated spaces. Manuscripta Math. 31 (1980), 425-438.

14. Lambrinos, P. Th. The bounded open topology on function spaces. Manuscripta Math. 36 (1981), 47-66.

15. Lambrinos, P. Th., and B. Papadopoulos. The (strong) Isbell topology and (weakly) continuous lattices. This volume (Chap. 11).

16. Lawson, J., and B. Madison. Comparisons of notions of weak Hausdorffness. Topol. Proc. Memphis State Univ. Confer. 1975 (S. P. Franklin and B. V. Smith Thomas, eds.). Marcel Dekker, New York, 1976, 207-215.

17. Myers, S. B. Equicontinuous sets of mappings. Ann. of Math. 47 (1946), 496-502.

18. Noble, N. Ascoli theorems and the exponential map. Trans. Amer. Math. Soc. 143 (1969), 393-411.

19. Poppe, H. Compactness in general function spaces. VEB Deutscher Verlag der Wissenschaften, Berlin, 1974.

20. Schwarz, F. Topological continuous convergence. To appear, Manuscripta Math., 1984.

21. Weston, J. D. A generalization of Ascoli's theorem. Mathematika 6 (1959), 19-24.

22. Wilansky, A. Topology for Analysis. Xerox College Publishing, Lexington, Massachusetts; Toronto, 1970.

23. Wilker, P. Adjoint product and hom functors in general topology. Pacific J. Math. 34 (1970), 269-283.

24. Wyler, O. Convenient categories for topology. Gen. Topol. Appl. 3 (1973), 225-242.

11

The (Strong) Isbell Topology and (Weakly) Continuous Lattices *

PANOS TH. LAMBRINOS and BASIL PAPADOPOULOS

School of Engineering
Democritus University of Thrace
Xanthi, Greece

1. INTRODUCTION

It is very well known that the category Top fails to have decently behaved function spaces; i.e., it fails to be cartesian closed, and so it is not convenient for several applications. Given any category $\mathcal{A}$ with finite products, an $\mathcal{A}$-object Y is exponential in $\mathcal{A}$ if the functor $\cdot \times Y: \mathcal{A} \to \mathcal{A}$ has a right adjoint. $\mathcal{A}$ is cartesian closed provided each of its objects is exponential in $\mathcal{A}$ [6]. In particular, an arbitrary space Y is exponential in Top if for every space Z there is a (unique) topology t on the set $\mathcal{C}(Y,Z)$ of continuous functions on Y to Z, such that for every space X the exponential function $E_{XYZ}: \mathcal{C}(X \times Y, Z) \to \mathcal{C}(X,\mathcal{C}_t(Y,Z))$, where $E_{XYZ}(f)(x)(y) = \hat{f}(x)(y) = \hat{f}(x,y)$ for all x and y, is a bijection.

Perhaps the most familiar example of a cartesian closed subcategory of Top is the category CHG of compact Hausdorff-generated spaces, i.e., the category of all quotients of disjoint unions of compact Hausdorff spaces.

The problem of determining the existence of the largest or of a maximal cartesian closed coreflective subcategory of Top is still open (see [6], p. 20). For instance, it is not even known whether the cartesian closed category BLCG of basic locally compact-generated spaces is strictly broader than CHG ([6], problem 5). In what follows, separation axioms are never implied unless explicitly stated; thus, e.g., regular or compact does not imply T_2. A space is basic locally compact if each point has a neighborhood basis consisting of compact sets.

---

*Dedicated to Professor Nicolas Oeconomides of the Aristotle University of Thessaloniki on the occasion of his twentieth Professorial anniversary.

The above is part of the motivation for the search for "reasonable" topologies on function spaces whose domains Y are objects of particular subcategories $\mathcal{A}$ of Top. The study of such reasonable function space topologies is also interesting in its own right. Indeed, the point of view of the present paper is mostly purely topological.

In Section 2, first we review the well-known fact that the compact-open topology is the appropriate topology to use on function spaces whose domains are basic locally compact. Next, we recall that the exponential objects in Top are precisely those spaces Y whose lattice of open sets $\mathcal{O}(Y)$ is continuous, and we investigate some properties of the corresponding appropriate function space topology for such domains, namely, the Isbell topology $\mathcal{T}_{is}$.

This leads naturally to the introduction of the strong Isbell topology $\mathcal{T}_{is}*$ in Section 3, which is defined via the strong (Day-Kelly-) Scott topology for complete lattices. The strong Isbell topology $\mathcal{T}_{is}*$ enjoys some nice properties on function spaces whose domains Y are such that $\mathcal{O}(Y)$ satisfies the naturally arising new lattice theoretic property of being weakly continuous, the further study of which might perhaps be desirable.

2. CORECOMPACTNESS AND THE ISBELL FUNCTION SPACE TOPOLOGY

Given arbitrary spaces Y and Z, let $\mathcal{C}_t(Y,Z)$ denote the set $\mathcal{C}(Y,Z)$ of continuous functions on Y to Z equipped with some topology t. The topology t is said to be splitting on $\mathcal{C}(Y,Z)$ whenever for every space X the continuity of a function $f: X \times Y \to Z$ implies that of its adjoint function $\hat{f}: X \to \mathcal{C}_t(Y,Z)$ where $\hat{f}(x)(y) = f(x,y)$ for all x and y, i.e., if the exponential injection $E_{XYZ}: \mathcal{C}(X \times Y, Z) \to \mathcal{C}(X,\mathcal{C}_t(Y,Z))$, where $E_{XYZ}(f) = \hat{f}$, is well defined [4]. If for every X the continuity of $\hat{f}: X \to \mathcal{C}_t(Y,Z)$ implies that of $f: X \times Y \to Z$, then t is called jointly continuous on $\mathcal{C}(Y,Z)$; equivalently, t is jointly continuous on $\mathcal{C}(Y,Z)$ if the evaluation function $e: \mathcal{C}_t(Y,Z) \times Y \to Z$ is continuous, where $e(g,y) = g(y)$ [12].

There exists at most one topology t on $\mathcal{C}(Y,Z)$ that is both splitting and jointly continuous; such a topology is the finest splitting and the coarsest jointly continuous topology on $\mathcal{C}(Y,Z)$.

A space Y is exponential in Top if for every space Z there is a splitting jointly continuous topology on $\mathcal{C}(Y,Z)$. Schwarz [22] has observed that the coarsest jointly continuous topology on $\mathcal{C}(Y,Z)$, if it exists, is also (the finest) splitting.

There is always a finest splitting topology, but in general there is no coarsest jointly continuous topology on $\mathcal{C}(Y,Z)$; consider for instance the set $\mathcal{C}(\mathbb{Q},\mathbb{R})$ of the continuous real-valued functions on the rationals. More generally, let Y be completely regular and let Z be a T_0-space containing a nondegenerate path; then there is no coarsest jointly continuous topology on $\mathcal{C}(Y,Z)$ iff Y fails to be locally compact ([9], p. 156). Hence, there are many spaces which are not exponential in Top.

Recall that the sets of the form $(K,P) = \{f \in \mathcal{C}(Y,Z): K \subset f^{-1}(P)\}$, where $P \in \mathcal{O}(Z)$ and K runs through the compact sets $\mathcal{K}(Y)$ in Y, generate the compact-open topology $\mathcal{T}_{co}$ on $\mathcal{C}(Y,Z)$, which is always splitting.

$\mathcal{T}_{co}$ is very well behaved on $\mathcal{C}(Y,Z)$ whenever the domain Y is basic locally compact; specifically, in this case $\mathcal{T}_{co}$ is (the coarsest) jointly continuous for arbitrary Z ([1], p. 156; [27], p. 280). For further relevant properties of $\mathcal{T}_{co}$ see [18], [11]. Notice that, in the absence of basic local compactness in the domain, $\mathcal{T}_{co}$ may fail badly to be the appropriate function space topology to use. Consider, for instance, the set $\mathcal{C}(\mathbb{R}^*,\mathbb{R})$, where $\mathbb{R}^*$ is the real line $\mathbb{R}$ retopologized by taking for each point $x \in \mathbb{R}$ a (finer) neighborhood system of the form $\{\{x\} \cup (V \cap \mathbb{Q}): V \in \mathcal{N}(x)\}$, where $\mathcal{N}(x)$ are the usual neighborhoods of x in $\mathbb{R}$ [15]. $\mathcal{T}_{co}$ is not jointly continuous on $\mathcal{C}(\mathbb{R}^*,\mathbb{R})$, though there does exist a (splitting) coarsest jointly continuous topology, namely, the bounded open topology $\mathcal{T}_{eo}$(see [16] and below). In fact, more generally, given any Urysohn (i.e., functionally separated) space Y and a T_0-space Z containing a nondegenerate path, $\mathcal{T}_{co}$ fails to be jointly continuous on $\mathcal{C}(Y,Z)$ iff the domain Y fails to be locally compact [17]. On the other hand, let Y be arbitrary and let $\overline{2}$ denote the Sierpinski space $\{0,1\}$; then the joint continuity of $\mathcal{T}_{co}$ on $(Y,\overline{2})$ implies that Y is basic locally compact [17]. Thus, the basic locally compact spaces are precisely those exponential objects Y in Top for which the compact-open topology is jointly continuous on $\mathcal{C}(Y,Z)$ for arbitrary Z.

Consider the following class of spaces which clearly contains the class of basic locally compact spaces. Let $B \subset A$ be subsets in an arbitrary space Y. Then B is bounded in A if every (directed) open cover of A contains finitely many members covering B ([10], p. 49); B is bounded in Y iff B is bounded in A for every $A \supset \overline{B}$ ([14]; see also [13] and below). A space Y is called corecompact if for every point $y \in Y$ and each open set V containing y there is some open set W bounded in V containing y. Ward ([26], 1968) introduced these spaces under the name quasi-locally compact.

Subsequently they were investigated under various names by Day and Kelly [3] (spaces with property C), Day [2] (Ω-compact), Isbell[10] (semi-locally bounded), Hofmann [7] (CL-spaces), and by Hofmann and Lawson [8] (corecompact), who showed that a sober space is corecompact iff it is basic locally compact. Isbell [10] and Hofmann and Lawson ([8], section 7) have exhibited examples of second countable corecompact spaces in which every compact subset has empty interior, thus showing that in the absence of sobriety a corecompact space may fail badly to be basic locally compact, or even weakly locally compact (viz., each point has some compact neighborhood).

On the other hand, it is not known whether every corecompact space is necessarily a k-space, i.e., whether it has the weak topology with respect to its compact subsets.

Day [2] showed that the category CORCG (= corecompact-generated spaces) of all quotients of corecompact spaces is cartesian closed. It is not clear, however, whether CORCG is strictly broader than BLCG ([6], problem 6).

The class of corecompact spaces is very nice; Day and Kelly [3] showed that Y is corecompact iff for every quotient map $q: X \to Z$, $q \times 1_Y: X \times Y \to Z \times Y$ is also quotient. Wyler ([29], p. 227) noted that Y is corecompact iff it is exponential in Top. In order to describe explicitly the appropriate function-space topology for corecompact domains, recall first that for an arbitrary space Y a subset $H \subset \mathcal{O}(Y)$ is open in the Day-Kelly-Scott topology $\Omega(Y)$ if (i) H contains every open set containing some member of H, and (ii) for every collection of open sets whose union belongs to H there are finitely many members of it whose union also belongs to H [3].

Consider the topology $\mathcal{T}_{is}$ on $\mathcal{C}(Y,Z)$ ([16], p. 56) generated by the sets of the form $(H,P) = \{f \in \mathcal{C}(Y,Z): f^{-1}(P) \in H\}$, where $H \in \Omega(Y)$ and $P \in \mathcal{O}(Z)$, which is always splitting and finer than the compact-open topology $\mathcal{T}_{co}$ [note that for each compact set K in Y, the filter $H(K) = \{V \in \mathcal{O}(Y): K \subset V\} \in \Omega(Y)$, and $(K,P) = (H(K),P)$]. We are indebted to F. Schwarz for bringing to our attention Isbell's earlier equivalent definition of $\mathcal{T}_{is}$ as the initial topology on $\mathcal{C}(Y,Z)$ with respect to the functions τ_P: $\mathcal{C}(Y,Z) \to \Omega(Y)$, where $\tau_P(f) = f^{-1}(P)$ and $P \in \mathcal{O}(Z)$ ([11], p. 323). Thus $\mathcal{T}_{is}$ should be called the Isbell topology instead of the Scott topology as in [16],[19],[20]. Especially so, since Schwarz and Weck show in [24], 2.10, that for a complete lattice Z, the lattice-theoretically defined Scott topology $\mathcal{T}_\sigma$ on $\mathcal{C}(Y,\sigma(Z))$ ([5], p. 99) does not always coincide with $\mathcal{T}_{is}$; though, of course, always $\mathcal{C}_{is}(Y,\overline{2}) = \mathcal{C}_\sigma(Y,\overline{2}) \cong \Omega(Y)$, where $\overline{2}$ denotes the Sierpinski space $\{0,1\}$. They also point out that $\mathcal{C}_\sigma(Y,Z) = \mathcal{C}_{is}(Y,Z)$ whenever Y is corecompact and Z is an injective T_0-space ([24], 2.10), thus suggesting the Isbell topology as a reasonable extension of the Scott topology for non-T_0 Z.

Isbell noted ([11], 1.4) that $\mathcal{C}_{is}(\cdot,\cdot)$: $\underline{Top}^{op} \times \underline{Top} \to \underline{Top}$ is bifunctorial. Clearly, the restrictions $\mathcal{C}_{is}(\cdot,\cdot)$: $T_i\underline{Top}^{op} \times T_i\underline{Top} \to T_i\underline{Top}$, where i = 0, 1, 2, are also bifunctors since the Isbell topology is finer than the pointwise topology.

Consider the induced covariant functor $\mathcal{C}_{is}(Y,\cdot)$ for fixed Y, the induced contravariant functor $\mathcal{C}_{is}(\cdot,Z)$ for fixed Z, and the usual composition operation T: $\mathcal{C}_{is}(X,Y) \times \mathcal{C}_{is}(Y,Z) \to \mathcal{C}_{is}(X,Z)$, where $T(f,g) = g\circ f$. Since T is clearly separately continuous, one wonders whether it is also continuous under certain conditions. Observe first that in view of [4], p. 259, T: $\mathcal{C}_{co}(X,Y) \times \mathcal{C}_{co}(Y,Z) \to \mathcal{C}_{co}(X,Z)$ is continuous for arbitrary X and Z, provided Y is basic locally compact. As expected, a corresponding result holds with respect to the Isbell topology, namely:

2.1 PROPOSITION If Y is corecompact, then the usual composition operation T: $\mathcal{C}_{is}(X,Y) \times \mathcal{C}_{is}(Y,Z) \to \mathcal{C}_{is}(X,Z)$ is continuous for arbitrary X and Z; also, T: $\mathcal{C}_{co}(X,Y) \times \mathcal{C}_{is}(Y,Z) \to \mathcal{C}_{co}(X,Z)$ is continuous.

Proof: Let (H,P) be a subbasic open neighborhood in $\mathcal{C}_{is}(X,Z)$ of the image $T(f,g) = g\circ f$ of a point $(f,g) \in \mathcal{C}(X,Y) \times \mathcal{C}(Y,Z)$. For each point $y \in g^{-1}(P) \in \mathcal{O}(Y)$ there is some open set V_y bounded in $g^{-1}(P)$ containing y. Consequently, $g^{-1}(P) = \cup\{V_y: y \in g^{-1}(P)\}$ and $\cup f^{-1}(V_y) = f^{-1}(g^{-1}(P)) \in H$ implies the existence of finitely many sets V_{y_λ} whose union $V = \cup V_{y_\lambda}$ is bounded in $g^{-1}(P)$ and such that $f^{-1}(V) = \cup f^{-1}(V_{y_\lambda}) \in H$.

Recall now ([5], II.1.10(i); cf. also [10], 2.8), according to which in a corecompact space Y, the sets $H(W) = \{W' \in \mathcal{O}(Y)$: W is bounded in $W'\}$, $W \in \mathcal{O}(Y)$, are a basis for the Day-Kelly-Scott topology $\Omega(Y)$. Thus, the set $(H,V) \times (H(V),P)$ is an open neighborhood of (f,g) in $\mathcal{C}_{is}(X,Y) \times \mathcal{C}_{is}(Y,Z)$, and it is contained in $T^{-1}(H,P)$, since for every pair $(p,q) \in (H,V) \times (H(V),P)$, $p^{-1}(V) \in H$ and $V \subset q^{-1}(P) \in H(V)$ clearly imply that $(q\circ p)^{-1}(P) = p^{-1}(q^{-1}(P)) \in H$. ■

In the special case where X is a singleton, clearly $\mathcal{C}_{is}(X,T) \cong T$; and so one gets as an immediate corollary of the preceding the implication (i) ⇒ (v) in the following well-known characterization theorem for corecompactness, which also motivates the characterization 3.4 of local boundedness. Recall that in a complete lattice L, x is way below y, or x is bounded in y ([11], p. 326), denoted $x \ll y$, iff for directed subsets $D \subset L$ the relation $y \leq \sup D$ always implies the existence of a $d \in D$ with $x \leq d$ ([5], p. 38). Recall also that a complete lattice L is continuous if for each $x \in L$, $x = \sup\{u \in L: u \ll x\}$ ([5], p. 41).

2.2 THEOREM For an arbitrary space Y the following are equivalent:

(i) Y is corecompact.

(ii) $\mathcal{O}(Y)$ is a continuous lattice.

(iii) Y has the Day-Kelly property (C), i.e., for every open set V containing a point $y \in Y$ there is an open set H_y in the Day-Kelly-Scott topology $\Omega(Y)$ such that $V \in H_y$ and $\bigcap\{W: W \in H_y\}$ is a neighborhood of y in Y.

(iv) The evaluation function $e: \mathcal{C}_{is}(Y,\overline{2}) \times Y \to \overline{2}$ is continuous.

(v) For every space Z the evaluation $e: \mathcal{C}_{is}(Y,Z) \times Y \to Z$ is continuous.

(vi) Y is exponential in <u>Top</u>, i.e., for every space Z there is a (unique) topology, namely, $\mathcal{T}_{is}$, on $\mathcal{C}(Y,Z)$ such that for every space X the exponential function $E_{XYZ}: \mathcal{C}(X \times Y,Z) \to \mathcal{C}(X,\mathcal{C}_{is}(Y,Z))$ is a bijection.

Proof: For the equivalence of (i), (ii), and (iii), see [8], 4.2. The above imply the equivalence of (i), (v), (vi), and it is easy to see that (iv) $\Leftrightarrow$ (iii). ∎

For alternate proofs and for more characterizations, see [23], 3.3, [24], 2.16, and [5], II.4.10,V.5.10.

The Isbell topology enjoys pleasant functorial properties, as the following proposition indicates. Recall first that a space is called a KR space [18] if all its compact subsets are regular subspaces. The class of KR spaces contains all T_2- and all regular spaces, though a basic locally compact space need not be a KR space.

2.3 PROPOSITION (i) For any fixed space Y the induced covariant functor $\mathcal{C}_{is}(Y,\cdot)$: <u>Top</u> → <u>Top</u> preserves homotopies of (a) arbitrary functions if Y is corecompact, (b) functions having corecompact domain, for arbitrary Y, (c) functions having KR range if Y is weakly locally compact.

(ii) For any fixed space Z the induced contravariant functor $\mathcal{C}_{is}(\cdot,Z)$: <u>Top</u>^{op} → <u>Top</u> preserves homotopies of (a) functions having corecompact range, for arbitrary Z, (b) functions having corecompact domain, for arbitrary Z, (c) functions having weakly locally compact domain, if Z is a KR space.

Proof: (i) Let $H: Z_1 \times I \to Z_2$ be a homotopy for $g_1, g_2: Z_1 \to Z_2$; we need find a homotopy $\Phi: \mathcal{C}_{is}(Y,Z_1) \times I \to \mathcal{C}_{is}(Y,Z_2)$ for the pair $g_{1+} = \mathcal{C}(1_Y, g_1), g_{2+} = \mathcal{C}(1_Y, g_2): \mathcal{C}_{is}(Y,Z_1) \to \mathcal{C}_{is}(Y,Z_2)$. The adjoint function $\hat{H}: Z_1 \to \mathcal{C}_{is}(I,Z_2) = \mathcal{C}_{co}(I,Z_2)$ is continuous, and so is its induced function $\hat{H}_+ = \mathcal{C}(1_Y,\hat{H}): \mathcal{C}_{is}(Y,Z_1) \to \mathcal{C}_{is}(Y, \mathcal{C}_{is}(I,Z_2))$.

(a) If Y is corecompact, then by Proposition 2.15 below, $\mathcal{C}_{is}(Y, \mathcal{C}_{is}(I, Z_2)) \cong \mathcal{C}_{is}(Y \times I,Z_2)$ and $\mathcal{C}_{is}(I \times Y,Z_2) \cong \mathcal{C}_{is}(I, \mathcal{C}_{is}(Y,Z_2))$. Observe also that $\mathcal{C}_{is}(Y \times I,Z_2) \cong \mathcal{C}_{is}(I \times Y,Z_2)$. Consequently, the continuous composition $h \circ \hat{H}_+ = \hat{\Phi}: \mathcal{C}_{is}(Y,Z_1) \to \mathcal{C}_{is}(I, \mathcal{C}_{is}(Y,Z_2))$ of the natural homeomorphism $h: \mathcal{C}_{is}(Y, \mathcal{C}_{is}(I,Z_2)) \to \mathcal{C}_{is}(I, \mathcal{C}_{is}(Y,Z_2))$ with $\hat{H}_+$ has a continuous adjoint $\Phi: \mathcal{C}_{is}(Y,Z_1) \times I \to \mathcal{C}_{is}(Y,Z_2)$ which is the required homotopy for g_{1+}, g_{2+}.

(b) Consider the continuous adjoint $\hat{H}^*: I \to \mathcal{C}_{is}(Z_1,Z_2)$ of $H^*: I \times Z_1 \to Z_2$, where $H^*(t,z) = H(z,t)$ for all $(z,t) \in Z_1 \times I$. If the domain Z_1 of g_1, g_2 is corecompact, then the continuity of the composition operation $T: \mathcal{C}_{is}(Y,Z_1) \times \mathcal{C}_{is}(Z_1,Z_2) \to \mathcal{C}_{is}(Y,Z_2)$ follows from Proposition 2.1 above. Therefore, the continuous composition $\Phi = T \circ (1 \times \hat{H}^*): \mathcal{C}_{is}(Y,Z_1) \times I \to \mathcal{C}_{is}(Y,Z_2)$ is the required homotopy.

(c) Since the Isbell topology $\mathcal{T}_{is}$ is always finer than $\mathcal{T}_{co}$, the composition $1 \circ \hat{H}_+: \mathcal{C}_{is}(Y,Z_1) \to \mathcal{C}_{is}(Y, \mathcal{C}_{is}(I,Z_2)) \to \mathcal{C}_{co}(Y, \mathcal{C}_{co}(I,Z_2))$ is continuous. Also, if the range Z_2 is a KR space, then the main theorem 2.1 in [18] implies the existence of a natural homeomorphism $h: \mathcal{C}_{co}(Y, \mathcal{C}_{co}(I, Z_2)) \to \mathcal{C}_{co}(I \times Y,Z_2)$. Since Y is weakly locally compact, the exponential function $E_{IYZ_2}: \mathcal{C}_{co}(I \times Y,Z_2) \to \mathcal{C}_{co}(I, \mathcal{C}_{co}(Y,Z_2))$ is a homeomorphism, and $\mathcal{T}_{co}$ is jointly continuous on $\mathcal{C}(Y,Z_2)$ by proposition 4.8 of [18]. Hence $\mathcal{T}_{co}$ coincides with $\mathcal{T}_{is}$ on $\mathcal{C}(Y,Z_2)$, because they are both splitting and jointly continuous. Thus, the continuous composition $\hat{\Phi}: E_{IYZ_2} \circ h \circ 1 \circ \hat{H}_+: \mathcal{C}_{is}(Y,Z_1) \to \mathcal{C}_{co}(I, \mathcal{C}_{co}(Y,Z_2))$ has a continuous adjoint $\Phi: \mathcal{C}_{is}(Y,Z_1) \times I \to \mathcal{C}_{is}(Y,Z_2)$, which is the required homotopy.

(ii) Let $H: Y_1 \times I \to Y_2$ be a homotopy for $f_1, f_2: Y_1 \to Y_2$; we need find a homotopy $\Phi: \mathcal{C}_{is}(Y_2,Z) \times I \to \mathcal{C}_{is}(Y_1,Z)$ for the pair $f_1^+ = \mathcal{C}(f_1, 1_Z)$, $f_2^+ = \mathcal{C}(f_2, 1_Z): \mathcal{C}_{is}(Y_2,Z) \to \mathcal{C}_{is}(Y_1,Z)$.

(a) If the range Y_2 of f_1, f_2 is corecompact, then the composition operation $T: \mathcal{C}_{is}(Y_1,Y_2) \times \mathcal{C}_{is}(Y_2,Z) \to \mathcal{C}_{is}(Y_1,Z)$ is continuous. Consider the composition $\Phi = T \circ (\hat{H}^* \times 1)$, where $\hat{H}^*$ is the adjoint of $H^*: I \times Y_1 \to Y_2$ with $H^*(t,y) = H(y,t)$. Then Φ is the required homotopy.

(b) Consider the continuous induced function $H^+ = \mathcal{C}(H, 1_Z): \mathcal{C}_{is}(Y_2, Z) \to \mathcal{C}_{is}(Y_1 \times I, Z)$. If the domain Y_1 of f_1, f_2 is corecompact, then by

Proposition 2.15 below, there is a natural homeomorphism h: $\mathcal{C}_{is}(Y_1 \times I, Z) \to \mathcal{C}_{is}(I, \mathcal{C}_{is}(Y_1, Z))$. Then $\hat{\mathcal{S}} = h \circ H^+$ is the adjoint of the required homotopy.

(c) If Z is KR and if the domain Y_1 is weakly locally compact, then the exponential function $E_{IY_1Z}: \mathcal{C}_{co}(I \times Y_1, Z) \to \mathcal{C}_{co}(I, \mathcal{C}_{co}(Y_1, Z))$ is a homeomorphism and $\mathcal{T}_{co} = \mathcal{T}_{is}$ on $\mathcal{C}(Y_1, Z)$ by 2.1 and 4.8 in [18]. Furthermore, the identity 1: $\mathcal{C}_{is}(I \times Y_1, Z) \to \mathcal{C}_{co}(I \times Y_1, Z)$ is continuous and $\mathcal{C}_{is}(Y_1 \times I, Z) \underset{h}{\cong} \mathcal{C}_{is}(I \times Y_1, Z)$. Consequently, the adjoint of the composition $E_{IY_1Z} \circ 1 \circ h \circ H^+$ is the desired homotopy. ■

Concerning the problem of identifying the homotopy classes of the set $\mathcal{C}(Y,Z)$, recall from [18] that for a k-space Y, they coincide with the path components in $\mathcal{C}_{co}(Y,Z)$, provided that either Z is a KR space or Y is an hl^* space (i.e., a space whose compact subsets are contained in basic locally compact subspaces). Note that the class of hl^* (= Hausdorff-like*) spaces contains the basic locally compact and the KR spaces.

2.4 PROPOSITION Given arbitrary spaces Y and Z, if $f,g \in \mathcal{C}(Y,Z)$ are in the same homotopy class, then they belong to the same path component in $\mathcal{C}_{is}(Y,Z)$. Conversely, given f,g belonging to the same path component in $\mathcal{C}_{is}(Y,Z)$, they are homotopic if either (i) Y is corecompact or (ii) Z is KR and Y is a k-space, or (iii) Y is an hl^* k-space.

Proof: Let H: $Y \times I \to Z$ be a homotopy for the pair f,g. Then the adjoint $\hat{H}^*: I \to \mathcal{C}_{is}(Y,Z)$ where $H^*(t,y) = H(y,t)$ is a continuous path from f to g. Conversely, if Y is corecompact, then by Proposition 2.2(vi) above, given a path $\varphi: I \to \mathcal{C}_{is}(Y,Z)$ from f to g, there is a homotopy H: $Y \times I \to Z$ for the pair f,g, where $\hat{H}^* = \varphi$. If either (ii) or (iii) is satisfied, then the well-defined exponential injection $E_{IYZ}: \mathcal{C}(I \times Y,Z) \to \mathcal{C}(I, \mathcal{C}_{is}(Y,Z)) \subset \mathcal{C}(I, \mathcal{C}_{co}(Y,Z))$ is a bijection onto $\mathcal{C}(I, \mathcal{C}_{co}(Y,Z))$ by 4.9 (iii) or (iv*)(c) of [18], because $I \times Y$ is a k-space. Therefore, to each path in $\mathcal{C}_{is}(Y,Z)$ from f to g corresponds a homotopy for f,g.

Let us now compare the Isbell topology $\mathcal{T}_{is}$ on $\mathcal{C}(Y,Z)$ to other topologies, such as the compact-open $\mathcal{T}_{co}$; the bounded open $\mathcal{T}_{eo}$, which is generated by the sets $(\bar{B},P) = \{f \in \mathcal{C}(Y,Z): \bar{B} \subset f^{-1}(P)\}$, where $P \in \mathcal{O}(Z)$ and B runs through the collection $\mathcal{B}(Y)$ of the bounded sets in Y [16], and to the bounding topology $\mathcal{T}_{bg}$, defined by Isbell [10], which is generated by the sets $\langle B,P\rangle = \{f \in \mathcal{C}(Y,Z)$: B is bounded in $f^{-1}(P)\}$, where $P \in \mathcal{O}(P)$ and $B \in \mathcal{B}(Y)$.

Both $\mathcal{T}_{is}$ and $\mathcal{T}_{bg}$ are clearly finer than $\mathcal{T}_{co}$ on any $\mathcal{C}(Y,Z)$. Furthermore, recall from [16] that $\mathcal{T}_{eo}$ is finer than $\mathcal{T}_{co}$ if either the range Z is

T_2 or regular, or if the domain Y is T_2; in particular, if the domain Y is regular, then $\mathcal{T}_{eo} = \mathcal{T}_{co}$ on $\mathcal{C}(Y,Z)$ for arbitrary Z. Proposition 2.6 below shows that $\mathcal{T}_{eo}$ coincides with $\mathcal{T}_{bg}$ quite often. We will need first a lemma, part (i) of which is Isbell's 2.4 in [10]. Recall from [14] that B is bounded in Y iff B is bounded in the subspace $\overline{B} \subset Y$; or so to speak, that boundedness in Y of B is not an inherent property of the subspace B itself (like compactness); it is in fact a property of the relative subspace $\overline{B}$ in Y.

2.5 LEMMA Let B be a bounded subset in a space $Y \supset A \supset B$.

(i) If Y is T_2, then B is bounded in A iff $\overline{B} \subset A$.

(ii) If Y is regular, then B is bounded in A iff $\overline{B} \subset \text{sat}(A) = \cap\{W \in \mathcal{O}(Y): A \subset W\}$.

2.6 PROPOSITION (i) If the domain Y is either T_2 or regular, then $\mathcal{T}_{eo} = \mathcal{T}_{bg}$ on $\mathcal{C}(Y,Z)$ for arbitrary Z.

(ii) If the range Z is either T_2 or regular, then $\mathcal{T}_{eo} = \mathcal{T}_{bg}$ on $\mathcal{C}(Y,Z)$ for arbitrary Y.

Proof: Observe first that for any bounded set B in Y, $(\overline{B},P) \subset \langle B,P\rangle$ without any assumptions on Y or on Z.

(i) Let $f \in \langle B,P\rangle$, i.e., B is bounded in $f^{-1}(P)$. Then, if Y is T_2 or regular, the preceding lemma implies that $f \in (\overline{B},P)$. So the defining subbasic elements of the two topologies coincide.

(ii) Again, let $f \in \langle B,P\rangle \subset \mathcal{C}_{bg}(Y,Z)$. B bounded in $f^{-1}(P)$ implies that the continuous image f(B) is bounded in P. Thus, if the range Z is T_2 or regular, Lemma 2.5 above implies that $f(\overline{B}) \subset \overline{f(B)} \subset P$. Therefore, again $\langle B,P\rangle = (\overline{B},P)$ for all $B \in \mathcal{B}(Y), P \in \mathcal{O}(Z)$, and so $\mathcal{T}_{eo} = \mathcal{T}_{bg}$. ∎

Recall that the class of weakly locally compact spaces contains properly the class of basic locally compact spaces. However, every weakly locally compact hl^*-space is also basic locally compact. As mentioned above, the class of corecompact spaces is not contained in the class of weakly locally compact spaces. However, both these classes are contained properly in the class of locally bounded spaces introduced in [15]. A space is <u>locally bounded</u> if each point has a neighborhood bounded in the space; equivalently, if each point has an (open) neighborhood basis consisting of sets bounded in the space. There are even T_2 locally bounded spaces that are not corecompact, hence neither weakly nor basic locally compact. The space $\mathbb{R}^*$ mentioned above is such an example, and the Katetov extension kN of the discrete countable set N is another (see [15,16]).

However, regular locally bounded spaces are necessarily basic locally compact. We will make use of the following exponential law for $\mathcal{T}_{eo}$.

2.7 THEOREM [16] Given arbitrary Y and regular Z, the bounded open topology $\mathcal{T}_{eo}$ is splitting on $\mathcal{C}(Y,Z)$. Furthermore, if, in addition, Y is locally bounded, then $\mathcal{T}_{eo}$ is also (the coarsest) jointly continuous on $\mathcal{C}(Y,Z)$; and, in fact, for every space X the function spaces $\mathcal{C}_{eo}(X \times Y, Z)$ and $\mathcal{C}_{eo}(X, \mathcal{C}_{eo}(Y,Z))$ are homeomorphic.

It is obvious from the above that if the domain Y is basic locally compact, then $\mathcal{T}_{is} = \mathcal{T}_{co}$ on $\mathcal{C}(Y,Z)$ for arbitrary Z. Isbell's 2.11, [10] (cf. also [24], 2.19), shows that in general, basic local compactness of Y cannot be relaxed to corecompactness. The following shows that such a relaxation is indeed possible at the cost of imposing some kind of additional restrictions either on the range Z or on the domain Y.

2.8 PROPOSITION Let Y and Z be arbitrary spaces. Consider the topologies $\mathcal{T}_{co}$, $\mathcal{T}_{is}$, $\mathcal{T}_{eo}$, and $\mathcal{T}_{bg}$ on $\mathcal{C}(Y,Z)$.

(i) If Y is basic locally compact, then $\mathcal{T}_{co} = \mathcal{T}_{is} = \mathcal{T}_{bg}$.

(ii) If Y is corecompact, then $\mathcal{T}_{is} = \mathcal{T}_{bg} \supset \mathcal{T}_{co}$.

(iii) If Y is a weakly locally compact corecompact space and if Z is a KR-space, then $\mathcal{T}_{co} = \mathcal{T}_{is} = \mathcal{T}_{bg}$.

(iv) If Y is corecompact, $\mathcal{C}_{co}(Y,Z) \times Y$ is a k-space, and if Z is a KR-space, then $\mathcal{T}_{co} = \mathcal{T}_{is} = \mathcal{T}_{bg}$.

(v) If Y is a corecompact hl^*-space and if $\mathcal{C}_{co}(Y,Z) \times Y$ is a k-space, then $\mathcal{T}_{co} = \mathcal{T}_{is} = \mathcal{T}_{bg}$.

(vi) If Y is basic locally compact and Z is T_2 or regular, then $\mathcal{T}_{co} = \mathcal{T}_{is} = \mathcal{T}_{bg} = \mathcal{T}_{eo}$.

(vii) If Y is corecompact and Z is T_2 or regular, then $\mathcal{T}_{is} = \mathcal{T}_{bg} = \mathcal{T}_{eo} \supset \mathcal{T}_{co}$.

(viii) If Y is weakly locally compact, then Z regular implies $\mathcal{T}_{co} = \mathcal{T}_{is} = \mathcal{T}_{bg} = \mathcal{T}_{eo}$; whereas Z T_2 implies $\mathcal{T}_{co} = \mathcal{T}_{is} \subset \mathcal{T}_{bg} = \mathcal{T}_{eo}$.

(ix) If Y is locally bounded, then Z regular implies $\mathcal{T}_{co} \subset \mathcal{T}_{is} \subset \mathcal{T}_{bg} = \mathcal{T}_{eo}$, and, in fact, $\mathcal{T}_{eo} \neq \mathcal{T}_{co}$ in general; whereas Z T_2 implies $\mathcal{T}_{co} \subset \mathcal{T}_{bg} = \mathcal{T}_{eo}$.

(x) If Y is regular locally bounded or T_2 corecompact, then for any Z, $\mathcal{T}_{co} = \mathcal{T}_{is} = \mathcal{T}_{bg} = \mathcal{T}_{eo}$.

Proof: (i) $\mathcal{T}_{co} \subset \mathcal{T}_{is}$ always. Since $\mathcal{T}_{co}$ is jointly continuous, $\mathcal{T}_{co} \supset \mathcal{T}_{is}$. Also, $\mathcal{T}_{is} = \mathcal{T}_{bg}$ by (ii) below.

(ii) If $f \in \langle B,P\rangle$, i.e., B is bounded in $f^{-1}(P) \in \mathcal{O}(Y)$, then since Y is corecompact, the interpolation property ([10], 2.8, or [5], I.1.18) guarantees the existence of an open set V bounded in $f^{-1}(P)$ such that $B \subset V$ and $H(V) = \{W \in \mathcal{O}(Y)\colon V \text{ is bounded in } W\} \in \Omega(Y)$. Consequently, $f \in (H(V),P) \subset \langle B,P\rangle$, and so $\mathcal{T}_{bg} \subset \mathcal{T}_{is}$. Conversely, if $f \in (H,P)$, i.e., $f^{-1}(P) \in H \in \Omega(Y)$, then by II.1.10.i of [5], there is an open bounded set $B \in H$ such that $f^{-1}(P) \in H(B) \subset H$, since Y is corecompact. Hence, $f \in \langle B,P\rangle \subset (H,P)$ and so $\mathcal{T}_{is} \subset \mathcal{T}_{bg}$.

(iii), (iv), and (v) follow from (ii), and from the fact that both $\mathcal{T}_{is}$ and $\mathcal{T}_{co}$ are splitting and jointly continuous, by Theorem 2.2(vi) above and by the exponential laws 4.9, (iv*b) of [18], respectively, and so they coincide.

(vi) and (vii) are clear in view of (i), (ii), and Proposition 2.6(ii) above.

(viii) If Z is regular and Y is weakly locally compact, then the exponential laws 4.9 [18], and Theorem 2.7 above imply that $\mathcal{T}_{co} = \mathcal{T}_{eo}$, because they are both splitting and jointly continuous [see also [22], 2.17(2)]. Since a splitting topology is contained in every jointly continuous topology, we get $\mathcal{T}_{is} \subset \mathcal{T}_{co}$, i.e., $\mathcal{T}_{co} = \mathcal{T}_{is}$.

If Z is T_2, then again $\mathcal{T}_{co}$ is jointly continuous, by 4.9 [18], and thus $\mathcal{T}_{co} = \mathcal{T}_{is}$; whereas $\mathcal{T}_{eo} = \mathcal{T}_{bg}$ by Proposition 2.6(ii).

(ix) If Z is regular, then by Theorem 2.7 above, the jointly continuous $\mathcal{T}_{eo} \supset \mathcal{T}_{is} \supset \mathcal{T}_{co}$. Examples such as the sets $\mathcal{C}(\mathbb{R}^*,\mathbb{R})$ or $\mathcal{C}(kN,\mathbb{R})$ mentioned above show that $\mathcal{T}_{eo}$ may be strictly finer than $\mathcal{T}_{co}$ in such situations; notice that $\mathcal{T}_{co}$ reduces to the pointwise topology on $\mathcal{C}(kN,\mathbb{R})$ [16]. The case Z is T_2 as well as (x) is clear, in view of Proposition 2.6(ii) and (i) above. ∎

Isbell has shown ([10],2.8) that his bifunctorial deep bounding function space topology $\mathcal{T}_{db}$ coincides with the bounding topology $\mathcal{T}_{bg}$ if the domain is corecompact; hence, by Proposition 2.8(ii), $\mathcal{T}_{db} = \mathcal{T}_{is}$ in this case. He also showed ([10],2.5) that $\mathcal{T}_{db} = \mathcal{T}_{co}$ if the domain is T_2. In fact, by Lemma 2.5(ii) above, a similar argument shows that if the domain is regular, then $\mathcal{T}_{db} = \mathcal{T}_{co}$ also; notice that if $A \subset A'$ in Y, then A is bounded in A′ iff the saturation sat A of A is bounded in A′. Hence, $\mathcal{T}_{db} = \mathcal{T}_{co} = \mathcal{T}_{eo} = \mathcal{T}_{bg}$ in this case. For some further results in [10] the reader should also consult [11], according to Isbell's own recommendation.

In view of Propositions 2.1 and 2.8, we get the following:

2.9 COROLLARY If Y and Z are such that one of the conditions 2.8 (i, iii, iv, v) is satisfied, then both T: $\mathcal{C}_{is}(X,Y) \times \mathcal{C}_{co}(Y,Z) \to \mathcal{C}_{is}(X,Z)$ and T: $\mathcal{C}_{co}(X,Y) \times \mathcal{C}_{co}(Y,Z) \to \mathcal{C}_{co}(X,Z)$ are continuous for arbitrary X.

Recall that the compact-open topology is regular (resp., uniformizable) iff the range is such ([12]. 7.4, 7.11); the same is true for the bounded-open topology ([16], 2.1.iii). Since the Isbell topology is generally well behaved, one is motivated to state the following.

2.10 CONJECTURE The Isbell topology is regular (resp., uniformizable) if the range space is regular (resp., uniformizable).

By Proposition 2.8 (vii) and (viii) the conjecture is true at least in case the domain is corecompact or weakly locally compact. Whether the conjecture remains valid for locally bounded domain is not known. Observe that in 2.8(ix) above, it is not known whether $\mathcal{T}_{eo} = \mathcal{T}_{is}$ or $\mathcal{T}_{is} = \mathcal{T}_{co}$ under these conditions. In the latter situation, there is a close relationship between $\mathcal{T}_{is}$ and the Scott topology given in Proposition 2.12 below, which is motivated by proposition 2.10 in [24]. The following lemma can be shown using standard techniques, and it may be useful toward proving the preceding conjecture. Recall that a quasi-uniformity satisfies the usual uniformity axioms with the possible exception of the symmetry axiom.

2.11 LEMMA Let $\mathcal{U}$ be a compatible quasi-uniformity on a space Z. If B is bounded in $A \in \mathcal{O}(Z)$, then there exists an entourage $V \in \mathcal{U}$ such that $V(B) \subset A$.

Recall that a space Z is injective (in Top) if for any subspace X′ of a space X, every continuous function g: $X' \to Z$ can be extended to a continuous function G: $X \to Z$ ([5],II.3).

2.12 PROPOSITION Let Y be a locally bounded space. Then

(i) $\mathcal{C}_{co}(Y,\cdot)$ preserves injective regular spaces.

(ii) If Z is an injective T_3-space, then the bounded-open topology $\mathcal{T}_{eo}$ agrees with the Scott topology $\sigma(\mathcal{C}(Y,Z))$, which is determined by the pointwise order induced on $\mathcal{C}(Y,Z)$ by the specialization order of Z.

Proof: (i) Let Z be an injective regular space, let X′ be a subspace of a space X, and let g: $X' \to \mathcal{C}_{eo}(Y,Z)$ be a continuous function. By the exponential law 2.7, there is a unique continuous function $\bar{g}$: $X \times Y \to Z$ whose adjoint $\hat{\bar{g}}$ is the given function g. Since Z is injective, $\bar{g}$

extends continuously to $G: X \times Y \to Z$. Then again, Theorem 2.7 guarantees the continuity of the adjoint $\hat{G}: X \to \mathcal{C}_{eo}(Y,Z)$, which is the required continuous extension of the given function g.

(ii) If Z is a T_0 regular injective space, then $\mathcal{C}_{eo}(Y,Z)$ is also such by the preceding. Thus, by theorem II.3.8 [5], $\mathcal{C}(Y,Z)$ is a continuous lattice in the specialization order of $\mathcal{T}_{eo}$, which coincides with the Scott topology of this lattice. The assertion follows from the coincidence of the specialization order of $\mathcal{T}_{eo}$ with the pointwise order induced on $\mathcal{C}(Y,Z)$ by the specialization order of Z. Indeed, let $f \leq g$ in the pointwise order, i.e., for each $y \in Y$, $f(y) \in \{g(y)\}^-$. Then, for every $P \in \mathcal{O}(Z)$, $f^{-1}(P) \subset g^{-1}(P)$. Therefore, every basic open neighborhood of f in $\mathcal{C}_{eo}(Y,Z)$ also contains g, i.e., f belongs to the $\mathcal{T}_{eo}$-closure of $\{g\}$. Conversely, let $P \in \mathcal{O}(Z)$ and $y \in f^{-1}(P)$. Since Z is regular, let $Q_y \subset \bar{Q}_y \subset P$ be an open neighborhood of f(y) in Z. Since Y is locally bounded, let $B_y \subset f^{-1}(Q_y)$ be a bounded (in Y) neighborhood of y. Then $\bar{B}_y \subset f^{-1}(\bar{Q}_y) \subset f^{-1}(P)$, and so $g \in (\bar{B}_y,P)$ also. Therefore, the arbitrary $y \in g^{-1}(P)$, i.e., $f(y) \in \{g(y)\}^-$, or $f \leq g$ in the pointwise order. ∎

It is well known that corecompactness is finitely productive, but as far as we know, there is no general product theorem for corecompactness recorded in the literature. The next result provides such a theorem, motivated by the corresponding product theorems for local compactness and local boundedness [15].

2.13 THEOREM The nonempty product ΠY_i of a family of spaces Y_i, $i \in I$, is corecompact iff each coordinate space Y_i is corecompact and all but finitely many of them are compact.

Proof: Necessity follows from the observation that, in general, open continuous surjections preserve corecompactness, and from proposition 4 [15] asserting that if a product space contains a bounded subset with nonempty interior, then all but finitely many of its coordinate spaces are compact.

Conversely, let $V = \bigcap_k pr_{i_k}^{-1}(V_k)$ be a basic open neighborhood of a given point $y = (y_i)$ in ΠY_i, where, for each $i_k \in F$, $V_k \in \mathcal{O}(Y_{i_k})$, F is finite, and $pr_{i_k}: \Pi Y_i \to Y_{i_k}$ denotes the projection. For each $i_k \in F$, pick an open neighborhood W_k of y_{i_k} bounded in V_k. In case finitely many of the spaces X_i, $i \in I - F$, say X_{i_λ}, $i_\lambda \in F'$, fail to be compact, choose for each $i_\lambda \in F'$ a bounded (in X_{i_λ}) open neighborhood W_λ of y_{i_λ}, since every X_i is locally bounded.

We claim that the basic neighborhood $W = \cap \{pr_i^{-1}(W_i): i \in F \cup F'\}$ of y is bounded in the given set V. To verify this, use the ultrafilter characterization of corecompactness ([26], or [8],4.2.2). So let $\mathcal{U}$ be an ultrafilter on ΠY_i with W as a member. Then for each $i_k \in F$, the ultrafilter basis $pr_{i_k}(\mathcal{U})$ converges to a point a_{i_k} in V_k, and for each $i \in I - F$, $pr_i(\mathcal{U})$ converges to a point a_i in Y_i; notice that a subset B is bounded in a space X iff every ultrafilter on X with B as a member converges in X ([13], corollary 2.1). Consequently, $\mathcal{U}$ converges to the point $a = (a_i) \in V$ as required. ■

Consider the well-defined exponential injection $E_{XYZ}: \mathcal{C}_{is}(X \times Y,Z) \to \mathcal{C}_{is}(X, \mathcal{C}_{is}(Y,Z))$ for arbitrary X,Y, and Z.

In view of certain applications such as Proposition 2.3 above, we need to know conditions under which E_{XYZ} is continuous, or an embedding, or even a homeomorphism. For instance, with respect to the compact-open topology $\mathcal{T}_{co}$, we know that if X satisfies Wilker's condition (D), then E_{XYZ} is continuous; also, that if X x Y satisfies (D), then E_{XYZ} is continuous; also, if X x Y satisfies (D), then actually E_{XYZ} is an embedding [18]. In this respect, Jackson-type subbasis lemmas ([9],V.1.3,3.8, and [18]) for $\mathcal{C}_{co}(X \times Y, Z)$ and $\mathcal{C}_{co}(X, \mathcal{C}_{co}(X,Z))$ are the standard approach to such questions. A result of this type is given next. Observe first that at least basic locally compact spaces satisfy the following variation of Wilker's condition (A). A space Y satisfies <u>condition (A*)</u> if given $H \in \Omega(Y)$ and open sets A_1, A_2, $A_1 \cup A_2 \in H$ implies the existence of filters $H_i \in \Omega(Y)$, $A_i \in H_i$, $i = 1,2$, such that $H_1 \cap H_2 \subset H$.

2.14 SUBBASIS LEMMA Let X satisfy condition (A^*) and let $\mathcal{S}$ be a subbasis for the topology $\mathcal{O}(T)$ of an arbitrary space T. Then the sets (H,R), where $R \in \mathcal{S}$ and $H \in \Omega(X)$, form a subbasis for the Isbell topology $\mathcal{T}_{is}$ on $\mathcal{C}(X,T)$.

Proof: Let $f \in (H,P)$, $P \in \mathcal{O}(T)$, and $H \in \Omega(X)$, and let $P = \cup Q_i$, $i \in I$, where each $Q_i = \cap R_{ij}$, $j \in F_i$, each F_i is finite, and $R_{ij} \in \mathcal{S}$ for all $j \in F_i$, $i \in I$. Since $\cup f^{-}(Q_i) \in H$, there is a finite union $\cup f^{-1}(Q_{i_\lambda}) \in H$. By condition (A^*) there are finitely many filters $H_\lambda \in \Omega(X)$, $f^{-1}(Q_{i_\lambda}) \in H_\lambda$, and $\cap H_\lambda \subset H$. Consequently, $f \in \cap_\lambda \cap_j (H_\lambda, R_{i_\lambda j})$ for all λ, $j \in F_{i_\lambda}$. On the other hand, for any $g \in \mathcal{C}(X,T)$ such that for each λ, $g^{-1}(R_{i_\lambda j}) \in H_\lambda$ for all $j \in F_{i_\lambda}$, we get $g^{-1}(P) \supset g^{-1}(Q_{i_\lambda}) = \cap_j g^{-1}(R_{i_\lambda j}) \in H_\lambda$ for each λ, and so $g^{-1}(P) \in H$, thus completing the proof. ■

Whether the above implies the continuity of the exponential function is not completely clear. However, the following holds.

2.15 PROPOSITION If both X and Y are corecompact, then for arbitrary Z the function spaces $\mathcal{C}_{is}(X \times Y, Z)$ and $\mathcal{C}_{is}(X, \mathcal{C}_{is}(Y,Z))$ are homeomorphic.

Proof: By Theorem 2.2(vi), the exponential function $E_{XYZ}: \mathcal{C}(X \times Y,Z) \to \mathcal{C}(X, \mathcal{C}_{is}(Y,Z))$ is a bijection. Since $X \times Y$ is corecompact, the evaluation $e: \mathcal{C}_{is}(X \times Y, Z) \times X \times Y \to Z$ is continuous. Hence, its adjoint $\hat{e}: \mathcal{C}_{is}(X \times Y, Z) \times X \to \mathcal{C}_{is}(Y,Z)$ is continuous as well as $\hat{\hat{e}}: \mathcal{C}_{is}(X \times Y, Z) \to \mathcal{C}_{is}(X, \mathcal{C}_{is}(Y,Z))$; notice that $\hat{\hat{e}} = E_{XYZ}$. On the other hand, the evaluation $e_1: \mathcal{C}_{is}(X, \mathcal{C}_{is}(Y,Z)) \times X \to \mathcal{C}_{is}(Y,Z)$ is continuous. Then observe that E^{-1}_{XYZ} is the continuous adjoint $\hat{h}$ of the continuous composition $h = e_2 \circ (e_1 \times 1_Y)$: $\mathcal{C}_{is}(X, \mathcal{C}_{is}(Y,Z)) \times X \times Y \to \mathcal{C}_{is}(Y,Z) \times Y \xrightarrow{e_2} Z$. In fact, this argument is categorical in nature (see [6], p. 15; and [25, p. 141). ■

As an application of the Subbasis Lemma 2.14, we get the following result on the natural bijection $h: \mathcal{C}(X,Y \times Z) \to \mathcal{C}(X,Y) \times \mathcal{C}(X,Z)$, where for each continuous function $f: X \to Y \times Z$, $h(f) = (\pi_Y \circ f, \pi_Z \circ f)$, $\pi_Y: Y \times Z \to Y$ and $\pi_Z: Y \times Z \to Z$ are the usual projections, and for each pair $(p,q) \in \mathcal{C}(X,Y) \times \mathcal{C}(X,Z)$ there is a unique function $\langle p,q \rangle \in \mathcal{C}(X,Y \times Z)$ [defined for each $x \in X$ by $\langle p,q \rangle(x) = (p(x),q(x))$] such that $h(\langle p,q \rangle) = (p,q)$.

2.16 PROPOSITION (i) The natural bijection $h: \mathcal{C}_{is}(X, Y \times Z) \to \mathcal{C}_{is}(X,Y) \times \mathcal{C}_{is}(X,Z)$ is always continuous.
(ii) Furthermore, h above is a homeomorphism if the space X satisfies the condition (A^*).

Proof: (i) The continuity of the induced functions $\pi_{Y+} = \mathcal{C}_{is}(X, Y \times Z) \to \mathcal{C}_{is}(X,Y)$ and $\pi_{Z+}: \mathcal{C}_{is}(X, Y \times Z) \to \mathcal{C}_{is}(X,Z)$ implies directly the continuity of the product function $h = \langle \pi_{Y+}, \pi_{Z+} \rangle$ by the (categorical) definition of products in <u>Top</u>.

(ii) If X satisfies the condition (A^*), then Lemma 2.14 implies that the collection of all sets $(H, P \times R)$, where $H \in \Omega(X)$, $P \in \mathcal{O}(Y)$, and $R \in \mathcal{O}(Z)$, generates the Isbell topology $\mathcal{T}_{is}$ on $\mathcal{C}(X, Y \times Z)$; and in fact, we can restrict attention to those sets $H \in \Omega(X)$ that are filters since they are a basis for $\Omega(X)$. Then the assertion is established if we check that each $h((H, P \times R)) = (H,P) \times (H,R)$. Indeed, for each $f \in (H, P \times R)$, $(\pi_Y \circ f)^{-1}(P) \cap (\pi_Z \circ f)^{-1}(R) = f^{-1}(P \times R) \in H$ implies $h(f) \in (H,P) \times (H,R)$. On the other hand, for each pair $(p,q) \in (H,P) \times (H,R)$, $\langle p,q \rangle^{-1}(P \times R) = p^{-1}(P) \cap q^{-1}(R)$ is a member of the filter H and so $(p,q) = h(\langle p,q \rangle) \in h(H, P \times R)$.

2.17 PROPOSITION Let Y be a second countable corecompact space. Then the Isbell topology $\mathcal{T}_{is}$ on $\mathcal{C}(Y,Z)$ is second countable iff Z is second countable.

Proof: Necessity follows clearly from the fact that any space Z is canonically embedded in $\mathcal{C}_{is}(Y,Z)$. Conversely, let $\mathcal{V}$ and $\mathcal{W}$ be countable bases for $\mathcal{O}(Y)$ and $\mathcal{O}(Z)$, respectively, and let $\mathcal{W}^*$ denote the set of finite unions of members of $\mathcal{W}$. We claim that the sets $(H(V),W^*)$, where $V \in \mathcal{V}$ and $W \in \mathcal{W}^*$, form the required countable subbasis for $\mathcal{C}_{is}(Y,Z)$. Indeed, let (H,P) be a subbasic open set containing a point $f \in \mathcal{C}_{is}(Y,Z)$. Since Y is corecompact, $f^{-1}(P) = \cup V_n$, $n \in N$, with each $V_n \in \mathcal{V}$ bounded in $f^{-1}(P)$; thus, there are finitely many V_{n_k}, $k \in F$, such that $\cup V_{n_k} \in H$. Observe that each $H(V_{n_k}) \in \Omega(Y)$ and that clearly $f \in \cap(H(V_{n_k}),P) \subset (H,P)$. Furthermore, $P = \cup W_n$, $n \in N$, $W_n \in \mathcal{W}$. Hence, for each $k \in F$ there is a finite union $\cup W_{n_\lambda} = W^*_k \in \mathcal{W}^*$, $\lambda \in F_k$, such that $f^{-1}(W^*_k) = \cup f^{-1}(W_{n_\lambda}) \in H(V_{n_k})$. Therefore, $f \in \cap(H(V_{n_k}), W^*_k) \subset \cap(H(V_{n_k}),P) \subset (H,P)$, thus establishing the claim. ∎

If, in addition to the conditions of Proposition 2.17, Z is also T_3, then $\mathcal{T}_{is} = \mathcal{T}_{eo}$, and so it is metrizable by proposition 13, [17].

3. LOCAL BOUNDEDNESS AND THE STRONG ISBELL FUNCTION SPACE TOPOLOGY

The following new concepts arise very naturally in view of the ideas presented in the preceding section. Throughout this section L denotes a complete lattice, though some of the results remain valid in a wider lattice theoretic context.

3.1 DEFINITION (i) A complete lattice L is <u>weakly continuous</u> if for each $x \in L$, $x = \sup\{b \in L: b \leq x \text{ and } b \ll 1\}$, i.e., if each element in L is the supremum of its smaller elements that are bounded in L.

(ii) The open sets of the <u>strong Scott topology</u> $\sigma^*(L)$ on L are those subsets $V \subset L$ satisfying the following two conditions: (a) $V = \uparrow V$, and (b) for every directed $D \subset L$, $\sup D = 1$ implies $D \cap V \neq \emptyset$. For any space Y denote by $\Omega^*(Y)$ the lattice $\mathcal{O}(Y)$ equipped with the strong (Day-Kelly-) Scott topology $\sigma^*(\mathcal{O}(Y))$.

(iii) Given arbitrary spaces Y and Z, the <u>strong Isbell topology</u> $\mathcal{T}_{is}*$ is the initial topology on $\mathcal{C}(Y,Z)$ with respect to the functions τ_p: $\mathcal{C}(Y,Z) \to \Omega^*(Y)$, where $\tau_p(f) = f^{-1}(P)$ and $P \in \mathcal{O}(Z)$. Thus, the sets $(H^*,P) = \{f \in$

$\mathcal{C}(Y,Z)$: $f^{-1}(P) \in H^*\}$, where $H^* \in \Omega^*(Y)$ and $P \in \mathcal{O}(Z)$ form a subbasis for $\mathcal{T}_{is^*}$.

3.2 REMARKS (i) Every continuous lattice L is clearly weakly continuous, since for every x,y in L, $y \ll x$ implies $y \le x$ and $y \ll 1$. The class of weakly continuous lattices is strictly broader than the class of continuous lattices. Indeed, observe first that a space Y is locally bounded iff the lattice $\mathcal{O}(Y)$ is weakly continuous. On the other hand, we have already mentioned examples of (even T_2) locally bounded spaces, such as $\mathbb{R}^*$ and kN, that are not corecompact.

(ii) The strong Scott topology $\sigma^*(L)$ is always finer than the Scott topology $\sigma(L)$, since every Scott-open set trivially contains the top element 1. In particular, the strong Isbell topology $\mathcal{T}_{is^*}$ is always finer than the Isbell topology $\mathcal{T}_{is}$. In fact, both the strong Isbell topology and the strong Scott topology may be strictly finer than the Isbell topology or the Scott topology on a lattice function space, as the following shows.

(iii) For every space Y, the strong Isbell topology and the strong Scott topology agree on the lattice $\mathcal{C}(Y,\overline{2})$, since it is straightforward to see that each of the spaces $\mathcal{C}_{is^*}(Y,\overline{2})$ and $\mathcal{C}_{\sigma^*}(Y,\overline{2})$ is homeomorphic to $\Omega^*(Y)$. Theorem 3.4(viii) below and Theorem 2.2(iv) above clearly imply that if Y is any locally bounded noncorecompact space, then on $\mathcal{C}(Y,\overline{2})$, $\mathcal{T}_{is^*}$ [$= \sigma^*(\mathcal{C}(Y,\overline{2}))$] is strictly finer than $\mathcal{T}_{is}$ [$= \sigma(\mathcal{C}(Y,\overline{2}))$].

(iv) Though $\sigma^*(L)$ and $\sigma(L)$ are different in general, they do present some similarities, as follows (recall [5], II.1.4, 1.13).

(a) A proper subset $A \subset L$ is strong Scott-closed iff $A = \downarrow A$, and there is no directed subset in A whose supremum reaches the top.

(b) For every $x \in L$, the strong Scott closure $\{x\}^{-*} = \downarrow x = \{x\}^-$ (= the Scott closure of $\{x\}$).

(c) $\sigma^*(L)$ is a T_0-topology.

(d) Every upper set is the intersection of its strong Scott-open neighborhoods.

(e) $\sigma^*(L)$ is always a compact topology. Since every subset in a compact space is trivially bounded, $\sigma^*(L)$ is also locally bounded always. Each bounded element $b \ll 1$ in L clearly has a smallest compact neighborhood $\uparrow b$ in $\sigma^*(L)$.

(f) If L is weakly continuous, then each Scott-open set $A \subset L$ is the union of strong Scott-open compact sets, namely, $A = \bigcup\{\uparrow b: b \ll 1, b \in A\}$. Indeed, the set of the bounded elements in L is directed since $b_1 \ll 1$ and

$b_2 \ll 1 \Rightarrow b_1 \vee b_2 \ll 1$. Therefore, for each $x = \sup\{b: b \leq x, b \ll 1\} \in A$, there is some $b_x \in A$, $b_x \leq x$, $b_x \ll 1$, and so $x \in \uparrow b_x \subset A$.

The strong Isbell topology $\mathcal{T}_{is^*}$ defines a bifunctor $\mathcal{C}_{is^*}(\cdot,\cdot)$: $\underline{\text{Top}}^{op} \times \underline{\text{Top}} \to \underline{\text{Top}}$ and by restriction, bifunctors $\mathcal{C}_{is}(\cdot,\cdot)$: $\underline{T_i\text{-Top}}^{op} \times \underline{T_i\text{-Top}} \to \underline{T_i\text{-Top}}$, $i = 0,1,2$. To verify this, one needs only to check that for any open $H^* \in \Omega^*(Y_2)$ and continuous function $f: Y_2 \to Y_1$, the set $\tilde{H}^* = \{V \in \mathcal{O}(Y_1): f^{-1}(V) \in H^*\} \in \Omega^*(Y_1)$.

Consider also the b-topology $\mathcal{T}_b$ on $\mathcal{C}(Y,Z)$, which is generated by the sets $(B,P) = \{f \in \mathcal{C}(Y,Z): B \subset f^{-1}(P)\}$, where $P \in \mathcal{O}(Z)$ and B runs through the bounded subsets $\mathcal{B}(Y)$ in Y. Since for any continuous function $f: Y_2 \to Y_1$, $B \in \mathcal{B}(Y_2)$ implies $f(B) \in \mathcal{B}(Y_1)$, the b-topology defines bifunctors similar to the preceding ones. Observe that for any $B \in \mathcal{B}(Y)$, the set $H^*(B) = \{V \in \mathcal{O}(Y): B \subset V\} \in \Omega^*(Y)$. Thus clearly, $\mathcal{T}_{co} \subset \mathcal{T}_b \subset \mathcal{T}_{is^*} \supset \mathcal{T}_{is} \supset \mathcal{T}_{co}$ always. There is not much known about the comparison of the preceding topologies. However, one can see that if Z (or Y) is regular, then $\mathcal{T}_{eo} \subset \mathcal{T}_b$; and also if the domain Y is locally bounded, then $\mathcal{T}_{is} \subset \mathcal{T}_b$. The results of Section 2 are clearly leading to several questions about $\mathcal{T}_b$ and $\mathcal{T}_{is^*}$. The following statements are, in some sense, the analogs of Proposition 2.1 and Theorem 2.2, respectively.

3.3 PROPOSITION Let Y be a locally bounded space and let Z,X be arbitrary. Then

(i) The usual composition operation T: $\mathcal{C}_{is}(X,Y) \times \mathcal{C}_b(Y,Z) \to \mathcal{C}_{is}(X,Z)$ is continuous.
(ii) T: $\mathcal{C}_{co}(X,Y) \times \mathcal{C}_b(Y,Z) \to \mathcal{C}_{co}(X,Z)$ is continuous.
(iii) T: $\mathcal{C}_{is}(X,Y) \times \mathcal{C}_{is^*}(Y,Z) \to \mathcal{C}_{is}(X,Z)$ is continuous.
(iv) T: $\mathcal{C}_{co}(X,Y) \times \mathcal{C}_{is^*}(Y,Z) \to \mathcal{C}_{co}(X,Z)$ is continuous.
(v) If, in addition, Y is T_2 and Z is regular, then T: $\mathcal{C}_b(X,Y) \times \mathcal{C}_{eo}(Y,Z) \to \mathcal{C}_{eo}(X,Z)$ is also continuous.

Proof: (i) and (ii) can be shown by slight modifications of the proof of Proposition 2.1; (iii) and (iv) are corollaries of (i),(ii), respectively.

(v) Let $(\bar{B},P)$ be a subbasic open [in $\mathcal{C}_{eo}(X,Z)$] neighborhood of the image $T(f,g) = g \circ f$ of a point (f,g) in the domain. Recall from [16], p. 52, that continuous images of closures of bounded sets are compact if the range is regular. Hence, there is an open set $Q \in \mathcal{O}(Z)$ such that $g(f(\bar{B})) \subset Q \subset \bar{Q} \subset P$. Also, by [16], p. 52, $f(\bar{B}) = \overline{f(B)}$, because Y is T_2 and $f(\bar{B})$

is an almost compact (or generalized H-closed) set. Since Y is locally bounded, $g^{-1}(Q) = \cup R_i$, $i \in I$, for suitably chosen open bounded subsets R_i in Y. Pick a finite union $R = \cup R_{i_k}$ covering the bounded subset f(B), and observe that $f(B) \subset R \subset \bar{R} \subset g^{-1}(\bar{Q}) \subset g^{-1}(P)$. Therefore, the open set (B,R) $\times$ ($\bar{R}$,P) in $\mathcal{C}_b(X,Y) \times \mathcal{C}_{eo}(Y,Z)$ is a neighborhood of (f,g) contained in $T^{-1}(\bar{B},P)$, because for any pair (f_1,g_1) in the domain, $B \subset f_1^{-1}(R)$ and $\bar{R} \subset g_1^{-1}(P)$ imply $\bar{B} \subset f_1^{-1}(\bar{R}) \subset f_1^{-1}(g_1^{-1}(P))$. ∎

For arbitrary Y and Z, perhaps neither $\mathcal{T}_b$ nor $\mathcal{T}_{is^*}$ is splitting on $\mathcal{C}(Y,Z)$. At least, in case the domain Y is locally bounded, they are both reasonably well behaved, according to the following characterization theorem, which can be proved using the methods employed in Section 2.

3.4 THEOREM For any space Y the following are equivalent.

(i) Y is locally bounded.

(ii) The lattice $\mathcal{O}(Y)$ is weakly continuous.

(iii) The space Y has property (C^*), i.e., for any open neighborhood V of $y \in Y$ there is an open set H_y^* in the strong Day-Kelly-Scott topology $\Omega^*(Y)$ such that $V \in H_y^*$ and $\cap\{W: W \in H_y^*\}$ is a neighborhood of y in Y.

(iv) The set $T = \{(y,V): y \in V \in \mathcal{O}(Y)\}$ is open in $Y \times \Omega^*(Y)$.

(v) For every Z, the evaluation e: $\mathcal{C}_b(Y,Z) \times Y \to Z$ is continuous.

(vi) The evaluation e: $\mathcal{C}_b(Y,\bar{2}) \times Y \to \bar{2}$ is continuous.

(vii) For every Z, the evaluation e: $\mathcal{C}_{is^*}(Y,Z) \times Y \to Z$ is continuous.

(viii) The evaluation e: $\mathcal{C}_{is^*}(Y,\bar{2}) \times Y \to \bar{2}$ is continuous.

REFERENCES

1. Brown, R. Elements of Modern Topology. McGraw-Hill, London, 1968.

2. Day, B. A reflection theorem for closed categories. J. Pure Appl. Algebra 2 (1972), 1-11.

3. Day, B., and G. M. Kelly. On topological quotient maps preserved by pullbacks or products. Proc. Camb. Phil. Soc. 67 (1970), 553-558.

4. Dugundji, J. Topology. Allyn and Bacon, Boston, 1966.

5. Gierz, G., K. H. Hofmann, K. Keimel, J. D. Lawson, M. Mislove, and D. S. Scott. A Compendium of Continuous Lattices. Springer-Verlag, Berlin-Heidelberg-New York, 1980.

6. Herrlich, H. Categorical topology 1971-1981. Proc. Fifth Prague Topol. Symp. 1981 (General Topology and Its Relations to Modern Analysis and Algebra V, J. Novak, ed.) Heldermann Verlag, Berlin, 1982, 279-383.

7. Hofmann, K. H. Continuous lattices, topology and topological algebra. Topology Proceedings 2 (1977), 179-212.

8. Hofmann, K. H. and J. D. Lawson. The spectral theory of distributive continuous lattices. Trans. Amer. Math. Soc. 246 (1978), 285-310.

9. Hu, S. T. Elements of General Topology. Holden-Day, San Francisco-London-Amsterdam, 1964.

10. Isbell, J. R. Meet-continuous lattices. Symposia Mathematica 16 (Convegni del Marzo 1973 e del Gennaio 1974, Roma: INDAM), Academic Press, London-New York, 1975, 41-54.

11. Isbell, J. R. Function spaces and adjoints. Math. Scand. 36 (1975), 317-339.

12. Kelley, J. L. General Topology. Springer-Verlag, Berlin-Heidelberg-New York, 1955.

13. Lambrinos, P. Th. A topological notion of boundedness. Manuscr. Math. 10 (1973), 289-296.

14. Lambrinos, P. Th. Subsets (m,n)-bounded in a topological space. Mathematica Balkanica 4.70 (1974), 391-397.

15. Lambrinos, P. Th. Locally bounded spaces. Proc. Edinburgh Math. Soc. 19, Series II (1975), 321-325.

16. Lambrinos, P. Th. The bounded-open topology on function spaces. Manuscr. Math. 36 (1981), 47-66.

17. Lambrinos, P. Th. Some properties of the domains of certain function spaces. Submitted. A preliminary report appeared in Abstracts Amer. Math. Soc. No. 3, 2 (1981), 786-54-68.

18. Lambrinos, P. Th. On the exponential law for function spaces equipped with the compact-open topology. These Proceedings.

19. Lambrinos, P. Th. Weakly continuous lattices and the strong Scott topology on function spaces. Preliminary report, Abstracts Amer. Math. Soc. No. 5, 2 (1981), 788-54-92.

20. Lambrinos, P. Th. On the Scott function space topology. Preliminary report, Abstracts Amer. Math. Soc. No. 5, 3 (1982), 796-54-282.

21. Rice, M. D. Function spaces and Cartesian-closed categories. Preprint, George Mason University, Fairfax, Virginia, 1979.

22. Schwarz, F. Topological continuous convergence. To appear, Manuscripta Math., 1984.

23. Schwarz, F. Exponential objects in categories of (pre)topological spaces and their natural function spaces. C. R. Math. Rep. Acad. Sci. Canada 4 (1982), 321-326.

24. Schwarz, F. and S. Weck. Scott topology, Isbell topology, and continuous convergence. This volume, Chap. 15.

25. Steenrod, N. A convenient category of topological spaces. Michigan J. Math. 14 (1967), 133-152.

26. Ward, A. J. Problem in Proceedings of the International Symposium on Topology and Its Applications, Herceg-Novi 1968. D. R. Kurepa, ed. Savez Matematičara, Beograd 1969, p. 352.

27. Wilansky, A. Topology for Analysis. Xerox College Publishing, Lexington, Massachusetts and Toronto, 1970.

28. Wilker, P. Adjoint product and hom functors in general topology. Pacific J. Math. 34 (1970), 269-283.

29. Wyler, O. Convenient categories for topology. Gen. Top. Appl. 3 (1973), 225-242.

12
Obtaining the T_0-Essential Hull

JIMMIE D. LAWSON*
Louisiana State University
Baton Rouge, Louisiana

ABSTRACT

A general construction for the essential hull of a T_0-space X is given. If X is embedded in any injective space Y (= a continuous lattice with the Scott topology), then the essential hull of X is given by first taking the continuous lattice $M \subseteq L$ generated by X, then taking all sups of subsets of X in M, and endowing the latter with the relative topology from Y.

INTRODUCTION

Banaschewski [1] introduced the notion of an essential extension of a T_0-space X. An essential extension is an embedding i: X → Y with the property that for any f: Y → Z, if fi is an embedding, then so also is f. (We work entirely in the category of T_0-spaces and continuous mappings.) We also say that the embedding i is essential and speak of Y as an essential extension (with the embedding understood).

Banaschewski showed that the space X always has a largest essential extension X, the essential hull of X, and gave an explicit construction for the essential hull in the filter lattice of $\mathcal{O}(X)$, where $\mathcal{O}(X)$ denotes the lattice of open sets of X.

A space is called essentially complete if it is its own essential hull. The essential completion, essentially complete spaces, spaces with injective essential completions, and related topics have been explored in some detail in the work of Hoffmann (see, e.g., [4,5,6]) and more recently by Hofmann and Mislove ([7,8,9]).

*The author gratefully acknowledges the support of NSF grant MCS-7900295.

The purpose of this paper is to give a general construction for the essential hull of a space. This construction bears resemblance to a standard method of realizing the Dedekind-MacNeille completion of a poset. Recall that if a poset P is embedded (as an ordered set) in a complete lattice L, then the Dedekind-MacNeille completion of P can be obtained by first forming L_1, the set of all infs of subsets of P (equivalently the lattice order generated by P) and then forming L_2, the set of all sups of subsets of P in the complete lattice L_1. The inclusion of P into L_2 is then well known to be the Dedekind-MacNeille completion of P [2].

Now let X be a T_0-space that is topologically embedded in an injective space M, i.e., a continuous lattice equipped with its Scott topology. (This embedding will also be an order embedding for the specialization order defined by $x \leq y$ iff $x \in \overline{\{y\}}$.) There are a variety of ways of obtaining such an embedding. Any T_0-space can be embedded in a product of 2's by using the characteristic maps of open sets. In [4] and [10] it is shown that any continuous poset endowed with its Scott topology can be embedded as the set of coprimes of a completely distributive (hence continuous) lattice. Banaschewski's construction actually embeds a space X in the lattice of filters of $\mathcal{O}(X)$; the lattice of filters is algebraic, hence continuous.

With X embedded in M, we first form M_1, the continuous lattice generated by X. This is obtained by adding all infs of nonempty subsets of X, then all sups of directed subsets of this set, and then finally adjoining 1 if necessary to obtain a lattice. Next we form M_2, the set of all sups of subsets of X in the complete lattice M_1. Endow M_2 with the relative topology from M. Then we prove that the inclusion of X into M_2 gives the essential hull of X.

This construction was suggested to me by the work of Hofmann and Mislove [9], who carried it out for a special case.

THE MAIN RESULT

In the following we lean rather heavily on some of the basic theory of continuous lattices. The reader is referred to [3] for needed definitions, terminology, results, etc. In particular, we employ the Scott topology on continuous lattices (giving rise to the injective T_0-spaces) and the CL- or λ-topology, a refinement of the Scott topology making the lattice into a compact Hausdorff topological semilattice with respect to the new topology.

1. DEFINITION Let $\emptyset \neq A \subseteq L$, a continuous lattice. The meet-semilattice generated by A is denoted $\langle A \rangle$. The CL-subobject generated by A is $\overline{\langle A \rangle}^{\lambda} \cup \{\sup A\}$ (where $\overline{\quad}^{\lambda}$ denotes the closure in the CL- or λ-topology). (Note that it may or may not be the case that $\sup A \in \overline{\langle A \rangle}^{\lambda}$.) We say A <u>generates</u> L if L is the CL-subobject generated by A.

2. REMARK The CL-subobject generated by A is obtained by closing up $\langle A \rangle$ under directed infs, then directed sups, then adding sup A (if necessary) [3, theorem 1.11, p. 147].

3. LEMMA Let X generate L, where L is a continuous lattice. If U is Scott open, then $\inf(U \cap X) = \inf(U)$.

Proof: If $U = \emptyset$ or $U = \{1\}$, then $\inf(U \cap X) = 1 = \inf(U)$. Otherwise, $\inf U = \inf(U \setminus \{1\}) = \inf(U \setminus \{1\} \cap \langle X \rangle)$ (since $U \setminus \{1\}$ is λ-open and $\langle X \rangle$ is λ-dense in $L \setminus \{1\}$) $= \inf(U \cap \langle X \rangle) = \inf(U \cap X)$ (since $U = \uparrow U$). ∎

Let $i: X \to Y$ be an embedding of T_0-spaces. Then $i_*: \mathcal{O}(Y) \to \mathcal{O}(X)$ defined by $V \to i^{-1}(V)$ is a lower adjoint with upper adjoint $i^*: \mathcal{O}(X) \to \mathcal{O}(Y)$ defined by $i^*(U) = U^* = \bigcup\{V \in \mathcal{O}(Y): U = i^{-1}(V)\}$, i.e., U^* is the largest open set of Y such that $i^{-1}(U^*) = U$. Recall from [1] that the embedding i is strict if $\{U^*: U \in \mathcal{O}(X)\}$ forms a basis for the topology of Y.

4. PROPOSITION Let L be a continuous lattice generated by X. Assume that L is equipped with the Scott topology and X with the subspace topology. Then the inclusion $i: X \to L$ is strict.

Proof: Let $x \in V$ where V is Scott open. Then there exists $z \ll x$ such that $z \in V$. Let $U = (\twoheaduparrow z \cap X)^*$. Then $U \cap X = \twoheaduparrow z \cap X$. Thus $\inf U = \inf(U \cap X)$ (by Lemma 3) $= \inf(\twoheaduparrow z \cap X) \geq z$. Thus we have $x \in U$ (by definition of i^*) $\subseteq \uparrow z \subseteq V$. Since x and V were arbitrary, i is strict. ∎

We recall further from [1] that an embedding $i: X \to Y$ is (i) essential if given any continuous $f: Y \to Z$ in <u>Top_0</u> such that if fi is an embedding, then f is an embedding; (ii) superstrict if given any collection $\mathcal{B} \subseteq \mathcal{O}(Y)$ closed under finite intersections such that $\mathcal{B}|X$ is a basis for X, then $\mathcal{B}$ itself is a basis for Y. By Proposition 1 of [1], i is superstrict iff it is essential. We shall need only the easily verifiable implication that superstrict implies essential.

5. LEMMA Let L be a T_0-space satisfying:

(i) The associated partial order $(L,\leq)$ is a complete lattice.
(ii) The topology on L is contained in the Scott topology of $(|L|,\leq)$.
(iii) $\vee: L \times L \to L$ sending $(x,y) \to x \vee y$ is (jointly) continuous.

Let X be a subspace of L such that $j: X \to L$ is strict. Let $Y = \{\sup A: A \subseteq X\}$ be endowed with the relative topology from L. Then $i: X \to Y$ is essential.

6. NOTE By Proposition 1.1 of [5] the T_0-spaces satisfying (i), (ii), and (iii) are precisely the essentially complete T_0-spaces. A continuous lattice endowed with the Scott topology satisfies those conditions.

Proof of Lemma 5: Let $\mathcal{B}$ be a collection of open subsets of Y closed under finite intersection such that $\mathcal{B}|X$ is a basis for X. If we show this implies that $\mathcal{B}$ is a basis for Y, then $i: X \to Y$ will be superstrict and hence essential.

Let $y \in Y$ and let V be an open set (in L) containing y. If $y = 0$, then $V = L$, so $V \cap Y = Y$, which is the intersection of the empty collection and hence in $\mathcal{B}$.

If $y \neq 0$, then $y = \sup A$ where $\emptyset \neq A \subseteq X$. Since by (ii) V is Scott open, there exists $F = \{x_1,\ldots,x_n\} \subseteq A$ such that $\sup F \in V$. By continuity of sup, there exist open sets V_i such that $x_i \in V_i$ for $1 \leq i \leq n$ and $\bigcap_{i=1}^{n} V_i \subseteq V$. Since $j: X \to L$ is strict, there exist sets U_i, $i = 1,\ldots,n$, open in X, such that $x_i \in j^*(U_i) \subseteq V_i$. Then $x_i \in j^*(U_i) \cap X = U_i$. Since $\mathcal{B}|X$ is a basis, there exist $B_i \in \mathcal{B}$, $1 \leq i \leq n$ such that $x_i \in B_i \cap X \subseteq U_i$. Then $B = \bigcap_{i=1}^{n} B_i \in \mathcal{B}$ by hypothesis. Furthermore, $x_i \leq y$ for all i implies $y \in B_i$, $1 \leq i \leq n$; thus $y \in B$. Finally, $B_i \subseteq i^*(B_i \cap X) \subseteq j^*(B_i \cap X) \subseteq j^*(U_i) \subseteq V_i$ for $1 \leq i \leq n$. Thus $y \in B = \bigcap_{i=1}^{n} B_i \subseteq V$. ∎

7. THEOREM Let X be a subspace of a continuous lattice L endowed with the Scott topology. Let L_1 be the CL-subobject generated by X and let $\lambda X = \{\sup A: A \subseteq X\}$, where sups are taken in the complete lattice L_1. Then the inclusion of X into λX endowed with the relative topology gives the essential hull of X.

Proof: Since nets in L_1 have the same lim inf in L_1 and L, by [3, theorem II-1.8], the relative topology of L_1 is the Scott topology. By Proposition 4, the inclusion from X into L_1 is strict. By Lemma 5 and Note 6, the inclusion from X into λX is essential. Now λX still satisfies (i), (ii), and (iii) of Lemma 5, and hence is essentially complete by Proposition 1.1 of [5]. Thus, λX is the essential hull of X. ∎

8. COROLLARY If a subset X of a continuous lattice L generates and order cogenerates, then ΣL, L equipped with the Scott topology, is the essential hull of X equipped with the subspace topology. In particular, X has an injective hull.

9. COROLLARY (Hoffmann [4]) A continuous poset endowed with its Scott topology has an injective hull.

Proof: As remarked earlier, a continuous poset P can be embedded as the set of coprimes of a completely distributive lattice L (see [4] or [9]). This is also a topological embedding, since the Scott topology is the upper topology in a completely distributive lattice. Let L_1 be the CL-subobject generated by P in L. Since P cogenerates L, it also cogenerates L_1. Hence by Theorem 7, L_1 is the essential hull of P. Hence P has an injective hull.

REFERENCES

1. Banaschewski, B. Essential extensions of T_0-spaces. General Top. and Appl. 7 (1977), 233-246.

2. Banaschewski, B., and Bruns, G. Categorical characterizations of the MacNeille completion. Arch. Math. (Basel) 18 (1967), 369-377.

3. Gierz, G., Hofmann, K. H., Keimel, K., Lawson, J. D., Mislove, M., and Scott, D. S. A Compendium of Continuous Lattices. Springer-Verlag, Berlin-Heidelberg-New York, 1980.

4. Hoffmann, R.-E. Continuous posets, prime spectra of completely distributive complete lattices, and Hausdorff compactifications. In Continuous Lattices, Lecture Notes in Mathematics 871. Springer-Verlag, Berlin-Heidelberg-New York, 1981, 159-208.

5. Hoffmann, R.-E. Essentially complete T_0-spaces. Manuscripta Math. 27 (1979), 401-432.

6. Hoffmann, R.-E. Essentially complete T_0-spaces II, Math. Zeit. 179 (1982), 73-90.

7. Hofmann, K. H. The order theoretical aspects of the essential hull of a topological T_0-space, preprint, Technische Hochschule, Darmstadt.

8. Hofmann, K. H. Complete distributivity and the essential hull of a T_0-space, this volume, Chap. 6.

9. Hofmann, K. H., and Mislove, M. A continuous poset whose compactification is not a continuous poset. SCS memo 5/28/82.

10. Lawson, J. D. The duality of continuous posets. Houston J. Math. 5 (1979), 357-386.

13

The Local Approach to Programming Language Theory

ARESKI NAIT ABDALLAH*

University of Waterloo
Waterloo, Ontario, Canada

1.0 INTRODUCTION

The need for a unified model theory for program semantics arises from the variety and the apparent nonequivalence of the models that have been proposed so far: continuous lattice [17], complete ordered magma [13], complete partial order (cpo) [20], type structure over a domain [18], "collection d'algorithmes" [16], etc. All these models describe one feature or another of program semantics. Incorporating features from the different models, we have started the development of a new theory which provides a unified conceptual frame for the systematic study of all these models, and the discovery of some new ones. This theory we call <u>bundle theory</u>.

Bundle theory is based upon the fundamental notion of <u>convergence</u>. This notion is present in all the aforementioned structures. Indeed, a continuous lattice is a complete lattice in which every element x is the least upper bound (the limit) of all the elements which are way below x. In a complete ordered magma, every element x is the least upper bound (limit) of the finite trees smaller than x. A cpo is a poset with a bottom element in which every ascending chain converges. In a "collection d'algorithmes" every element is the union (limit) of the atomic elements it contains, etc.

Our formal definition of convergence is given by that of a bundle [10]. This notion is related to, but does not coincide with, other notions that have been defined and studied elsewhere for mathematical reasons: sheaves

*Current affiliation: McMaster University, Hamilton, Ontario, Canada

[8], convergence spaces [4], approximating orders [7], completely distributive algebraic lattices [6]. We illustrate in this paper how the notion of a bundle provides a description and classification of all the program semantics models known so far; new models that were considered only after our study of bundles was started [1,15] also fit nicely into this frame.

2.0 PREBUNDLES, BUNDLES, SAGITTAL BUNDLES

Let U be a von Neumann universe, $X \in U$ be a set, and

$$F(X) = \{f: I \to X \mid I \in U\}$$

be the set of families of elements of X. If $x \in X$, a family $\sigma \in F(X)$ whose only element is x will be denoted by $\underline{x}$. Thus $\underline{x}$ means all <u>constant</u> functions with value x. We also consider

$\lim: F(X) \to X$ a partial function called <u>limit</u>

$\beta: X \to P(F(X))$ a total function called <u>fibre</u>

where P denotes the powerset operation.

DEFINITION: A <u>prebundle</u> is a triple $(X, \lim, \beta)$ with $\lim: F(X) \to X$ and $\beta: X \to P(F(X))$ as above.

Let $\sim$ be an equivalence relation over X; then $\sim$ induces an equivalence relation $\sim^*$ on F(X) defined by:

for all $\sigma,\tau \in F(X)$

$$\sigma \sim^* \tau \Leftrightarrow \mathrm{Dom}(\sigma) = \mathrm{Dom}(\tau) \text{ and } \forall\, i \in \mathrm{Dom}(\sigma) \quad \sigma(i) \sim \tau(i)$$

An equivalence relation $\sim$ over the set X equipped with the prebundle $(X, \lim, \beta)$ is <u>regular</u> iff

for all $\sigma,\tau \in F(X)$

$$\sigma \sim^* \tau \Rightarrow \lim(\sigma) \sim \lim(\tau)$$

DEFINITION: Let $(X,\lim_X,\beta_X)$ and $(Y,\lim_Y,\beta_Y)$ be prebundles. A function $f: X \to Y$ is <u>strongly regular</u> at $x \in X$ if and only if

$$\forall\, \sigma \in \beta(x) \quad f(\sigma) = (f \circ \sigma) \in \beta(f(x))$$

PROPOSITION 1: The class $\{(X,\lim,\beta): X \in U\}$ of prebundles whose carrier is in U, and such that $\forall\, x \in X \ \lim \underline{x} = x$, together with the strongly regular functions, form a category with products, coproducts, quotients, equalizers, and coequalizers. This category will be denoted by <u>Prb</u>.

Proof: (i) Product: Let

$$(X,\lim,\beta) \times (X',\lim',\beta') = (X \times X', \lim \times \lim', \beta \times \beta')$$

where X × X is defined as usual and

$$(\lim \times \lim')(\sigma,\sigma') = (\lim(\sigma), \lim'(\sigma'))$$
$$(\lim \times \lim')(\underline{x},\sigma') = (x,\lim'(\sigma'))$$
$$(\lim \times \lim')(\sigma,\underline{x}') = (\lim(\sigma),x')$$

Note that we define lim × lim′ from $F(X) \times F(X') \subseteq F(X \times X')$ into X × X′.

$$(\beta \times \beta')(x,x') = (\{\underline{x}\} \cup \beta(x)) \times (\{\underline{x}'\} \cup \beta'(x')) - \alpha(x,x')$$

with $\alpha(x,x') = \begin{cases} \emptyset & \text{if } \underline{x} \in \beta(x) \text{ and } \underline{x}' \in \beta'(x') \\ \{(\underline{x},\underline{x}')\} & \text{otherwise} \end{cases}$

(ii) Coproduct (also called sum): Let

$$(X,\lim,\beta) + (X',\lim,\beta') = (X + X', \lim + \lim', \beta + \beta')$$

where $X + X' = X \underline{\cup} X'$ is the disjoint union of X and X′; lim + lim′ is the canonical extension of lim and lim′ to F(X) + F(X′); and β + β′ is the canonical extension of β and β′ to X + X′.

(Note that the closure by product and coproduct extends to any number of prebundles.)

(iii) Quotient: Let ~ be an equivalence relation over X, and (X,lim, β) be a prebundle. Define

$$\forall\, x \in X \quad [x] = \{y \in X\colon x \sim y\} \in X/\sim$$
$$\forall\, \sigma \in F(X) \quad [\sigma] = (\lambda i \in \text{Dom}(\sigma) \cdot [\sigma(i)]) \in F(X/\sim)$$

The function

$$\beta/\sim\colon [x] \to \{[\sigma]\colon \sigma \sim^* \tau, \qquad \tau \in \beta(u),\ u \in [x]\}$$

is always well defined.

The function

$$\lim/\sim\colon [\sigma] \to [\lim(\sigma)]$$

is well defined if ~ is regular, i.e.,

$$\sigma \sim^* \tau \Rightarrow \lim(\sigma) \sim \lim(\tau)$$

Thus, if the equivalence relation ~ is regular, then prebundle (X/~, lim/~, β/~) is well defined, and is the quotient of (X,lim,β) under ~.

(iv) Equalizers: Let X and Y be two prebundles and let

$$X \underset{g}{\overset{f}{\rightrightarrows}} Y \quad \text{with } f,g \text{ strongly regular}$$

Let $Z^* = \{x \in X\colon f(x) = g(x)\}$. Let Z be the largest subset E of Z^* such that the trace on E of the prebundle structure of X defines a prebundle on E, and such that the injection $i: E \to X$ commutes with the limits, i.e.,

$$\forall \sigma \in F(E) \quad (i \circ \lim_E)(\sigma) = (\lim_X \circ\, i)(\sigma)$$

Then the injection $e: Z \to X$ is strongly regular and $Z \overset{e}{\to} X \overset{f}{\underset{g}{\rightrightarrows}} Y$ supplies an equalizer of f,g. Indeed, let $W \overset{h}{\to} X \overset{f}{\underset{g}{\rightrightarrows}} Y$ be such that $f \circ h = g \circ h$. We have $h(W) \subseteq X$. Since h is strongly regular, h(W) is a sub-prebundle of X and $h(W) \subseteq Z^* \Rightarrow h(W) \subseteq Z$ is a sub-prebundle. Whence a unique arrow k: $W \to Z$ such that the diagram

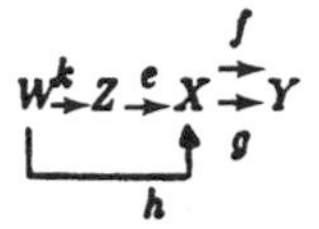

commutes.

(v) Coequalizers: Under the same conditions, let

$$X \overset{f}{\underset{g}{\rightrightarrows}} Y$$

Let ~ be the coarsest equivalence relation on Y such that $\forall \sigma \in X$ $f(a) \sim g(a)$. Define $Z = Y/\sim$. The projection $e: Y \to Y/\sim = Z$ is a coequalizer of f,g in <u>Ens</u>. It is now sufficient to supply Z with the quotient prebundle since ~ is regular.

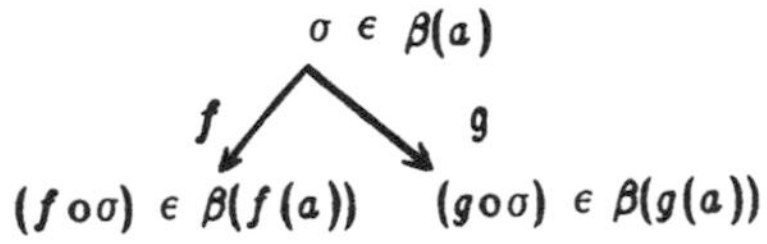

By definition we shall have $(f \circ \sigma) \sim^* (g \circ \sigma)$, and also $f(a) \sim g(a)$. ∎

DEFINITION: If (X, lim, β) is a prebundle, then its <u>kernel</u> is the set

$$N(X) = \{u \in X: \exists x \in X \ \ \exists \sigma \in \beta(X) \ \ \exists i \in \mathrm{Dom}\ (\sigma) \quad u = \sigma(i)\}$$

An element $x \in X$ is <u>rational</u> iff

$$\forall \sigma \in \beta(x) \ \ \exists\ i \in \mathrm{Dom}(\sigma) \quad x = \sigma(i)$$

The prebundle (X, lim, β) is <u>elementary</u> iff its kernel has only rational elements.

The idea of a kernel is that of reconstructing all the elements of a given space by starting with only a restricted subset of privileged elements and taking limits of those (cf. definition of bundles). The terminology for these privileged elements varies from author to author: atomic [16], compact [20], finite trees [13], etc.

A <u>presentation</u> of a prebundle (X, lim, β) is given by an <u>enumeration of the kernel</u>, which is a surjective mapping

$$a: \mathbb{N} \to N(X) \qquad n \to a_n$$

and for any $x \in X$, $\sigma \in \beta(x)$, $\sigma: I \to N(X)$, a recursive enumeration of the set $\{a^{-1}(\sigma(i)): i \in I\}$ of indices of the elements of the family σ.

EXAMPLES: (i) Let $X = P(\mathbb{N})$ where $\mathbb{N}$ is the set of natural numbers, $\lim = \cup$, $\beta(x) = \{y: y \subseteq x\}$ where each y is indexed by itself. Then $\emptyset$ and the singletons are rational and the prebundle (X,lim,β) is elementary. The kernel is $\{\emptyset\} \cup \{\{n\}: n \in \mathbb{N}\}$ and is enumerable, but the prebundle itself has no presentation.

(ii) Let X be the set of all Böhm trees [2] ordered by the prefix order, monotonically extended by "$\Omega \leq$ any other element." A natural prebundle structure for X is given by lim = least upper bound, and $\beta(x) = \{y \text{ finite}: y \leq x\}$. The kernel contains exactly all the finite Böhm trees. Its complement corresponds to those λ-expressions which are without a normal form, but with a head normal form. The kernel is enumerable, and the prebundle has a presentation.

DEFINITION: Let $(X, \lim_X, \beta_X)$ and $(Y, \lim_Y, \beta_Y)$ be prebundles. A function $f, X \to Y$ is <u>regular</u> at $x \in X$ if and only if

$$\forall \sigma \in \beta(x) \quad f(x) = \lim_Y(f(\sigma))$$

where $f(\sigma)$ is defined as $f(\sigma) = f \circ \sigma$.

LEMMA 2 (Point Reconstruction Property) Let (X, lim, β) be a prebundle. Then the following are equivalent:

(i) $\forall x \in X \ \forall \sigma \in \beta(x) \quad x = \lim(\sigma)$

(ii) The identity function $\lambda x \cdot x: X \to X$ is regular.

LEMMA 3 (Constant Family Property) Let (X, lim, β) be a prebundle. Then the following are equivalent:

(i) For every element $x \in X$, for every constant family $(\lambda i.c) \in \beta(x)$ the limit $\lim(\lambda i.c)$ of this family is

$\lim(\lambda i.c) = c = x$

(ii) For any $c \in X$, for any prebundle $(Y, \lim^*, \beta^*)$, the constant function $\lambda y \in Y \cdot c$ is regular.

LEMMA 4 (Fubini Property) Let $(X, \lim, \beta)$ be a prebundle such that

$$\forall \sigma: I \times J \to X \quad \lim_i \lim_j \sigma(i,j) \text{ exists} \Rightarrow \lim_i \lim_j \sigma(i,j) = \lim_{i,j} \sigma(i,j)$$

Then for any two prebundle Y,V and any function F: $Y \times V \to X$, f is regular from $Y \times V$ into X if and only if f is regular for each of its variables.

Proof: Suppose $f: Y \times V \to X$ is regular, and let $g_{v_0} = \lambda y \cdot f(y, v_0)$, $\sigma_1 \in \beta(y)$. Then

$$\begin{aligned} g_{v_0}(y) &= f(\lim \sigma_1, v_0) \\ &= f(\lim(\sigma_1, \underline{v_0})) \\ &= \lim f(\sigma_1, \underline{v_0}) \qquad \text{f regular} \\ &= \lim f(\sigma_1, v_0) = \lim_{g_{v_0}} (\sigma_1) \end{aligned}$$

Thus g_{v_0} is regular. We have the same for $g_{y_0} = \lambda v \cdot f(y_0, v)$.

Reciprocally, suppose g_{v_0} and g_{y_0} are regular. Then if $\sigma = (\sigma_1, \sigma_2) \in \beta(y, v)$ with $\sigma_1 \in \beta(y) \cup \{\underline{y}\}$, $\sigma_2 \in \beta(v) \cup \{\underline{v}\}$, we have

$$\begin{aligned} f(y, v) &= f(\lim \sigma) = f(\lim \sigma_1, \lim \sigma_2) \\ &= \lim_{\sigma_1} f(\sigma_1, \lim \sigma_2) \qquad \text{regularity with respect to y} \\ &= \lim_{\sigma_2} \lim_{\sigma_1} f(\sigma_1, \sigma_2) \qquad \text{regularity with respect to v} \\ &= \lim_{\sigma_1, \sigma_2} f(\sigma_1, \sigma_2) \qquad \text{condition of the lemma} \\ &= \lim f(\sigma) \end{aligned}$$

Thus f is regular. ∎

Summarizing the preceding statements, we define:

DEFINITION: A prebundle $(X, \lim, \beta)$ is a <u>bundle</u> if and only if:

(i) $\forall x \in X \; \forall \sigma \in \beta(x) \quad x = \lim(\sigma)$

(ii) $\forall \sigma \in F(X) \quad \sigma = \lambda i \cdot c, \; c \in X \Rightarrow \lim(\sigma) = c$

(iii) $\forall \sigma: I \times J \to X \quad \lim_i \lim_j \sigma(i,j) \text{ exists} \Rightarrow \lim_i \lim_j \sigma(i,j) = \lim_{i,j} \sigma(i,j)$

If $(X, \lim, \beta)$ is a bundle and $\sigma \in \beta(x)$, then we say that σ converges toward x.

The class of bundles, together with the strongly regular functions, form a full subcategory of the category Prb of prebundles. This subcategory, called Fais, has products, coproducts, quotients, equalizers, and coequalizers.

If X and Y are (carriers of) two bundles, we define

$$[X \to Y] = \{f: X \to Y \mid f \text{ is regular}\}$$

An interesting question is to investigate possible bundle structures for $[X \to Y]$.

PROPOSITION 5 Assume X, Y are two given bundles and suppose $[X \to Y]$ is given a bundle structure. Then the following are equivalent:

(i) $\forall f \in [X \to Y]\ \forall \sigma \in \beta(f)\ \forall x \in X\ \ (\lim_{[X \to Y]} \sigma)(x) = \lim_Y (\sigma(x))$

(ii) The evaluation function Ev: $[X \to Y] \times Y \to Y$, Ev $(f,x) = f(x)$ is regular in its first variable.

Furthermore, regularity of the evaluation function Ev in its second variable is automatic.

Proof: (i) $\Rightarrow$ (ii) By Lemma 4, it is sufficient to show that Ev is regular for each variable. Ev $(f_0, \cdot): x \to f_0(x)$ is always regular since f_0 is regular. For Ev $(\cdot, x_0): f \to f(x_0)$ let $\sigma \in \beta(f)$. Then

$$\begin{aligned} \text{Ev}\ (f, x_0) &= \text{Ev}\ (\lim \sigma, x_0) = (\lim_{[X \to Y]} \sigma)(x_0) \\ &= \lim_Y (\sigma(x_0)) \qquad \text{by hypothesis} \\ &= \lim \text{Ev}\ (\sigma, x_0) \end{aligned}$$

Thus Ev $(\cdot, x_0)$ is regular.

(ii) $\Rightarrow$ (i): Since Ev is regular, it is regular with respect to its first variable, i.e.,

$$\forall f \in [X \to Y]\ \forall \sigma \in \beta(f)\ \forall x \in X$$
$$\text{Ev}\ (f, x) = (\lim_{[X \to Y]} \sigma)(x) = \lim_Y \text{Ev}\ (\sigma, x) = \lim(\sigma(x))$$

which is (i). ∎

DEFINITION: Any bundle structure for $[X \to Y]$ verifying one of the assertions of Proposition 5 will be called <u>evaluation faithful.</u>

There always exists a coarsest evaluation faithful bundle over $[X \to Y]$ which is defined by:

(i) $(\lim_{[X \to Y]} \sigma) = \lambda x \in X \cdot \lim_Y \sigma(x)$ whenever defined.

(ii) $\forall f \in [X \to Y] \ \ \beta(f) = \{\sigma \in F([X \to Y]) : f = \lim_{[X \to Y]} (\sigma)\}$.

In particular, $\forall f \in [X \to Y] \ \ \underline{f} \in \beta(f)$. This bundle will be called the <u>simple convergence bundle</u>.

As for the abstraction operation, we may remark that if X,Y,Z are bundles, and if $[X \times Y \to Z]$, $[Y \to Z]$, and $[X \to [Y \to Z]]$ are given bundle structures, then we have the following:

(i) $g_f : X \to [Y \to Z]$ defined by $g_f(x) = \lambda y \cdot f(x,y)$ is strongly regular if and only if

$$\forall x \in X \ \ \forall \sigma \in \beta(x) \ \ \lambda y \cdot f(\sigma, y) \in \beta(\lambda y \cdot f(x,y))$$

(ii) $L: [X \times Y \to Z] \to [X \to [Y \to Z]]$ defined by $L(f) = \lambda x \cdot \lambda y \cdot f(x,y)$ is strongly regular if and only if

$$\forall f \in [X \times Y \to Z] \ \ \forall \Phi \in \beta(f) \ \ \lambda x \cdot \lambda y \cdot \Phi(x, y) \in \beta(\lambda x \cdot \lambda y \cdot f(x,y))$$

If the above two conditions (i) and (ii) are verified for any X,Y,Z, then we say that our functional bundles are <u>abstraction faithful</u>. We remark that simple convergence bundles are always abstraction faithful. Thus we see that the category <u>Fais</u> when canonically supplied with abstraction faithful and evaluation faithful functional bundles (containing only <u>strongly regular</u> functions) is a <u>cartesian closed category</u>.

Many of the notions described so far easily generalize to categories. If U is a von Neumann universe, a <u>small</u> category is any category whose classes of objects and of arrows are elements of U. A <u>diagram</u> in a category X is a functor $\sigma: I \to X$, where I is a small category. For any small category X, the <u>set</u> (or <u>discrete category</u>) F(X) <u>of diagrams</u> of X is

$$F(X) = \{\sigma: I \to X \mid I \text{ is small}\}$$

For any diagram σ in X, we have the usual notions of <u>limit</u> and <u>colimit</u> of σ, whenever they exist. This gives two partial functions

$$\lim, \text{colim}: F(X) \to Ob(X)$$

where Ob(X) is the class of objects of X; lim and colim are even <u>functors</u> when X is complete. The isomorphism of two objects x, y of X will be denoted by x = y.

DEFINITION: A <u>sagittal bundle</u> is a triple (X, λ, β) such that:

(i) X is a small category.

(ii) λ: F(X) → Ob(X) is a partial function, $\lambda \in \{\lim, \mathrm{colim}\}$.

(iii) β: Ob(X) →P(F(X)) is a total function.

(iv) $\forall x \in Ob(X)\ \ \forall \sigma \in \beta(x)\ \ x = \lambda(\sigma)$.

Classically, we have that for any diagram σ: I × J → X, if $(\forall j)\ \sigma_j$: I × {j} → X and $x_j = \lambda(\sigma_j)$, and if the diagram τ: j → x_j is such that $\lambda(\tau) = y$, then $\lambda(\sigma) = y$ (Fubini property).

Similarly, the set of diagrams $\beta(x)$ will be called the <u>fibre</u> of x.

As an example, the category <u>Fais</u> of all bundles, with the strongly regular functions as arrows, is such that every diagram has a limit and a colimit because <u>Fais</u> has equalizers, products, coequalizers, and coproducts. Ordered bundles (see below) may be seen as special cases of sagittal bundles by deducing arrows from the partial order in the usual way. Then the limits (resp., colimits) are simply the least upper bounds (resp., greatest lower bounds).

The idea behind the notion of sagittal bundles is that of formalizing the notion of approximation used in solving recursive domain equations [19].

If (X,λ,β) and (X',λ',β') are two sagittal bundles, then a functor f: X → Y is <u>regular</u> iff

$$\forall x \in Ob(X)\ \ \forall \sigma \in \beta(x)\ \ f(x) = \lambda'(f(\sigma))$$

(where = denotes an isomorphism).

3. BUNDLES IN PROGRAM SEMANTICS

3.1 Ordered Bundles:

If X ∈ U is a partially ordered set, then $\vee$: S → $\vee$ S (resp., $\wedge$: S → $\wedge$ S) is the partial function that takes least upper bounds (resp., greatest lower bounds) of subsets of X, whenever they exist.

A bundle $(X, \lim, \beta)$ is <u>ordered</u> iff X is a poset and $\lim \in \{\vee, \wedge\}$. A <u>monic ordered bundle</u> (mob) over a poset X is a couple $\langle \lim, s\rangle$ where $\lim \in \{\vee, \wedge\}$ and s: X → P(X) are such that

$\forall\ x \in X\quad x = \lim(s)x))$

For any $x \in X$, the set $s(x)$ is called the spectrum of x. Indeed, in an elementary mob we have

$\forall\ x \in X\quad \forall\ u \in s(x)\quad u \in s(u)$

For example, any algebraic cpo defines an elementary mob by $\lim = \vee$ and $s(x) = \{y \le x: y \text{ compact}\}$.

The following three notions of approximation, interpolation, and stability play an important role in the study of regular functions over ordered bundles.

Approximation

Let $(X, \lim, \beta)$ be a bundle; let $e, x \in X$ and $\sigma \in F(X)$; we say that

(i) e σ- approximates x, denoted $e <_\sigma x$, iff $\sigma \in \beta(x)$ and $e \in \text{Image}(\sigma)$.

(ii) e weakly approximates x, denoted $e <^* x$, iff $e <_\sigma x$ for some $\sigma \in F(X)$.

(iii) e strongly approximates x, denoted $e < x$, iff $\forall\ \sigma \in \beta(x)$, $e <_\sigma x$.

In particular, for every rational u, $u < u$.

Interpolation Properties

$(X, \lim, \beta)$ has the interpolation property at $e \in X$ iff

$\forall\ x \in X\quad \forall\ \sigma \in \beta(x)\quad e <_\sigma x \Leftrightarrow \exists u\quad e < u\quad u <_\sigma x$

The bundle $(X, \lim, \beta)$ has the global interpolation property at $e \in X$ iff

$\forall\ x \in X\quad e < x \Leftrightarrow \exists u\quad e < u < x.$

We notice here that every element $x \in X$ that has an interpolation property (whether global or not) is an element of the kernel.

Stability

The ordered bundle $(X, \vee, \beta)$ is stable at $x \in X$ iff

$\forall\ e,y \in X\quad e \le x \le y \Rightarrow (e <^* x \Leftrightarrow e <^* y)$

Every Galois connection is stable [10]. Stability allows us to obtain interpolation properties over certain mob. Stability in bundles is also related to Berry's stable functions [3] in the following way: if X and Y are two mob with X stable, and f: $X \to Y$ is stable in Berry's sense at $x \in X$, then f is stable on the principal lowerset $\downarrow x$.

Threshold functions intervene in the study of continuous function spaces over continuous lattices, and of semantic domains with data types as objects [17,16,18]. They are of two kinds:

(i) Upper threshold functions (utf):

$\forall\ a \in X\ \ \forall\ m,b \in Y,\ m \geq b$

$$\langle a,b\rangle_m(x) = \begin{cases} b & \text{if } x \leq a \\ m & \text{otherwise} \end{cases}$$

(ii) Lower threshold functions (ltf):

$\forall\ a \in X\ \ \forall\ b \in Y\ \ \forall\ m \in N(Y),\ m \leq b$

$$[a,b]_m\ x = \begin{cases} b & \text{if } x \prec a \\ m & \text{otherwise} \end{cases}$$

where $\prec$ denotes strong approximation.

PROPOSITION 6 (i) The threshold function $\langle a,b\rangle_m$: $X \to Y$ is regular for any two ordered bundles $(X, \vee, \beta)$ and $(Y, \vee, \beta')$.

(ii) The threshold function $\langle a,b\rangle_m$ is regular from $(X, \wedge, \beta)$ into $(Y, \wedge, \beta')$ at $x \in X$ iff a is <u>finite</u> for the fibre of x, namely, iff

$$a \leq x \Leftrightarrow \forall\ \sigma \in \beta(x)\quad a \prec_\sigma x$$

(iii) The lower threshold function $[a,b]_m$ is regular from $(X, \vee, \beta)$ into $(Y, \vee, \beta')$ at point x if bundle X has the interpolation property at x.

3.2 Ordered bundles in program semantics:

Recursive Procedure Calls

We consider two such bundles: the complete partial order (cpo) (in general), and the complete ordered free magma (= free algebra) M(F,V). Recursive procedure calls amount to fixed-point equations

$$\exists ?\ f\quad f(x) = \tau(f)(x) \tag{*}$$

where τ is some functional.

<u>cpo</u> [20]: The bundle structure of a cpo is given by $(X,\vee,\beta)$, where

$\forall\ x \in X\ \ \beta(x) = \{\sigma \text{ directed family: } x = \vee\ \sigma\}$

Every cpo is <u>stable</u>. The cpo that is usually taken is the <u>flat</u> cpo:

$\mathbb{N}_\perp$ =

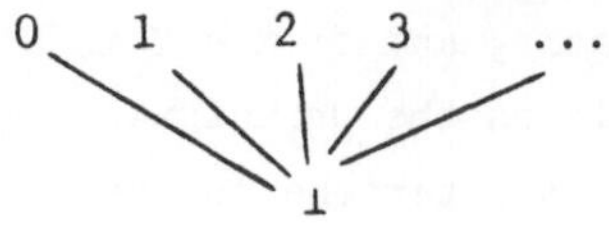

and some of the cpo it generates by taking regular function spaces. Regularity in this case corresponds to Scott continuity. The functional τ in (*) is assumed to be continuous and the least solution of the equation is given by $\vee_{n \in \mathbb{N}} \tau^n(\lambda z \cdot \perp)$ [20].

Free magma: A special case of the above is the complete free magma M(F,V), whose elements are trees. M(F,V) is an elementary bundle whose kernel is the enumerable set of finite trees [11], and which has a presentation.

Type Checking

If

F(n) ← if n = 0 then 0 else 3*F(n - 1) + 1

is a given program, one may want to check whether the function f computed by this program maps even integers into even integers, and odd integers into odd integers. Using threshold function, this can be expressed as:

Is f less defined than $\langle \text{even}, \text{even} \rangle_{\mathbb{N}} \wedge \langle \text{odd}, \text{odd} \rangle_{\mathbb{N}}$?

Here odd and even belong to the bundle

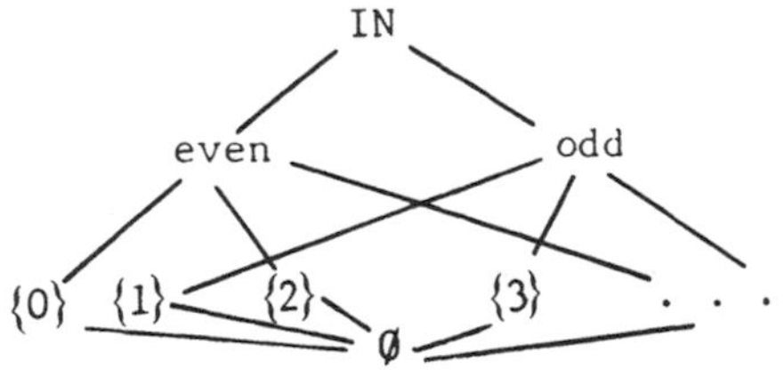

or to any bundle containing this bundle as a subbundle.

A type structure over a domain [18] is presented as a cpo X of subsets of a cpo D with

(i) $\{\downarrow d : d \in D\} \subseteq X$

(ii) $D \in X$

(iii) X is closed under infinite intersection.

(iv) $\forall\, x \in X \;\; \downarrow x = x$, and x contains the lub of its directed subsets.

Although the poset X is presented in [18] as a cpo, it is really used as an elementary mob $\langle \vee, s \rangle$ over X with $\forall x \quad s(x) = \{\downarrow d\colon d \in x\}$. The use of this bundle also solves some technical problems encountered in [18] (see [11]).

Recursive Programs with Type Declarations

The semantic of typed program

$$F\alpha\beta\colon G(x) \leftarrow \tau[G](x)$$

(if input is of type α, output is of type β), is given by using the following bundle X [10]:

(i) X is an elementary mob; $\lim = \vee$.

(ii) X is a semilattice with a top element.

(iii) $\forall a \in X \;\; \forall S \in X \;\; \vee S$ exists $\rightarrow a \wedge (\vee S) = \vee \{a \wedge s\colon s \in S\}$.

This structure is called a convenient domain. Any convenient domain that is a complete lattice is a complete Heyting algebra.

In a convenient domain X, function $\lambda x \cdot a \wedge x$ is regular for every a. If α (resp., β) denotes $a \in X$ (resp., $b \in X$), and if τ is a functional over $[X \rightarrow X]$, we define its transform $N_{a,b}(\tau)$ [resp., $N^*_{a,b}(\tau)$] by

$$N_{a,b}(\tau) = \lambda f \cdot \lambda x \cdot \langle a, b \wedge \tau(f)(x) \rangle_T(x)$$

[resp., $N^*_{a,b}(\tau) = \lambda f \cdot \lambda x \cdot \langle a, \tau(f)(x) \rangle_b(x)$]. Depending on the type-checking mechanism chosen, the function computed by the above typed program is then given by the greatest fixed point of one of these transforms.

Reflexive Domains with Types as Objects

One needs a domain A verifying

$$A = D + [A \rightarrow A]$$

where D contains the ground data types (e.g., integer, real, boolean, ...) as elements, as in the above type-checking example and $[A \rightarrow A]$ contains the functional data types [11,16]. The answer we give in [9,10] uses the following double bundle structure over A:

A has an elementary mob structure $\langle \wedge, s^* \rangle$ (upper bundle).

A has an elementary mob structure $\langle \vee, s_* \rangle$ (lower bundle).

where

$s^*(x) = \{y \in A: y \gg y \gg x\}$

and $\gg$ is the way below relation for the opposite order:

$y \gg x \Leftrightarrow \forall$ filter $F \subseteq A \quad y \geq \wedge F \Rightarrow \exists\, f \in F \quad x \geq f$

Under some general conditions, and up to a slight change in the upper bundle, this solution has been shown to be unique.

3.3 Metric Bundles

In this case, the bundle structure is given by the topology associated with the distance: the bundle theoretic limit is defined by the topological limit, and

$$\beta(x) = \{(y_n)_{n\in\mathbb{N}} : \lim_n y_n = x\}$$

These bundles have been applied to recursive procedure calls [14] and to the study of concurrent processes [1].

Recursive Procedure Calls

The algebra of finitary and infinitary terms M(F,V) is complete for the distance:

$$d(x,y) = \begin{cases} 0 & \text{if } x = y \\ \dfrac{1}{2^{\min\{n:\ \pi_n(x) \neq \pi_n(y)\}}} & \text{otherwise} \end{cases}$$

where $\pi_n(y)$ denotes the cut at height n of tree y.

The functional τ in the equation

$$\exists?\, f \quad f(x) = \tau(f)(x)$$

is shown to be contracting in most cases, and the solution of the equation is the unique fixed point of τ [14].

Concurrency

The domains of processes [1]

$$P = \{p_0\} \cup 2^{A \times P}, \text{ and } P = \{p_0\} \cup \{\Sigma \to 2^{\Sigma \times P}\}$$

are constructed by using metric completion and Hausdorff distance.

3.4 Other Bundles

For the study of concurrency and nondeterminism, the authors in [14] advocate the use of the following structure, which we call the <u>Kuratowski-Painlevé convergence "bundle"</u> (KP).

Let X be a T_2 topological space. For any $\{a_i\}_{i\in\mathbb{N}} \subseteq P(X)$, define

$LI_i(a_i) = \{x \in X:$ every open neighborhood of x meets $a_n, n \geq N$ for some $N \in \mathbb{N}\}$

$LS_i(a_i) = \{x \in X:$ every open neighborhood of x meets infinitely many $a_n\}$

If $LI_i(a_i) = LS_i(a_i)$, then this set is called $\lim_i(a_i)$ (KP-limit of $\{a_n\}_{n\in\mathbb{N}}$).

This defines over $2^X = \{u \subseteq X:$ u is closed and nonempty$\}$ the following "bundle"

$$\lim: F(2^X) \to 2^X$$
$$\{a_n\} \to \lim_n a_n \text{ whenever defined}$$
$$\beta: 2^X \to P(F(2^X))$$
$$u \to \{\{a_n\}_{n\in\mathbb{N}}: u = \lim a_n\}$$

with only the weak Fubini property: if $\lim_i \lim_j a_{ij}$ and $\lim_{i,j} a_{ij}$ both exist, then they are equal.

The fact that this construction does not give a bundle is due to the dissymmetrical role played by LI and LS. The limit inferior LI always defines a bundle over 2^X and has the Fubini property. The limit superior LS, on the other hand, only yields an inclusion $LS_i\ (LS_j\ a_{ij}) \subseteq LS_{i,j}\ a_{ij}$.

In the setting of [14], X is in fact a compact space and, in this case, the KP-limit coincides with the limit in the sense of Hausdorff distance (as shown in [5], p. 172), which brings us back to a construction already mentioned. Both LS and LI, however, are of interest in program semantics. See, for example, [12].

3.5 Operations on Bundles

The use of bundle products and coproducts is well known [17]. In [15], we find an application of the notion of a bundle quotient. If A is an alphabet, then A^* is the free monoid (the set of all possible finite sequences) generated by A, and A^∞ is the set of all sequences, finite or infinite, of elements of A. If A^∞ is ordered by the prefix order, then it is canonically supplied with an elementary ordered bundle.

$$\lim = \vee \qquad s(x) = \{y \text{ finite}: y \leq x\}$$

The set of two-way finite words ${}^\infty A^\infty$ is defined as the quotient of $A^\infty \times A^\infty$ under the equivalence relation

$(\alpha,\beta) \sim (\alpha',\beta')$ iff $\exists\, u \in A^*$ with

$(\beta = u\beta'$ and $\alpha' = u^R\alpha)$ or $(\beta' = u\beta$ and $\alpha = u^R\alpha')$

The underlying construction is that of a bundle quotient, and the required property is the <u>regularity</u> of ~. Since elements of A^∞ are obtained as limits of increasing sequences (for the prefix order) of elements of A^*, one easily sees that if, for example,

$$\exists\, u \in A^*\ \sigma_2(i) = u\ \tau_2(i) \quad \text{and} \quad \tau_1(i) = u^R\sigma_1(i)$$

then, by extending our words both ways, i.e., going to the limit, we also have for the same u:

$$\lim_i \sigma_2(i) = u \lim_i \tau_2(i) \quad \text{and} \quad \lim_i \tau_1(i) = u^R \lim_i \sigma_1(i)$$

i.e., $\lim\,(\sigma_1,\sigma_2) \sim \lim\,(\tau_1,\tau_2)$. Thus ~ is regular, and the set ${}^\infty A^\infty$ is supplied with the quotient bundle. As a set, ${}^\infty A^\infty$ turns out to be ${}^\infty A^\infty = A^Z/_R$ where $Z = \{\ldots, -3,-2,-1,0,1,2,3,\ldots\}$ and x R y iff x equals y up to a translation (shift).

3.6 Sagittal Bundles

The notion of regular functor is a natural one in category theory. If <u>Set</u> is the category whose objects are the sets $X \in U$, and arrows are all functions, and if C is a small category, it is a well-known fact that any hom-functor

$$C(a,\cdot)\colon C \to \underline{Set}$$

preserves all limits.

Sagittal bundles and regular functors occur in [17,19] and are used for computing fixed points for certain regular functors, and their approximations as limits or colimits of diagrams. Functors occur in recursive domain equations that occur naturally in program semantics. Historically, the first such construction for program semantics use appears in [17], where the domain equation

$$D = [D \to D]$$

is solved by using projective diagrams in the category of continuous lattices, with continuous functions as arrows, and the functor $D \to [D \to D]$ which is shown to be regular for such diagrams. This is considered in a more general setting in [7], p. 224, where the following sagittal bundle $(X, \lim, \beta)$ is used, such that

(i) X is a small category in which every projective diagram has a limit (called procomplete in [7].

(ii) lim takes limits of diagrams.

(iii) $\beta(x) = \{\sigma : \sigma \text{ is a projective diagram; } \lim \sigma = x\}$.

The procontinuous functors of [7] are just the regular functors for this bundle.

As an example, let Inf↑ be the category whose objects are the complete lattices, and whose arrows are the Lawson-continuous functions. Then, if Inf↑ is equipped with the sagittal bundle (Inf↑, lim, β) as above, the "function space" functor

Funct: Inf↑ → Inf↑, D → [D → D]

is regular.

All these sagittal bundles structures constitute the essence of various λ-calculus models constructions [3,11,17], and play in the solution of recursive domain equations a role similar to the one played by ordinary bundles in recursive function equations.

The author gratefully acknowledges the referee's comments on a first version of this paper.

REFERENCES

1. J. de Bakker and J. C. Zucker. Denotational semantics of concurrency, 14th SIGACT Conference, San Francisco (May 1982).

2. H. Barendregt. The λ-Calculus: Its Syntax and Semantics, North Holland, Amsterdam (1981).

3. G. Berry. Modèles complètement adéquats et stables des λ-calculus typés. Thèse d'Etat, Paris (1979).

4. E. Binz. Continuous convergence on C(X), Springer LNM 469 (1975).

5. F. Hausdorff. Set Theory, Chelsea Publishing Co. (1962).

6. K. H. Hofmann. On completely distributive lattices, Third Conference on Continuous Lattices, Bremen (1982).

7. K. H. Hofmann, D. Scott, et al. A Compendium of Continuous Lattices, Springer-Verlag (1980).

8. D. Mumford. Introduction to Algebraic Geometry.

9. M. A. Nait Abdallah. Types and approximating calculi in programming language semantics. 3rd Workshop on Continuous Lattices, Riverside (March 1979).

10. M. A. Nait Abdallah. Faisceaux et sémantique des programmes, Thèse d'Etat, Paris (1980).

11. M. A. Nait Abdallah. Sort theory, University of Waterloo TR CS-82-19 (1982).

12. M. A. Nait Abdallah. Metric interpretation and greatest fixpoint semantics of logic programs, University of Waterloo TR CS-83-29 (1983).

13. M. Nivat. On the interpretation of recursive polyadic program schemes, Symp. Math. 15 (1975), pp. 255-281.

14. M. Nivat and A. Arnold. The metric space of infinite trees, IRIA report 300 (1978).

15. M. Nivat and D. Perrin. Ensembles reconnaissables de mots biinfinis, 14th SIGACT, San Francisco (May 1982).

16. L. Nolin. Algorithmes universels, RAIRO rouge 4 (1974), pp. 5-18.

17. D. Scott. Continuous lattices, Springer LNM 274 (1972), pp. 97-136.

18. A. Shamir and W. Wadge. Data types as objects, Springer LNCS 52 (1977), pp. 465-479.

19. M. Smyth and G. Plotkin. The category-theoretic solution of recursive domain equations, DAI report 60, Edinburgh (1978).

20. J. Vuillemin. Syntaxe, sémantique et axiomatique d'un langage de programmation simple, Thèse d'Etat, Paris (1974).

14

Algebraic Lattices as Dual Spaces of Distributive Lattices

HILARY A. PRIESTLEY
Mathematical Institute
University of Oxford
Oxford, England

1. INTRODUCTION

It is well known that the Lawson topology on an algebraic lattice is compact and zero dimensional. In fact, slightly more is true: an algebraic lattice A equipped with its natural order and Lawson topology is a compact, totally order-disconnected space; and, as such, is the dual space of a distributive lattice (isomorphic to the lattice of Lawson-clopen upper sets of A). We use the theory of continuous lattices to investigate the properties of distributive lattices arising in this fashion.

In Sec. 4 we consider in particular the bounded distributive lattices whose dual spaces are compact, zero-dimensional, topological lattices; these turn out to be precisely the catalytic ones. A more detailed account of this material appears in [16]. Thanks are due to, among others, Dr. B. A. Davey for helpful discussions on the relationship between catalyticity in the category of bounded distributive lattices and lattice objects in the dual category.

We use the notation and terminology of [10] throughout. The duality theory we require is summarized in Sec. 2. For additional information on distributive lattices we refer the reader to [3].

2. THE $\underset{\sim}{D}$-$\underset{\sim}{P}$ DUALITY FOR DISTRIBUTIVE LATTICES

The category $\underset{\sim}{D}$ has as objects the bounded distributive lattices. We denote the universal bounds of any $L \in \underset{\sim}{D}$ by 0,1 and assume $0 < 1$. Morphisms in $\underset{\sim}{D}$ are 0,1-preserving lattice homomorphisms. To each $L \in \underset{\sim}{D}$ we associate the hom-set $X_L = \underset{\sim}{D}(L,\underline{2})$ of $\underset{\sim}{D}$-morphisms from L onto the two-element chain $\underline{2}$. We regard $\underline{2}$ as an element of $\underset{\sim}{D}$ and also as an ordered topological space;

for the latter role it is given the natural order and discrete topology. As a subset of the product space $\underline{2}^L$, X_L inherits the pointwise order (denoted by $\le$) and a compact topology $\mathcal{T}$. The dual space of L is defined to be the ordered space $(X_L,\mathcal{T},\le)$. It is totally order-disconnected: if $x \nleq y$ in X_L, there exists a $\mathcal{T}$-clopen upper set V such that $x \in V$, $y \notin V$. We define $\underset{\sim}{P}$ to be the category whose objects are the compact, totally order-disconnected spaces (Priestley spaces) and whose morphisms are the continuous monotone maps. It was established in [14] that the categories $\underset{\sim}{D}$ and $\underset{\sim}{P}$ are dually equivalent. For an up-to-date survey of consequences of the $\underset{\sim}{D}$-$\underset{\sim}{P}$ duality see [17], and, for its connection with the spectral theory of continuous lattices, see [10], p.325.

2.1 THEOREM [14,15] (i) For each $L \in \underset{\sim}{D}$, $(X_L,\mathcal{T},\le) \in \underset{\sim}{P}$ and L is isomorphic to the lattice of $\mathcal{T}$-clopen upper sets of X_L under the mapping $a \mapsto \{x \in X_L \mid x(a) = 1\}$, $a \in L$.

(ii) For each $(P,\mathcal{T},\le) \in \underset{\sim}{P}$, the lattice of $\mathcal{T}$-clopen upper sets of P belongs to $\underset{\sim}{D}$ and has dual space homeomorphic and order isomorphic to $(P,\mathcal{T},\le)$.

(iii) There exists a 1-1 correspondence between $\underset{\sim}{D}$-morphisms $f \in \underset{\sim}{D}(L,M)$ and $\underset{\sim}{P}$-morphisms $\varphi \in \underset{\sim}{P}(X_M,X_L)$ given by

$$(\varphi(y))(a) = y(f(a)) \qquad a \in L,\ y \in X_M$$

Further, f is surjective if and only if φ is an order embedding and f is injective if and only if φ is surjective.

Rephrased in more formal categorical terms, the duality asserts that the contravariant functors $\underset{\sim}{D}(\cdot,\underline{2})$ and $\underset{\sim}{P}(\cdot,\underline{2})$ set up an adjunction between $\underset{\sim}{D}$ and $\underset{\sim}{P}$. The unit and counit maps ϵ_L and η_P ($L \in \underset{\sim}{D}$, $P \in \underset{\sim}{P}$) are given by evaluation and are isomorphisms.

We shall henceforth identify $L \in \underset{\sim}{D}$ with the lattice of $\mathcal{T}$-clopen upper sets of $(X_L,\mathcal{T},\le)$.

As an easy consequence of the duality we have (see [14,15]):

2.2 LEMMA Let $L \in \underset{\sim}{D}$ have dual space $(X_L,\mathcal{T},\le)$.

(i) The ideal lattice Id(L) of L is isomorphic to the lattice of $\mathcal{T}$-open upper sets of X_L, under the map $I \mapsto \bigcup\{a \mid a \in I\}$.

(ii) The filter lattice Filt(L) of L is isomorphic to the lattice of $\mathcal{T}$-open lower sets of X_L, under the map $F \mapsto \bigcup\{X_L \setminus a \mid a \in F\}$.

(iii) The congruence lattice Con(L) of L is isomorphic to the lattice of $\mathcal{T}$-open subsets of X_L, under the map that associates to an open set U the congruence θ given by $a \equiv b\ (\theta)$ if and only if $a \cap (X_L \setminus U) = b \cap (X_L \setminus U)$.

(iv) The lattice L^{op} has dual space $(X_L, \mathcal{T}, \geq)$.

We denote by J(L) the set of join-irreducible elements (excluding 0) of $L \in \mathbf{D}$ and by M(L) the set of meet-irreducible elements (excluding 1). Subsequently, these sets will always be assumed to carry the induced order from L.

If P is a finite poset and L its lattice of upper sets, then $P \cong J(L)^{op}$ (via the map $x \mapsto \uparrow x$) and we may identify $a \in L$ with $\{b \in J(L) \mid b \leq a\}$. This recovers Birkhoff's classic representation for finite distributive lattices. Similarly, $P \cong M(L)^{op}$ via the map $x \mapsto P \setminus \downarrow x$.

2.3 LEMMA (Adams [1]) Let $L \in \mathbf{D}$, $a \in L$. Then

(i) $a \in J(L)$ if and only if $a = \uparrow x$ for some $x \in X_L$.

(ii) $a \in M(L)$ if and only if $a = X_L \setminus \downarrow x$ for some $x \in X_L$.

We now consider how duality works in $\mathbf{H}^d$, the category of dual Heyting algebras. A bounded distributive lattice L is in $\mathbf{H}^d$ if and only if L^{op} is a Heyting algebra, as defined, for example, in [10], 0.3.7. Explicitly, $L \in \mathbf{H}^d$ if for each $a,b \in L$, the relative dual pseudocomplement

$$a \leftarrow b = \min \{c \in L \mid c \vee a \geq b\}$$

exists in L. Morphisms in $\mathbf{H}^d$ are those $\mathbf{D}$-morphisms that preserve $\leftarrow$. The first part of Lemma 2.4 is well known; the second part is due to Bowen [6].

2.4 LEMMA (i) Let $L \in \mathbf{D}$ have dual space $(X_L, \mathcal{T}, \leq)$. Then $L \in \mathbf{H}^d$ if and only if $U \in \mathcal{T}$ implies $\uparrow U \in \mathcal{T}$. If this condition holds

$$a \leftarrow b = \uparrow(b \setminus a) \qquad \forall\, a,b \in L$$

(ii) Let $L,M \in \mathbf{H}^d$ and let $f \in \mathbf{D}(L,M)$ have dual $\varphi \in \mathbf{P}(X_M, X_L)$. Then $f \in \mathbf{H}^d(L,M)$ if and only if $\varphi(\downarrow y) = \downarrow\varphi(y)\ \forall\, y \in X_M$.

NOTE If X_M, X_L are semilattices and $\varphi\colon X_M \to X_L$ is surjective and preserves finite infima, then φ satisfies $\varphi(\downarrow y) = \downarrow\varphi(y)$ for all $y \in X_M$, the condition appearing in Lemma 2.4(ii). Indeed, since φ is monotone, $\varphi(\downarrow y) \subseteq \downarrow\varphi(y)$ for every y. To prove the reverse inclusion, suppose $z \leq \varphi(y)$ for some $z \in X_L$. Given that φ is surjective, there is some $x \in X_M$ with $\varphi(x) = z$, and so, using the fact that φ preserves finite infima, we have $\varphi(x \wedge y) = \varphi(x) \wedge \varphi(y) = z$. Thus $\downarrow\varphi(y) \subseteq \varphi(\downarrow y)$ also holds.

3. ALGEBRAIC LATTICES AS PRIESTLEY SPACES

Let A be an algebraic lattice, K(A) its subset of compact elements (for definitions and an account of the basic properties of algebraic lattices, see [10],I.4). We shall need to consider the following topologies on A:

(a) The lower topology $\omega(A)$ has as closed subbase the sets $\uparrow x$ ($x \in A$).

(b) The Scott topology $\sigma(A)$ has as an open base the sets $\uparrow k$ ($k \in K(A)$) (see [10],II.1.15).

(c) The interval topology IV(A) has as closed subbase the sets $\uparrow x$, $\downarrow x$ ($x \in A$).

(d) The Lawson topology $\lambda(A)$ is the join of $\omega(A)$ and $\sigma(A)$.

The topology of $\lambda(A)$-open upper sets is $\sigma(A)$ and the topology of $\lambda(A)$-open lower sets is $\omega(A)$ ([10],III.1.6 and IV.3.28).

3.1 LEMMA (cf. [10],III.1.12) If A is an algebraic lattice, with natural order $\le$, then $(A,\lambda(A),\le) \in \underset{\sim}{P}$.

Proof: If $x \not\le y$ in A, there exists $k \in K(A)$ such that $k \le x$, $k \not\le y$. The set $\uparrow k$ is $\lambda(A)$-clopen and contains x but not y. This proves total order disconnectedness. Compactness is proved in [10],III.1.9. ∎

It usually happens that an ordered set, or, more particularly, an algebraic lattice, can be topologized in many ways to become a Priestley space and so functions as the dual space of many nonisomorphic lattices. Consider, for example, M_ω, a countably infinite antichain with universal bounds 0,1 adjoined. Both $\lambda(M_\omega)$ and $\lambda(M_\omega{}^{op})$ make M_ω a Priestley space: the associated lattices of clopen upper sets are not isomorphic. This nonuniqueness does not occur in an important special case which arises in Sec. 4: whenever $(P,\mathcal{T},\le) \in \underset{\sim}{P}$ and IV(P) is Hausdorff, we have $\mathcal{T} = IV(P)$. Conditions under which the interval topology is Hausdorff are given in [9] and [11].

3.2 PROPOSITION Let L be the lattice of $\lambda(A)$-clopen upper sets of an algebraic lattice A. Then

(i) The map $k \mapsto \uparrow k$ is an order isomorphism from the $\vee$-subsemilattice K(A) of A onto $J(L)^{op}$.

(ii) Every nonzero element of L is a finite join of elements in J(L).

(iii) $L \in \underset{\sim}{H}^d$.

Proof: A set $\uparrow k$ is $\lambda(A)$-clopen if and only if $k \in K(A)$, and every nonempty $\lambda(A)$-clopen upper set is a finite union of sets $\uparrow k$ ($k \in K(A)$). Properties (i) and (ii) now follow from Lemma 2.3.

Property (iii) is a consequence of Lemma 2.4(i) and [10],VI.1.13. More directly, one can use (ii): for a,b $\in$ L, $\bigvee \{c \in J(L) \mid c \leq b, c \not\leq a\}$ defines a $\leftarrow$ b. ∎

3.3 PROPOSITION Let L be the lattice of $\lambda(A)$-clopen upper sets of an algebraic lattice A. Then Id(L) is completely distributive and is isomorphic to the lattice of all lower sets of J(L).

Proof: Lemma 2.2 implies that Id(L) $\cong$ $\sigma(A)$, the $\lambda(A)$-open upper sets of A. By [10],I.1.14, Id(L) is completely distributive. It follows that Id(L) is isomorphic to the lattice of lower sets of the set of complete join irreducibles in $\sigma(A)$ (see, for example, [8], Proposition 1.1). A non-zero element of $\sigma(A)$ is a complete join irreducible if and only if it is of the form $\uparrow k$, with k compact. The result now follows from Proposition 3.2 (i). ∎

The filter lattice of the lattice L in Proposition 3.3 is isomorphic to $\omega(A)$, but in general this does not provide a useful description of Filt(L) analogous to that given of Id(L). In Proposition 3.5 we give descriptions of both Id(L) and Filt(L) in terms of projective limits.

We now consider morphisms. The fundamental theorem on compact zero-dimensional semilattices ([12],II.3, [10],VI.3.4, and VI.3.13) asserts that the two categories

$\underset{\sim}{AL}$: algebraic lattices and maps preserving directed sups and arbitrary infs

$\underset{\sim}{Z}$: compact zero-dimensional semilattices and continuous identity-preserving semilattice morphisms

are isomorphic. The isomorphism associates to each algebraic lattice A the topological semilattice $(A,\lambda(A),\wedge)$ and is the identity on morphisms. We shall henceforth treat algebraic lattices as lying in $\underset{\sim}{AL}$ or in $\underset{\sim}{Z}$, as expedient.

3.4 LEMMA Let A,B be algebraic lattices, and L,M their lattices of Lawson-clopen upper sets. Suppose the injective map f $\in \underset{\sim}{D}(L,M)$ has as dual the surjective map $\varphi \in \underset{\sim}{P}(B,A)$. Then statements (i) and (ii) below are equivalent and imply (iii).

(i) $f(J(L)) \subseteq J(M)$.

(ii) φ is a $\underset{\sim}{Z}$-morphism (i.e., φ preserves arbitrary infs).

(iii) f $\in \underset{\sim}{H}^d(L,M)$.

Proof: Note first that $f(a) = \varphi^{-1}(a)$ for all $a \in L$ [by Theorem 2.1 (iii)]. Proposition 3.2(i) implies that (i) holds if and only if, for each k in $K(A)$, $\varphi^{-1}(\uparrow k) = \uparrow k_\varphi$ for some $k_\varphi \in K(B)$.

Now suppose φ is a $\underset{\sim}{Z}$-morphism. Then φ has a lower adjoint $D(\varphi)\colon A \to B$ which maps $K(A)$ into $K(B)$ and which is given by

$$(D(\varphi))(x) = \min \varphi^{-1}(\underline{\uparrow} x) \qquad x \in A$$

(see [10],0.3 and IV.1.12). Hence (ii) implies (i).

Conversely, suppose that whenever $k \in K(A)$ there exists $k_\varphi \in K(B)$ such that $\varphi^{-1}(\underline{\uparrow} k) = \underline{\uparrow} k_\varphi$. Since A is algebraic, for each $x \in A$ we have

$$x = \wedge \{\, k \mid k \in K(A),\ k \le x\}$$

so

$$\varphi^{-1}(\uparrow x) = \underline{\uparrow}\sup\{k_\varphi \mid k \in K(A),\ k \le x\}$$

This shows that $\min \varphi^{-1}(\underline{\uparrow} x)$ exists for each $x \in A$. Hence φ possesses a lower adjoint and so preserves arbitrary infs.

That (i) and (ii) imply (iii) is a consequence of the note following Lemma 2.4. ∎

Surjectivity of φ is used crucially only to prove that (ii) implies (iii); the first part of the lemma holds with minor adaptations for non-surjective φ. Statement (iii) implies (i) and (ii) if we impose the additional (strong) restriction that $\varphi^{-1}(x)$ contain a unique minimal point for each x. This condition necessarily holds if (ii) is satisfied and φ is surjective, since then φ has a lower adjoint d, defined by $d(x) = \inf \varphi^{-1}(\downarrow x) = \min \varphi^{-1}(x)$ ([12], 0.3.7). A simple example of the failure of (iii) implies (ii) is obtained by taking A to be the three-element chain $\{0,a,1\}$, B the four-element Boolean algebra $\{0,b,c,1\}$ and defining $\varphi\colon B \to A$ by $\varphi(0) = 0$, $\varphi(1) = 1$, $\varphi(b) = \varphi(c) = a$. Then φ is surjective and $\varphi(\downarrow y) = \downarrow\varphi(y)$ for all y, but φ does not preserve infima.

We shall now make use of the profiniteness of objects in $\underset{\sim}{P}$ and in $\underset{\sim}{Z}$ (or in $\underset{\sim}{AL}$) and the fact that various functors preserve (strict) projective limits to prove some results about the ideal, filter, and congruence lattices of a bounded distributive lattice whose dual space is an algebraic lattice in its Lawson topology. We require the following functors:

(a) σ is the covariant functor on $\underset{\sim}{AL}$ given by:

(i) For each algebraic lattice A, $\sigma(A)$ is the lattice of Scott-open subsets of A.

(ii) For each $\underset{\sim}{AL}$-morphism $\varphi: B \to A$, $\sigma(\varphi): \sigma(B) \to \sigma(A)$ is defined by

$$(\sigma(\varphi))(b) = (D(\varphi))^{-1}(b)$$

for $b \in \sigma(B)$, where $D(\varphi): A \to B$ is the lower adjoint of φ.

(b) P is the covariant functor on $\underset{\sim}{AL}$ given by:

(i) For each algebraic lattice A, P(A) is the lattice of closed upper sets of A, ordered by reverse inclusion.

(ii) For each $\underset{\sim}{AL}$-morphism $\varphi: B \to A$, $P(\varphi): P(B) \to P(A)$ is defined by

$$(P(\varphi))(b) = \uparrow \varphi(b) \qquad b \in P(B)$$

(c) Γ is the covariant functor on $\underset{\sim}{AL}$ given by:

(i) For each algebraic lattice A, $\Gamma(A)$ is the lattice of Lawson-closed subsets of A, ordered by reverse inclusion.

(ii) For each $\underset{\sim}{AL}$-morphism $\varphi: B \to A$, $\Gamma(\varphi): \Gamma(B) \to \Gamma(A)$ is defined by

$$(\Gamma(\varphi))(b) = \varphi(b) \qquad (b \in \Gamma(B))$$

Each of σ, P, Γ takes $\underset{\sim}{AL}$ into $\underset{\sim}{AL}$ ([10].II.1.15, VI.3.28, and VI.3.8). Additionally, it is known that σ, P preserve $\underset{\sim}{AL}$-projective limits ([10],VI.3.20, VI.3.22) and that Γ (regarded as a functor on $\underset{\sim}{Z}$) preserves strict projective limits in $\underset{\sim}{Z}$ ([12], p. 15).

3.5 PROPOSITION Let $L \in \underset{\sim}{D}$ have dual space $(A, \lambda(A), \leq)$, where A is an algebraic lattice. Then

(i) A is the projective limit (in $\underset{\sim}{AL}$, in $\underset{\sim}{Z}$, and in $\underset{\sim}{P}$) of an inverse system $(A_\alpha, \varphi_{\alpha\beta})$ where each A_α is finite and the maps $\varphi_{\alpha\beta}: A_\beta \to A_\alpha$ are surjective and preserve infs.

(ii) L is the direct limit (in $\underset{\sim}{H}^d$ and in $\underset{\sim}{D}$) of a direct system $(L_\alpha, f_{\alpha\beta})$ where each L_α is finite and the maps $f_{\alpha\beta}: L_\alpha \to L_\beta$ are dual Heyting algebra embeddings.

(iii) Id(L) is the projective limit (in $\underset{\sim}{AL}$ and in $\underset{\sim}{Z}$) of an inverse system $(L_\alpha, \psi_{\alpha\beta})$ where $\psi_{\alpha\beta}: L_\beta \to L_\alpha$ is given by

$$\psi_{\alpha\beta}(b) = \bigvee\{a \in J(L_\alpha) \mid f_{\alpha\beta}(a) \leq b\} \qquad b \in L_\beta$$

(iv) Filt(L) is the projective limit (in $\underset{\sim}{AL}$ and in $\underset{\sim}{Z}$ of an inverse system $(L_\alpha, \xi_{\alpha\beta})$ where $\xi_{\alpha\beta}: L_\beta \to L_\alpha$ is given by

$$\xi_{\alpha\beta}(b) = \bigwedge\{a \in M(L_\alpha) \mid f_{\alpha\beta}(a) \geq b\} \qquad b \in L_\beta$$

(v) Con(L) is the projective limit (in $\underset{\sim}{AL}$ and in $\underset{\sim}{Z}$ of an inverse system $(\mathrm{Con}(L_\alpha), \zeta_{\alpha\beta})$ where, for $\theta \in \mathrm{Con}(L_\alpha)$, $a, b \in L_\alpha$,

$$a \equiv b\ (\zeta_{\alpha\beta}(\theta)) \Leftrightarrow f_{\alpha\beta}(a) \equiv f_{\alpha\beta}(b)\quad(\theta)$$

[In (iii)-(v), L_α and $f_{\alpha\beta}$ are as in (ii).]

Proof: We can obtain the algebraic lattice A as the projective limit (in $\underset{\sim}{Z}$ and in $\underset{\sim}{AL}$) of an inverse system $(A_\alpha, \varphi_{\alpha\beta})$ of finite lattices A_α and surjective inf-preserving maps $\varphi_{\alpha\beta}$ [12]. Concretely,

$$A \cong \{x \in \prod A_\alpha \mid \varphi_{\alpha\beta}\pi_\beta(x) = \pi_\alpha(x)\}$$

where $\pi_\gamma\colon \prod A_\alpha \to A_\gamma$ is the canonical projection. This is also how the projective limit is computed in $\underset{\sim}{P}$ (see [18]).

We get (ii), at least with L a direct limit in $\underset{\sim}{D}$, by dualizing and using Lemma 3.4. To see that we also have an $\underset{\sim}{H}^d$-limit, note that, because L can be regarded as the directed union of dual Heyting subalgebras L_α, the universal property for a colimit cone which certainly holds in $\underset{\sim}{D}$ actually holds in $\underset{\sim}{H}^d$.

To prove (iii), we observe first that $\sigma(A_\alpha) = L_\alpha$ and that

$$\mathrm{Id}(L) \cong \sigma(A) = \lim_{\leftarrow} \sigma(A_\alpha)$$

using maps $\sigma(\varphi_{\alpha\beta})$. Also, for any $b \in L_\beta$,

$$\begin{aligned}(\sigma(\varphi_{\alpha\beta}))(b) &= (D(\varphi_{\alpha\beta}))^{-1}(b)\\ &= \{x \in A_\alpha \mid \min\varphi_{\alpha\beta}{}^{-1}(\uparrow x) \in b\}\\ &= \bigcup\{\uparrow x \mid x \in A_\alpha, \varphi_{\alpha\beta}{}^{-1}(\uparrow x) \subseteq b\}\\ &= \bigvee\{a \in J(L_\alpha) \mid f_{\alpha\beta}(a) \le b\}\end{aligned}$$

Similarly, $P(A_\alpha) = L_\alpha$ and

$$\mathrm{Filt}(L) \cong P(A) = \lim_{\leftarrow} P(A_\alpha)$$

using maps $P(\varphi_{\alpha\beta})$. Also, for $b \in L_\beta$,

$$\begin{aligned}A_\alpha \setminus (P(\varphi_{\alpha\beta}))(b) &= A_\alpha \setminus \uparrow\varphi_{\alpha\beta}(b)\\ &= \{x \in A_\alpha \mid (\forall\, y \in b)\ x \not\ge \varphi_{\alpha\beta}(y)\}\\ &= \{x \in A_\alpha \mid b \cap \varphi_{\alpha\beta}{}^{-1}(\downarrow x) = \emptyset\}\end{aligned}$$

$$= \{x \in A_\alpha \mid b \subseteq \varphi_{\alpha\beta}^{-1}(A_\alpha \setminus \downarrow x)\}$$
$$= \bigcup\{\downarrow x \mid x \in A_\alpha,\ b \subseteq \varphi_{\alpha\beta}^{-1}(A_\alpha \setminus \downarrow x)\}$$

Hence,

$$(P(\varphi_{\alpha\beta}))(b) = \bigcap\{(A_\alpha \setminus \downarrow x) \mid x \in A_\alpha,\ b \subseteq \varphi_{\alpha\beta}^{-1}(A_\alpha \setminus \downarrow x)\}$$
$$= \bigwedge\{a \in M(L_\alpha) \mid f_{\alpha\beta}(a) \geq b\}$$

For the congruence lattice we have

$$\mathrm{Con}(L) \cong \Gamma(A) = \lim_{\leftarrow} \Gamma(A_\alpha) \cong \lim_{\leftarrow} \mathrm{Con}(L_\alpha)$$

It remains to identify the bonding maps $\zeta_{\alpha\beta}\colon \mathrm{Con}(L_\beta) \to \mathrm{Con}(L_\alpha)$. Suppose $\theta \in \mathrm{Con}(L_\beta)$ is identified with $b \in \Gamma(L_\beta)$, so that $c \equiv d\ (\theta)$ if and only if $c \cap b = d \cap b$ [see Lemma 2.2(iii)]. Then for $p,q \in L_\alpha$,

$$f_{\alpha\beta}(p) \equiv f_{\alpha\beta}(q) \Leftrightarrow \varphi_{\alpha\beta}^{-1}(p) \cap b = \varphi_{\alpha\beta}^{-1}(q) \cap b$$
$$\Leftrightarrow p \cap \varphi_{\alpha\beta}(b) = q \cap \varphi_{\alpha\beta}(b)$$
$$\Leftrightarrow p \equiv q\ (\zeta_{\alpha\beta}(\theta))$$

Note that, in Proposition 3.5, L is expressed as a direct limit of a system of finite lattices L_α, while Id(L), Filt(L) are expressed as inverse limits of the L_α. The "natural" functors Id, Filt are of course covariant (see, for example, [5]).

We can use the fact that the maps $f_{\alpha\beta}$ preserve join irreducibles (by Lemma 3.4) to obtain an alternative formula for the maps $\psi_{\alpha\beta}$. We identify each A_α with the isomorphic lattice $J(L_\alpha)^{op}$ (see the remarks following Lemma 2.2). Let $g_{\alpha\beta} = f_{\alpha\beta} \mid J(L_\alpha)$, so $g_{\alpha\beta}$ maps $J(L_\alpha)$ into $J(L_\beta)$. Then, for $a \in J(L_\alpha)$, $b \in J(L_\beta)$,

$$f_{\alpha\beta}(a) \leq b \Leftrightarrow g_{\alpha\beta}(a) \in \{c \in J(L_\beta) \mid c \leq b\} = b$$
$$\Leftrightarrow a \in g_{\alpha\beta}^{-1}(b)$$

Hence $\psi_{\alpha\beta}(b) = g_{\alpha\beta}^{-1}(b)$.

4. COMPACT ZERO-DIMENSIONAL TOPOLOGICAL LATTICES AND CATALYTICITY

Let A be a bialgebraic lattice (that is, assume that both A and A^{op} are algebraic). Then A becomes a Priestley space when given either of the topologies $\lambda(A)$, $\lambda(A^{op})$. If these topologies agree, A is said to be a linked bialgebraic lattice. Some important and well-known characterizations of such lattices are collected together in Theorem 4.1. For the proofs and for many equivalents of (v), see [10], VII.2.6,8,9; [12], III.1.53; and [8], Proposition 1.1.

4.1 THEOREM Let A be a complete lattice. Then the following are equivalent:

(i) A is a linked bialgebraic lattice.

(ii) A is bialgebraic and the interval topology IV(A) is Hausdorff.

(iii) A is a compact zero-dimensional topological lattice with respect to $\lambda(A)$.

(iv) A is a compact zero-dimensional topological lattice with respect to some topology.

If, in addition, A is distributive, (i)-(iv) are equivalent to

(v) A is isomorphic to the lattice of upper sets of some poset.

(vi) A is a completely distributive, algebraic lattice.

We can now characterize those bounded distributive lattices whose dual spaces are topological lattices. We recall ([2,13]) that L is said to be $\underset{\sim}{D}$-catalytic if $\underset{\sim}{D}(L,M)$ is a lattice (with respect to the pointwise order) for any $M \in \underset{\sim}{D}$.

4.2 THEOREM [16] Let $L \in \underset{\sim}{D}$. Then the following are equivalent:

(i) The dual space of L is a topological lattice.

(ii) L satisfies the following conditions:

(C1) Every nonzero element of L is a finite join of elements in J(L).

(C2) Every nonidentity element of L is a finite meet of elements in M(L).

(C3) J(L) is a $\wedge$-subsemilattice of L.

(C4) M(L) is a $\vee$-subsemilattice of L.

(C5) $1 \in J(L)$, $0 \in M(L)$.

(iii) L is $\underset{\sim}{D}$-catalytic.

(iv) For some poset Q, L is a retract of $F\underset{\sim}{D}(Q)$, the $\underset{\sim}{D}$-object freely generated by Q.

The equivalence of these statements can be proved in a variety of ways. Obviously (i) $\Rightarrow$ (ii) is a consequence of Theorem 4.1 and Proposition 3.2 [applied to L and L^{op}; see Lemma 2.2(iv)]. The equivalence of (ii) and (iii) was first established by Balbes [2]. The implications (ii), (iii) ⇒ (i), and (i) ⇒ (ii) are proved in [16]. A direct proof of (i) ⇔ (iii) proceeds along the following lines. For $L,M \in \underset{\sim}{D}$, each of $\underset{\sim}{D}(L,M)$, $\underset{\sim}{P}(X_M,X_L)$ carries a natural (pointwise) order; these hom-sets are in fact order isomorphic. If X_L is a topological lattice, $\underset{\sim}{D}(L,M) \cong \underset{\sim}{P}(X_M,X_L)$ is certainly a lattice. Conversely, if L is catalytic, then $X_L = \underset{\sim}{D}(L,\underline{2})$ is a lattice, and it remains to show that the lattice operations are continuous. The way to accomplish this is to utilize the lattice structure of $\underset{\sim}{P}(X_M,X_L)$ for some specific M. Details of this approach will be given in a forthcoming paper with Davey dealing with hom-set lattices more widely.

A $\underset{\sim}{D}$-object L is freely generated by a poset precisely when its dual space is a distributive topological lattice; in this case, L is freely generated by $J(L) \cap M(L)$ and has the ring of upper sets of $J(L) \cap M(L)$ (in the interval topology) as its dual space (see Theorem 4.1(v); a detailed proof appears in [16]. Since retracts of catalytic lattices are catalytic (see [13]), (iv) ⇒ (iii). In [16] it is shown that if A is a linked bialgebraic lattice, then A is a $\underset{\sim}{P}$-retract of a topological lattice $B \in \underset{\sim}{P}$; B is taken to be the lattice of Lawson-closed upper sets of A, ordered by reverse inclusion, and the maps $x \mapsto \uparrow x$ (from A to B) and $V \mapsto \inf V$ (from B to A) set up the retraction (see [10], p.289). A direct proof of (i) ⇔ (iv) has been obtained independently by G. Gierz.

We conclude with some comments on the elusive (weak) projectives in $\underset{\sim}{D}$: the retracts of free objects in $\underset{\sim}{D}$ ([3,4]). Balbes [2] has shown that every $\underset{\sim}{D}$-projective object is $\underset{\sim}{D}$-catalytic [this follows also from (iv) ⇒ (iii) in Theorem 4.2]. The converse fails in general (no uncountable chain is $\underset{\sim}{D}$-projective), but holds for countable lattices [2]. A finite distributive lattice is $\underset{\sim}{D}$-projective if and only if its dual space is a lattice.

Assume L is $\underset{\sim}{D}$-projective. Then all the attributes of lattices whose dual spaces are algebraic lattices in the Lawson topology are possessed by L and L^{op}. In particular, L is the direct limit in $\underset{\sim}{C}$ of finite projective lattices, where $\underset{\sim}{C}$ can be taken to be $\underset{\sim}{D}$, $\underset{\sim}{H}$, or $\underset{\sim}{H}^d$ ($\underset{\sim}{H}$ being the category of Heyting algebras). It is, however, in general not true that L is the direct limit in the category $\underset{\sim}{H} \cap \underset{\sim}{H}^d$ of double Heyting algebras of finite $\underset{\sim}{D}$-projectives. Clinkenbeard [7] constructs a countable linked bialgebraic lattice with no nontrivial finite quotients under lattice homomorphisms; this lattice is the dual space of a distributive lattice that is countable and catalytic, and hence projective, but that has no nontrivial finite double Heyting subalgebras.

A lattice L is $\underset{\sim}{D}$-projective if and only if its dual space X_L is a $\underset{\sim}{P}$-retract of $\underline{2}^{\kappa}$ for some κ. This certainly occurs if X_L is a $\underset{\sim}{Z}$-retract of some $\underline{2}^{\kappa}$, that is, if X_L is $\underset{\sim}{Z}$-injective. Such $\underset{\sim}{Z}$-objects are characterized in [12], III.3.4 as those distributive topological lattices in which $\downarrow k$ is finite for each compact element k. Thus included among the $\underset{\sim}{D}$-projectives are those $L \in \underset{\sim}{D}$ that are freely generated by $Q = J(L) \cap M(L)$ and are such that, for each finite subset F of Q, there are only finitely many subsets G of Q with $G \subseteq \uparrow F$. These observations amplify remarks made by Balbes and Horn [4] (before Theorem 10).

REFERENCES

1. M. E. Adams, The poset of prime ideals of a distributive lattice, Algebra Univ. 5 (1975), 141-142.

2. R. Balbes, Catalytic distributive lattices, Algebra Univ. 11 (1980), 334-340.

3. R. Balbes and Ph. Dwinger, Distributive Lattices, University of Missouri Press, Columbia, Missouri (1974).

4. R. Balbes and A. Horn, Projective distributive lattices, Pacific J. Math. 33 (1970), 273-279.

5. B. Banaschewski, Coherent frames, Lecture Notes in Mathematics 871, pp. 1-11, Springer-Verlag, Berlin-Heidelberg-New York (1981).

6. W. G. Bowen, Lattice Theory and Topology, D. Phil. thesis, University of Oxford (1981).

7. D. J. Clinkenbeard, Simple compact topological lattices, Algebra Univ. 9 (1979), 322-328.

8. B. A. Davey, On the lattice of subvarieties, Houston J. Math. 5 (1979), 183-192.

9. M. Erné, Separation axioms for interval topologies, Proc. Amer. Math. Soc. 79 (1980), 185-190.

10. G. Gierz, K. H. Hofmann, K. Keimel, J. D. Lawson, M. Mislove, and D. S. Scott, A Compendium of Continuous Lattices, Springer-Verlag, Berlin-Heidelberg-New York (1980).

11. G. Gierz and J. D. Lawson, Generalized continuous and hypercontinuous lattices, Rocky Mountain J. Math. 11 (1981), 271-296.

12. K. H. Hofmann, M. Mislove, and A. Stralka, The Pontryagin Duality of Compact 0-Dimensional Semilattices and Its Applications, Lecture Notes in Mathematics 396, Springer-Verlag, Berlin-Heidelberg-New York (1974).

13. T. G. Kucera and B. Sands, Lattices of lattice homomorphisms, Algebra Univ. 8 (1978), 180-190.

14. H. A. Priestley, Representation of distributive lattices by means of ordered Stone spaces, Bull. London Math. Soc. 2 (1970), 186-190.

15. H. A. Priestley, Ordered topological spaces and the representation of distributive lattices, Proc. London Math. Soc. (3) 24 (1972), 507-530.

16. H. A. Priestley, Catalytic distributive lattices and compact zero-dimensional topological lattices, Algebra Univ. 19 (1984), 322-329.

17. H. A. Priestley, Ordered sets and duality for distributive lattices, in Ordres, descriptions et vôles (M. Pouzet and D. Richard, eds.), pp. 39-60, Annals of Discrete Mathematics, North-Holland, Amsterdam, 1984.

18. T. P. Speed, Profinite posets, Bull. Austral. Math. Soc. 6 (1972), 177-183.

19. B. A. Davey and H. A. Priestley, Lattices of homomorphisms, J. Austral. Math. Soc. A (to appear).

15

Scott Topology, Isbell Topology, and Continuous Convergence

FRIEDHELM SCHWARZ* and SIBYLLE WECK*

Mathematical Institute
Universität Hannover
Hannover, Federal Republic of Germany

1. INTRODUCTION

In his fundamental paper on continuous lattices, Scott conjectured [1972, p. 133]: "Concerning function spaces there ought to be some connections with the limit spaces of Cook and Fischer..., but these are rather vague ideas." Indeed, in many important cases – in particular, under his assumption that X and Y are continuous lattices – the Scott topology on Top(X,Y) coincides with the limitierung of continuous convergence c(X,Y), i.e., the connection is equality [Proposition 2.10, Theorem 2.16(1,4)]. It is the aim of this paper to compare the Scott topology and other related function space topologies with the limitierung of continuous convergence and to make the intrinsic connections to well-known results of J. Isbell and B. Banaschewski clearly visible.

The main point about the limitierung of continuous convergence is: whenever it is a topology, then it is the "right" function space topology [see, for example, Proposition 2.4 and Schwarz (1982b, 1983)].

In this context, it is reasonable to consider the Scott topology of a complete lattice in terms of filter convergence, i.e., as the topological modification of the Scott convergence. A first basic observation is that on the lattice Top(X,$\overline{2}$) (where $\overline{2}$ denotes the Sierpinski space), continuous convergence always coincides with the order-theoretically defined Scott convergence [Proposition 2.12(1)], hence with the Scott topology whenever c(X,$\overline{2}$) is a topology. While in that case, $\mathrm{Top}_c(X,\cdot)$: Top → Lim (limit spaces) is an internal hom-functor of Top, the process of endowing Top(X,Y)

*Current affiliation: University of Toledo, Toledo, Ohio

with the Scott topology cannot be made functorial [Remark 2.11(5)]. So we replace the Scott topology by a closely related topology, here called the Isbell topology is(X,Y), which is functorial and coincides in essential cases with the Scott topology (Proposition 2.10). The functor $\mathrm{Top}_c(X,\cdot)$ equals $\mathrm{Top}_{is}(X,\cdot)$ if and only if the open sets of X form a continuous lattice. Further equivalent conditions, e.g., that $c(X,\overline{2})$ is a topology, are contained in the characterization theorem 2.16.

Scott observed [1972, 3.3] that in case of continuous lattices X and Y, the Scott topology on Top(X,Y) coincides with the topology of pointwise convergence pw(X,Y). By an order-theoretical description of $pw(X,\overline{2})$, one obtains that the equality of the functors $\mathrm{Top}_c(X,\cdot)$ and $\mathrm{Top}_{pw}(X,\cdot)$ is equivalent to the lattice of open sets of X being hypercontinuous. More characterizations are collected in Theorem 3.4, which is the analog of Theorem 2.16 where the Isbell topology is replaced by the topology of pointwise convergence.

The main connections between the function space topologies considered are summed up in Diagram 3.8 at the end of this paper.

2. ISBELL TOPOLOGY AND CONTINUOUS CONVERGENCE

Our main working categories are the limit spaces and the topological spaces. For the convenience of the reader, we record first some basic notions and facts on convergence categories, in particular, concerning the limitierung of continuous convergence.

2.1 DEFINITION [cf. Choquet (1948), Kowalsky (1954), Fischer (1959), Kent (1964)]. For any set X let $\mathcal{P}(X)$ denote the power set of X and $\mathbb{F}(X)$ the set of all proper filters on X.

(1) A function $\mathbb{K}$: $X \to \mathcal{P}(\mathbb{F}(X))$ is called a convergence structure on X and $(X,\mathbb{K})$ a convergence space iff for each $x \in X$, $\mathbb{K}(x)$ is an upper set of $(\mathbb{F}(X),\subset)$ containing the point ultrafilter $\dot{x}$. [The set $\mathbb{K}(x)$ may be considered as the set of all filters converging to x.] Instead of $\mathfrak{F} \in \mathbb{K}(x)$, we often use the more illustrative notations $\mathfrak{F} \xrightarrow[(X,\mathbb{K})]{} x$ or $\mathfrak{F} \xrightarrow[\mathbb{K}]{} x$, or simply $\mathfrak{F} \to x$. A function f: $(X,\mathbb{K}) \to (Y,\mathbb{L})$ between convergence spaces is said to be continuous iff $\mathfrak{F} \to x$ implies $f(\mathfrak{F}) \to f(x)$ [where $f(\mathfrak{F})$ denotes the filter generated by $\{f(F) \mid F \in \mathfrak{F}\}$]. The category of convergence spaces and continuous maps is called FCo.

(2) A convergence structure $\mathbb{K}$ on X is called a

<u>Limitierung</u> iff for each $x \in X$, $\mathbb{K}(x)$ is a dual ideal of $(\mathbb{F}(X),\subset)$.

<u>Pseudotopology</u> iff $\mathfrak{F} \in \mathbb{K}(x)$ whenever $\mathcal{U} \in \mathbb{K}(x)$ for each ultrafilter $\mathcal{U} \supset \mathfrak{F}$.

<u>Pretopology</u> iff for each $x \in X$, $\mathbb{K}(x)$ is a principal dual ideal of $(\mathbb{F}(X),\subset)$.

The corresponding subcategories of FCo are denoted by Lim, Pst, and Prt, respectively.

Obviously, every topological space may be considered as a convergence space. In the chain Top $\subset$ Prt $\subset$ Pst $\subset$ Lim $\subset$ FCo, every category is topological and properly contained as a bireflective subcategory in the following ones. On the set of all convergence structures on X a partial order is given by

$$\mathbb{K} \leq \mathbb{L} \quad \text{iff} \quad 1_X\colon (X,\mathbb{K}) \to (X,\mathbb{L}) \text{ is continuous}$$

We say $\mathbb{K}$ is smaller than $\mathbb{L}$ or $\mathbb{L}$ is larger than $\mathbb{K}$ (notice that this terminology includes equality). Then the Top-reflection of a convergence space $(X,\mathbb{K})$ – often called the "topological modification of $(X,\mathbb{K})$" and here denoted by $(X,\mathbb{K})^{\mathrm{Top}} = (X,\mathbb{K}^{\mathrm{Top}})$ – is given by X, endowed with the smallest topology larger than $\mathbb{K}$.

2.2 DEFINITION [Cook and Fischer (1965, theorem 1)] For $X,Y \in$ Lim, the <u>limitierung of continuous convergence</u> $c(X,Y)$ on $\mathrm{Lim}(X,Y)$ is defined by

$$\mathcal{G} \xrightarrow[c(X,Y)]{} f \quad \text{iff} \quad \mathfrak{F} \xrightarrow[X]{} x \Rightarrow \mathcal{G}(\mathfrak{F}) = \mathrm{ev}(\mathcal{G} \times \mathfrak{F}) \xrightarrow[Y]{} f(x)$$

for $f \in \mathrm{Lim}(X,Y)$, $\mathcal{G} \in \mathbb{F}(\mathrm{Lim}(X,Y))$. [Here $\mathrm{ev}\colon \mathrm{Lim}(X,Y) \times X \to Y$, $(f,x) \mapsto f(x)$, denotes the usual evaluation map.] In case of $X,Y \in$ Top, this reduces to

$$\mathcal{G} \xrightarrow[c(X,Y)]{} f \quad \text{iff} \quad \forall\, x \in X \;\; \mathcal{G}(\mathcal{U}(x)) \supset \mathcal{U}(f(x))$$

[where $\mathcal{U}(z)$ denotes the neighborhood filter of z with respect to the topology in question]. Instead of $(\mathrm{Lim}(X,Y),c(X,Y))$, we use the notation $\mathrm{Lim}_c(X,Y)$.

In analogy to the abbreviation $\mathrm{Lim}_c(X,Y)$, we write $(\mathrm{FCo}(X,Y),r(X,Y)) = \mathrm{FCo}_r(X,Y)$ if $r(X,Y)$ is a convergence structure on $\mathrm{FCo}(X,Y)$ depending on

X and Y. In case that X,Y ∈ Top, we have FCo(X,Y) = Top(X,Y) and we use the notation $Top_r(X,Y)$. [Note that in general, r(X,Y) need not be a topology.]

The importance of the limitierung of continuous convergence within Lim is evident from the following well-known fact:

2.3 PROPOSITION [Cook and Fischer (1965, theorem 5)] Lim is cartesian closed; for each X ∈ Lim, the functor $Lim_c(X,\cdot)$ is right adjoint to • x X.

Moreover, the limitierung of continuous convergence plays an important role also in connection with topological questions: Top and Prt are not cartesian closed; in fact, in a certain sense there are only a few cartesian closed subcategories of Prt. Hence it becomes an interesting problem to characterize the exponentiable objects, i.e., those objects X such that • x X has a right adjoint. It turns out that whenever • x X has a right adjoint, it is given by $Lim_c(X,\cdot)$.

2.4 PROPOSITION Let A be an epireflective subcategory of Prt.

(1) A is cartesian closed iff A contains only indiscrete spaces (i.e., A = indiscrete spaces or A = indiscrete T_0-spaces [Schwarz (1982a, 3.3; 1982b, 3.4)].

(2) For X ∈ A, the following are equivalent [Schwarz (1983, 5.1)].
 (a) X is exponentiable in A.
 (b) $Lim_c(X,\cdot)$ is right adjoint to • x X.
 (c) For all Y ∈ A, $Lim_c(X,Y)$ ∈ A.

In Top, • x X has a right adjoint iff for all Y ∈ Top, there is a topology t on Top(X,Y) such that for all W ∈ Top, E: Top(W x X, Y) → Top(W,(Top(X,Y),t)), f ↦ f^* [where $f^*(w)(x) = f(w,x)$ for all w ∈ W, x ∈ X] is a bijection. This gives rise to the following definition 2.5.

NOTE: For the rest of this paper, we assume X,Y ∈ Top, if not specified otherwise.

2.5 DEFINITION [Dugundji (1966), XII-10.1]. A topology t on Top(X,Y) is called

(1) Splitting iff for all W ∈ Top, E: Top(W x X,Y) → Top(W,(Top(X,Y),t)) is a function (i.e., iff f^* is continuous whenever f is).

(2) Conjoining iff for all W ∈ Top, E^{-1}: Top(W,(Top(X,Y),t)) → Top(W x X,Y) is a function (i.e., iff f is continuous whenever f^* is).

Equivalently, t is conjoining iff the evaluation map ev: $(\mathrm{Top}(X,Y),t) \times X \to Y$ is continuous.

There is a nice characterization of splitting and conjoining topologies via continuous convergence [cf. Arens and Dugundji (1951,2.4,2.5)].

2.6 LEMMA Let t be a topology on Top(X,Y).

(1) t is conjoining iff $t \leq c(X,Y)$.

(2) t is splitting iff $c(X,Y) \leq t$.

Proof: (1) c(X,Y) is the largest limitierung on Lim(X,Y) = Top(X,Y) that renders the evaluation map continuous.

(2) Of course, t is splitting whenever $c(X,Y) \leq t$ (Proposition 2.3). Since every splitting topology on Top(X,Y) is larger than every conjoining one, we have for splitting t:

$$t \geq \sup_{\mathrm{Lim}} \{r \mid r \text{ conjoining topology on } \mathrm{Top}(X,Y)\}$$
$$= \sup_{\mathrm{Lim}} \{r \mid r \text{ topology on } \mathrm{Top}(X,Y) \text{ with } r \leq c(X,Y)\} = c(X,Y)$$

by (1) and Machado [1973, 1.2]. ∎

Let $\overline{2}$ denote Sierpinski space $(\{0,1\},\{\emptyset,\{1\},\{0,1\}\})$. By Proposition 2.3, $(\mathrm{Top}(X,h)\colon \mathrm{Top}_c(X,Y) \to \mathrm{Top}_c(X,\overline{2}) \mid h \in \mathrm{Top}(Y,\overline{2}))$ is an initial source in Lim. If $c(X,\overline{2})$ is a topology, this is an initial source in Top (because Top is bireflective in Lim). Analogously, initial sources in Top may be constructed by starting from an arbitrary topology on $\mathrm{Top}(X,\overline{2})$.

2.7 LEMMA Let t be a topology on $\mathrm{Top}(X,\overline{2})$. Let t(X,Y) be the initial topology on Top(X,Y) with respect to $\{\mathrm{Top}(X,h)\colon \mathrm{Top}(X,Y) \to (\mathrm{Top}(X,\overline{2}),t) \mid h \in \mathrm{Top}(Y,\overline{2})\}$.

(1) $t(X,\overline{2}) = t$.

(2) $\mathrm{Top}^t(X,\cdot)\colon \mathrm{Top} \to \mathrm{Top}$ with $\mathrm{Top}^t(X,Y) = (\mathrm{Top}(X,Y),t(X,Y))$ is a functor [Isbell (1975b,1.4)].

(3) If t is a T_0-topology, then $\mathrm{Top}^t(X,\cdot)$ preserves the T_0-property.

Proof: (1) is obvious.

(2): Let $Y_1,Y_2 \in \mathrm{Top}$, g: $Y_1 \to Y_2$ continuous. Then $\mathrm{Top}(X,h) \circ \mathrm{Top}(X,g) = \mathrm{Top}(X,h \circ g)\colon \mathrm{Top}^t(X,Y_1) \to \mathrm{Top}^t(X,\overline{2})$ is continuous for each $h \in \mathrm{Top}(Y_2,\overline{2})$.

(3): For any T_0-space Y, $\{h \mid h \in \mathrm{Top}(Y,\overline{2})\}$ separates points; hence $\{\mathrm{Top}(X,h)\colon \mathrm{Top}^t(X,Y) \to (\mathrm{Top}(X,\overline{2}),t) \mid h \in \mathrm{Top}(Y,\overline{2})\}$ is a point-separating initial source, i.e., $\mathrm{Top}^t(X,Y)$ is a T_0-space. ∎

Improving on ideas of Wilker (1970), Isbell (1975b, 1.4) pointed out the importance of the functors $\mathrm{Top}^t(X,\cdot)$: He showed that every right adjoint endofunctor of Top is of this form, and – identifying $\mathrm{Top}(X,\overline{2})$ with the lattice $\mathcal{O}(X)$ of open sets of X – he proved that $\mathrm{Top}^t(X,\cdot)$ is a right adjoint iff t is a <u>topological topology</u> on $\mathcal{O}(X)$, i.e., iff t makes arbitrary union and finite intersection continuous operations.

In the sequel, we will consider mainly one special case of the functor defined in Lemma 2.7, namely, that t is the Scott topology. We give here a definition of the Scott topology in terms of (filter) convergence [cf. <u>A Compendium of Continuous Lattices</u> (1980, II-1.1)]; this has the big advantage that later on we are able to compare Scott convergence and the Scott topology with continuous convergence (Proposition 2.12). For a definition of the Scott topology in terms of open or closed sets, cf. Scott (1972, p. 102) and the <u>Compendium</u> (1980, II-1.2,1.3,1.4). A first definition of the Scott topology on a particular complete lattice, that of open sets of a topological space, was given by Day and Kelly (1970, theorem 3) when they characterized those topological spaces X such that $\cdot \times X$ preserves quotient maps.

2.8 DEFINITION Let L be a complete lattice. Then

$$\mathfrak{F} \xrightarrow[s(L)]{} x \quad \text{iff} \quad x \le \lim\inf \mathfrak{F} = \sup\{\inf F \mid F \in \mathfrak{F}\}$$

defines a convergence structure s(L) on L, called the <u>Scott convergence</u> of L. Its topological modification $s(L)^{\mathrm{Top}}$ is called the <u>Scott topology</u> of L and denoted by $\sigma(L)$. For any T_0-space Y, the <u>specialization order</u> of Y defined by

$$x \le y \quad \text{iff} \quad x \in \mathrm{cl}\,\{y\} \qquad x,y \in Y$$

is a partial order on Y. In case Top(X,Y) endowed with the pointwise order (determined by the specialization order of Y) is a complete lattice, we denote its Scott convergence and Scott topology by s(X,Y) and $\sigma(X,Y)$, respectively.

We warn the reader that Scott convergence does not mean convergence in the Scott topology: In general, Scott convergence is not even a limitierung [Weck (1981, 2.4)]; it is a limitierung iff L is meet-continuous [Weck (1981, 2.5(b,d)], and it is a (pre)topology iff L is a continuous lattice [<u>Compendium</u> (1980, II-1.8); Weck (1981, 2.7(b,d)].

2.9 DEFINITION For the special case $t = \sigma(X,\overline{2})$, the topology $t(X,Y)$ formed as in Lemma 2.7 is denoted by $is(X,Y)$ and called the <u>Isbell topology</u> of $Top(X,Y)$. Instead of $Top^{\sigma(X,\overline{2})}(X,Y)$, we write $Top_{is}(X,Y)$.

A subbase of $is(X,Y)$ is given by $\{\langle H,V\rangle \mid H \in \sigma(\mathcal{O}(X))$ and $V \in \mathcal{O}(Y)\}$ where $\langle H,V\rangle = \{f \in Top(X,Y) \mid f^{-1}(V) \in H\}$, as was mentioned by Lambrinos (1981, p. 10).

By Lemma 2.7(1), the Scott topology of $Top(X,\overline{2})$ equals the Isbell topology. Proposition 2.10, which is an application of Isbell (1975b, 2.3), shows that these topologies coincide in many important cases, in particular, if X and Y are continuous lattices in their Scott topologies [cf. Scott (1972, 3.3)].

X is called <u>core compact</u> [Hofmann and Lawson (1978, 4.1)] (or quasi-locally compact [Ward (1969)], semilocally bounded [Isbell (1975a,b)], a CL-space [Hofmann (1977, section 1)] iff for every open set U and every point $x \in U$, there exists an open set V with $x \in V \subset U$ such that every open cover of U contains finitely many members covering V.

Recall that a topological space X is injective in Top iff its T_0-reflection X^{T_0Top} is injective in T_0-Top [Wyler (1977, proposition 2)]. In particular, the T_0-Top-injectives are just the Top-injective T_0-spaces.

2.10 PROPOSITION If X is a core compact space and Y is an injective T_0-space (i.e., a continuous lattice in its Scott topology [Scott (1972, 2.12)]), then $Top_\sigma(X,Y) = Top_{is}(X,Y)$.

Proof: By the first part of the proof of [Isbell (1975b, 2.3)], $Top_{is}(X,Y)$ is an injective T_0-space. Hence $Top(X,Y)$ is a continuous lattice in the specialization order determined by $is(X,Y)$, and $is(X,Y)$ is the Scott topology of this lattice. But the specialization order with respect to $is(X,Y)$ is just the pointwise order with respect to the specialization order of Y, because X is core compact. ∎

2.11 REMARK (1) Isbell showed that $Top_{is}(X,Y)$ is injective whenever X is core compact and Y is an injective topological space [Isbell (1975b, 2.3)]. Under these assumptions, Proposition 2.10 suggests the Isbell topology as a reasonable extension of the Scott topology on function spaces for non-T_0 Y.

(2) Let $\overline{1}$ denote the singleton topological space. Since $Top_{is}(\overline{1},\overline{2}) \cong \overline{2}$ and $(Top(\overline{1},h): Top_{is}(\overline{1},Y) \to Top_{is}(\overline{1},\overline{2}) \mid h \in Top(Y,\overline{2}))$ is an initial source in Top, we obtain: $Y \cong Top_{is}(\overline{1},Y)$.

(3) Let $Y = ([0,1],t)$ where t is the A(lexandroff)-discrete topology consisting of all upper sets of $[0,1]$. Then $\mathrm{Top}(\bar{1},Y)$ is a complete lattice (isomorphic to $[0,1]$). Since $Y \neq ([0,1],\sigma([0,1]))$, we have $\mathrm{Top}_\sigma(\bar{1},Y) \neq \mathrm{Top}_{is}(\bar{1},Y)$ by (2). This simple example shows that in general, Scott topology and Isbell topology on $\mathrm{Top}(X,Y)$ do not coincide. (So Lambrinos' (1981, p.10) former use of the terminology the "Scott topology" on $\mathrm{Top}(X,Y)$ for $is(X,Y)$ was contradictory; accordingly, he now also uses the name "Isbell topology" [Lambrinos and Papadopoulos (1985, text preceding 2.1)].)

(4) There are several attempts to extend the definition of the Scott topology to arbitrary posets [Scott (1972, p. 102), Erné (1981, Section 2), Weck (1981, Section 3)]. Of course, each of these definitions coincides with the usual one in case of complete lattices. Thus (3) shows that, in general, on function spaces none of these Scott topologies equals the Isbell topology.

(5) Consider once more the example in (3). Since $is(\bar{1},Y) \subsetneq \sigma(\bar{1},Y)$ and $is(\bar{1},Y)$ is the largest topology on $\mathrm{Top}(\bar{1},Y)$ that makes all maps $\mathrm{Top}(\bar{1},h)\colon \mathrm{Top}(\bar{1},Y) \to \mathrm{Top}_{is}(\bar{1},\bar{2})$, $h \in \mathrm{Top}(Y,\bar{2})$, continuous, there is a continuous $h\colon Y \to \bar{2}$ such that $\mathrm{Top}(\bar{1},h)\colon \mathrm{Top}_\sigma(\bar{1},Y) \to \mathrm{Top}_\sigma(\bar{1},\bar{2})$ is not continuous. Consequently, it is not possible to extend the definition of the Scott topology to posets in a way that $\mathrm{Top}_\sigma(X,\cdot)\colon T_0\mathrm{Top} \to T_0\mathrm{Top}$ becomes a functor.

Of course, the unsatisfactory behavior of the Scott topology on function spaces results from the obvious fact that different T_0-topologies on a set may determine the same specialization order, hence the same Scott topology on the induced poset.

We want now to compare the limitierung of continuous convergence with the Isbell topology, the Scott topology, and a few other well-known topologies on function spaces. [Connections of the Isbell topology to further function space topologies, such as the bounded-open topology, bounding topology, etc. – but not to continuous convergence – are investigated by Lambrinos and Papadopoulos (1984, Section 2).] Let us begin our considerations with the particular function space $\mathrm{Top}(X,\bar{2})$. The fact that in this case Scott convergence coincides with continuous convergence will play a key role subsequently.

2.12 PROPOSITION (1) $s(X,\bar{2}) = c(X,\bar{2})$.

(2) $s(X,\bar{2})$ is a pseudotopology.

(3) $\mathrm{Top}_{is}(X,\bar{2}) = \mathrm{Top}_\sigma(X,\bar{2}) = \mathrm{Top}_s(X,\bar{2})^{\mathrm{Top}} = \mathrm{Top}_c(X,\bar{2})^{\mathrm{Top}}$.

Proof: (1) Let $f \in \mathrm{Top}(X,\bar{2})$, $\mathcal{H}$ a filter on $\mathrm{Top}(X,\bar{2})$. Then $\mathcal{H} \xrightarrow[s(X,\bar{2})]{} f$ iff $f(x) = 1$ implies $(\sup\{\inf H \mid H \in \mathcal{H}\})(x) = 1$. Now $(\sup\{\inf H \mid H \in \mathcal{H}\})(x) = 1$ means that there is a member $H \in \mathcal{H}$ such that $x \in \operatorname{int}(\bigcap\{h^{-1}(1) \mid h \in H\})$, i.e., there are $H \in \mathcal{H}$ and $U \in \mathcal{U}(x)$ with $U \subset \bigcap\{h^{-1}(1) \mid h \in H\}$. Hence $\mathcal{H} \xrightarrow[s(X,\bar{2})]{} f$ iff $f(x) = 1$ implies the existence of $H \in \mathcal{H}$, $U \in \mathcal{U}(x)$ with $H(U) = \{1\}$. On the other hand, $\mathcal{H} \xrightarrow[c(X,\bar{2})]{} f$ iff for each $x \in X$, $\mathcal{H}(\mathcal{U}(x)) \supset \mathcal{U}(f(x))$, i.e., iff $f(x) = 1$ implies $\mathcal{H}(\mathcal{U}(x)) = \dot{1}$, and the latter means also that $f(x) = 1$ implies the existence of $H \in \mathcal{H}$, $U \in \mathcal{U}(x)$ with $H(U) = \{1\}$.

(2) Every topology is a pseudotopology, and $c(X,Y)$ is a pseudotopology whenever Y is [Machado (1973, 3.7)].

(3) Clear. ■

Erné (1982, p. 42) asks whether a complete lattice is already continuous whenever its Scott convergence is a pseudotopology. This is answered negatively by Proposition 2.12(2): counterexamples are given by the lattices $\mathcal{O}(X)$ where X is not core compact [cf. Theorem 2.16(1,2)].

Denote by $co(X,Y)$ the compact-open topology on $\mathrm{Top}(X,Y)$; a subbase of $co(X,Y)$ is given by $\{(K,V) \mid K \subset X$ compact and V open in $Y\}$ where $(K,V) = \{f \in \mathrm{Top}(X,Y) \mid f(K) \subset V\}$. (Note that in our terminology, compactness does not include the Hausdorff property.) The topology of pointwise convergence on $\mathrm{Top}(X,Y)$, i.e., the topology induced by the usual product topology, is denoted $pw(X,Y)$ and called the pointwise topology. It follows from Proposition 3.2(2) that $\mathrm{Top}_{pw}(X,Y) = \mathrm{Top}^t(X,Y)$ for $t = pw(X,\bar{2})$ in Lemma 2.7.

2.13 PROPOSITION (1) $c(X,Y) \le c(X,Y)^{\mathrm{Top}} \le is(X,Y) \le co(X,Y) \le pw(X,Y)$.

(2) $c(X,Y)^{\mathrm{Top}}$ is the smallest splitting topology on $\mathrm{Top}(X,Y)$ [Schwarz (1983, 4.9(1))].

(3) The Isbell topology, the compact-open topology, and the pointwise topology are always splitting.

Proof: (1) $(\mathrm{Top}(X,h)\colon \mathrm{Top}_c(X,Y)^{\mathrm{Top}} \to \mathrm{Top}_c(X,\bar{2})^{\mathrm{Top}} \mid h \in \mathrm{Top}(Y,\bar{2}))$ is a source in Top (Proposition 2.3); $(\mathrm{Top}(X,h)\colon \mathrm{Top}_{is}(X,Y) \to \mathrm{Top}_{is}(X,\bar{2}) \mid h \in \mathrm{Top}(Y,\bar{2}))$ is an initial source in Top. Since $c(X,\bar{2})^{\mathrm{Top}} = is(X,\bar{2})$ by Proposition 2.12(3), we obtain $c(X,Y)^{\mathrm{Top}} \le is(X,Y)$.

Let K be a compact subset of X, V open in Y. Define $H = \{U \in \mathcal{O}(X) \mid K \subset U\}$. Then $H \in \sigma(\mathcal{O}(X))$ by the Compendium (1980, II-1.2), and $(K,V) = \langle H,V\rangle$ (cf. text following Definition 2.9). Thus $is(X,Y) \le co(X,Y)$.

$\{(\{x\},V) \mid x \in X$ and $V \in \mathcal{O}(Y)\}$ constitutes a subbase for $pw(X,Y)$, so $co(X,Y) \le pw(X,Y)$.

(2) Lemma 2.6(2).

(3) By (1) above and Lemma 2.6(2). ∎

In the sequel, we will mainly investigate under what conditions continuous convergence coincides with one of the topologies of Proposition 2.13(1). Of course, first of all, it has to be a topology. In this connection, initially dense classes are of some importance. A subclass $\underline{D}$ of Top is called initially dense in Top iff for every $Y \in$ Top, there is an initial source $(f_i: Y \to D_i \mid i \in I)$ with all D_i in $\underline{D}$. The most important example is given by $\underline{D} = \{\overline{2}\}$; indeed, every initially dense class can be characterized by $\overline{2}$.

2.14 PROPOSITION A class $\underline{D} \subset$ Top is initially dense in Top iff there is a $D \in \underline{D}$ that contains $\overline{2}$ as a subspace. In particular, every nonsingleton complete lattice in its Scott topology, hence every nonsingleton injective T_0-space, constitutes an initially dense class in Top.

Proof: If $\underline{D}$ is initially dense in Top, then $(f: \overline{2} \to Y \mid Y \in \underline{D}$ and $f \in \text{Top}(\overline{2},Y))$ is an initial source. Hence there exist a space $D \in \underline{D}$ and a nonconstant function $f \in \text{Top}(\overline{2},D)$ such that the subspace $D_f = f(\{0,1\})$ of D is not indiscrete. Since D_f cannot be discrete, it is homeomorphic to $\overline{2}$. The converse implication is obvious. ∎

If, for an initially dense subclass of Top, the functor $\text{Top}_c(X,\cdot)$: Top → Lim is Top-valued, then it is an endofunctor of Top.

2.15 PROPOSITION [Schwarz (1983, 5.1)] Let $\underline{D}$ be an initially dense class in Top. Then the following conditions are equivalent:

(1) For all $Y \in \underline{D}$, $\text{Top}_c(X,Y) \in$ Top.

(2) For all $Y \in$ Top, $\text{Top}_c(X,Y) \in$ Top.

We are now able to give various characterizations of the spaces X with the property that the Isbell topology coincides with the limitierung of continuous convergence on Top(X,Y) for all $Y \in$ Top. The equivalence of conditions (2) and (3) is known from Schwarz (1983, 6.5); application of Proposition 2.12(1) provides a new proof which is much more transparent. Conditions (8)-(11) are based on Isbell (1975b, 2.3).

2.16 THEOREM Let $\underline{D}$ be an initially dense class in Top. Let I be a nonsingleton injective T_0-space, e.g., $I = \overline{2}$. The following conditions are equivalent:

(1) X is core compact.

(2) $\mathcal{O}(X)$ is a continuous lattice.

(3) For all $Y \in \underline{D}$, $Top_c(X,Y) \in Top$.

(4) For all $Y \in \underline{D}$, $Top_{is}(X,Y) = Top_c(X,Y)$.

(5) For all $Y \in \underline{D}$, is(X,Y) is conjoining, i.e., ev: $Top_{is}(X,Y) \times X \to Y$ is continuous.

(6) For all $W \in Top$, $Y \in \underline{D}$, the natural map E: $Top(W \times X,Y) \to Top(W,Top_{is}(X,Y)$ is a bijection.

(7) There is a conjoining topological topology on $Top(X,\overline{2}) \cong \mathcal{O}(X)$ [namely, $\sigma(X,\overline{2}) = is(X,\overline{2}) = c(X,\overline{2})$].

(8) $Top_{is}(X,\cdot)$ preserves Top-injectives.

(9) $Top_{is}(X,\cdot)$ preserves T_0Top-injectives.

(10) $Top_{is}(X,I)$ is injective in Top (T_0Top).

(11) $Top_\sigma(X,I)$ is injective in Top (T_0Top).

Proof: (1) $\Leftrightarrow$ (2). Hofmann and Lawson [1978, 4.2(1,3)].

(2) $\Leftrightarrow$ (3). By the Compendium (1980, II-1.8), $\mathcal{O}(X)$ is continuous iff the Scott convergence of $\mathcal{O}(X)$, i.e., that of $Top(X,\overline{2})$, is a topology. This means $Top_c(X,\overline{2}) \in Top$ by Proposition 2.12(1). Apply Proposition 2.15.

(3) $\Rightarrow$ (4). By (3) and Proposition 2.15, $Top_c(X,\overline{2}) \in Top$. Hence $Top_c(X,\overline{2}) = Top_{is}(X,\overline{2})$ by Proposition 2.12(3). Since is(X,Y) and c(X,Y) are initial with respect to $(Top(X,h): Top(X,Y) \to Top_{is}(X,\overline{2}) = Top_c(X,\overline{2}) \mid h \in Top(Y,2))$ (cf. text before Lemma 2.7), it follows that $Top_{is}(X,Y) = Top_c(X,Y)$.

(4) $\Rightarrow$ (3). Obvious.

(4) $\Rightarrow$ (6). Proposition 2.3.

(6) $\Rightarrow$ (5). Choose $W = Top_{is}(X,Y)$, $f^* = 1_W$.

(5) $\Rightarrow$ (4). By Proposition 2.13(3), is(X,Y) is splitting. Apply Lemma 2.6.

Since we have shown that (1)-(6) are equivalent for arbitrary $\underline{D}$, the equivalence of (7) may be proved in the following way: (6) with $\underline{D}$ = Top $\Rightarrow$ (7): By assumption, $Top_{is}(X,\cdot)$ is right adjoint to the functor $\cdot \times X$. Then Isbell (1975b, 1.4) implies that $is(X,\overline{2})$ is a topological topology. Obviously, $is(X,\overline{2})$ is conjoining [cf. proof of (6) $\Rightarrow$ (5)]. In the proof of Isbell (1975a, 2.2), it is shown that a topology on a complete lattice is coarser than the Scott topology whenever arbitrary joins are continuous. Thus every topological topology on $Top(X,\overline{2})$ is splitting [Proposition 2.12 (3), Lemma 2.6(2)]. Application of Lemma 2.6 and Proposition 2.13(3) yields the expression in brackets.

(7) $\Rightarrow$ (5) with $\underline{D} = \{\overline{2}\}$. Trivial.

(1) $\Rightarrow$ (8). Isbell (1975b, 2.3).

(8) ⇒ (9). Lemma 2.7(3) and Wyler (1977, proposition 2).

(9) ⇒ (10). Obvious.

(10) ⇒ (2). $\overline{2}$ is a subspace of I (cf. Proposition 2.14); hence, being an injective T_0-space, it is a retract of I [Scott (1972, 1.2,1.7)]. By Lemma 2.7(2), $Top_\sigma(X,\overline{2}) = Top_{is}(X,\overline{2})$ is a retract of the injective T_0-space $Top_{is}(X,I)$ and is, consequently, also an injective T_0-space [Scott (1972, 1.4)]. Thus $Top(X,\overline{2})$ is a continuous lattice in its specialization order which coincides with the pointwise order.

(1) ⇒ (11). Apply (10) above and Proposition 2.10.

(11) ⇒ (2). Top(X,I) and $Top(X^{T_0Top},I)$ – each endowed with the pointwise order – as well as $\mathcal{O}(X)$ and $\mathcal{O}(X^{T_0Top})$ are order isomorphic. Apply the Compendium (1980, II-4.7). ■

Further equivalent conditions for core compactness may be found in Day and Kelly (1970, theorem 3), Hofmann and Lawson (1978, 4.2,4.5), the Compendium (1980, II-4.10, V-5.10), and Schwarz (1982b) 3.3).

In Theorem 2.16, the core-compact spaces are characterized by properties of the Isbell topology. Lambrinos and Papadopoulos (1985, 3.4) obtained analogous characterizations of the larger class of locally bounded spaces [Lambrinos (1975, def. 1)] via the "strong Isbell topology." This function space topology arises from their "strong (Day-Kelly-) Scott topology" on $\mathcal{O}(X) \cong Top(X,\overline{2})$ by the initiality construction in Lemma 2.7.

From the result that for core-compact X, the Isbell topology on Top(X,Y) coincides with continuous convergence, one obtains at once:

2.17 COROLLARY [Lambrinos and Papadopoulos (1985, 2.15)] If W and X are core-compact spaces, then $E: Top_{is}(W \times X, Y) \cong Top_{is}(W,Top_{is}(X,Y))$ for all $Y \in Top$.

Proof: For core-compact W and X, W x X is also core compact. Hence $Top_{is}(W \times X, Y) = Top_c(W \times X, Y) \cong Top_c(W,Top_c(X,Y)) = Top_{is}(W,Top_{is}(X,Y))$ by Theorem 2.16(1,4) and Proposition 2.3. ■

2.18 COROLLARY [Lambrinos and Papadopoulos (1985, 2.4)] Homotopic functions in Top(X,Y) belong to the same path component of $Top_{is}(X,Y)$. In case of core-compact X, the converse holds also.

Proof: Let $f,g \in Top(X,Y)$. A homotopy $h: f \simeq g$ is a function $h \in Top([0,1] \times X, Y)$ such that $h^*(0) = h(0,\cdot) = f$, $h^*(1) = h(1,\cdot) = g$. Now $Top([0,1] \times X, Y) \cong Top([0,1], Top_c(X,Y)) \subset Top([0,1], Top_{is}(X,Y))$ by Propositions 2.3 and 2.13(1), and for core-compact X, the inclusion turns into equality by Theorem 2.16(1,4). ■

X is called <u>locally compact</u> iff every point $x \in X$ has a neighborhood base consisting of compact sets. Obviously, every locally compact space is core compact [but not conversely, as is shown by Isbell (1975a, 2.11) and Hofmann and Lawson (1978, Section 7)]. Hence, by Theorem 2.16, we know that for locally compact X, the limitierung of continuous convergence coincides with the Isbell topology. Moreover, it coincides with the compact-open topology whenever X is locally compact, and conversely.

2.19 PROPOSITION Let $\underline{D} \subset$ Top be initially dense in Top. The following conditions are equivalent:

(1) X is locally compact.

(2) For all $Y \in \underline{D}$, $\mathrm{Top}_c(X,Y) = \mathrm{Top}_{co}(X,Y)$ [$= \mathrm{Top}_{is}(X,Y)$].

(3) For all $Y \in \underline{D}$, co(X,Y) is conjoining.

Proof: (2) $\Leftrightarrow$ (3). Lemma 2.6(1), Proposition 2.13(1).

(1) $\Rightarrow$ (3). Well known, e.g., Fox (1945, proof of theorem 1).

(3) $\Rightarrow$ (1). $co(X,\overline{2})$ is conjoining by an initiality argument and the fact that $\mathrm{Top}_{co}(X,\cdot)$ is an internal hom-functor. A neighborhood base of $f \in \mathrm{Top}(X,\overline{2})$ in the compact-open topology is given by $\{(K,\{1\}) \mid K$ compact in X and $f(K) \subset \{1\}\}$. It is then easily seen that for each open neighborhood U of $x \in X$, there exist a neighborhood V of x and a compact set $K \subset U$ such that every open set containing K also contains V. Hence $V \cup K$ is a compact neighborhood of x contained in U. ∎

P. Lambrinos (1985, beginning of section 2) observed independently that (1) follows from (3). The implication (1) $\Rightarrow$ (2) holds, more generally, for locally compact limit spaces X [Schwarz (1982c, 2.7)].

As a corollary of Proposition 2.19, we obtain a short proof of a result of Keimel and Gierz (1982). Let L be a complete lattice. A function $f: X \to L$ is called <u>lower semicontinuous</u> iff $f(x) \leq \liminf f(\mathcal{U}(x))$ for each $x \in X$, i.e., iff $f \in \mathrm{FCo}(X,(L,s(L)))$. Endowed with the pointwise order, the set $\mathrm{FCo}(X,(L,s(L)))$ forms a complete lattice (it is closed under arbitrary joins).

2.20 COROLLARY [Keimel and Gierz (1982, (6))] If X is a locally compact topological space and L a continuous lattice, then the Scott topology on the set of all lower semicontinuous functions coincides with the compact-open topology: $\mathrm{Top}_\sigma(X,(L,\sigma(L))) = \mathrm{Top}_{co}(X,(L,\sigma(L)))$.

Proof: First note that $s(L) = \sigma(L)$ because L is a continuous lattice, hence $\mathrm{FCo}(X,(L,s(L))) = \mathrm{Top}(X,(L,\sigma(L)))$. Apply Propositions 2.10 and 2.19 (1,2). ∎

While in this section we gave conditions for the limitierung of continuous convergence to coincide with the Isbell topology or the compact-open topology, the following section is concerned with the same question in case of the pointwise topology.

3. POINTWISE TOPOLOGY AND CONTINUOUS CONVERGENCE

Scott (1972) considers function spaces Top(X,Y) where X and Y are continuous lattices in their Scott topologies, i.e., injective T_0-spaces, in which case Top(X,Y) in its pointwise topology turns out to be injective T_0 [Scott (1972, 3.3)]. Scott's conjecture that his theorem "might possibly be generalized" to arbitrary T_0-spaces X [Scott (1972, p. 113)] was disproved by Isbell (1975a, 2.12), who characterized these spaces X as the "locally finite-bottomed" ones.

In Theorem 2.16, we showed that is(X,Y) = c(X,Y) for all Y ∈ Top iff the functor • × X has a right adjoint [namely, $\mathrm{Top}_{is}(X,\cdot) = \mathrm{Top}_c(X,\cdot)$]. Analogously, in this section we will prove that the pointwise topology coincides with the limitierung of continuous convergence iff $\mathrm{Top}_{pw}(X,\cdot)$ is right adjoint to • × X (Theorem 3.4). The spaces X fulfilling the last condition are again the locally finite-bottomed ones, as has been demonstrated by Banaschewski [1977, proposition 6(4,7)]; in other words, the part of the core-compact spaces in Theorem 2.16 is played here by the proper subclass of locally finite-bottomed spaces. In the same way, one has to replace the condition "$\mathcal{O}(X)$ is a continuous lattice" by "$\mathcal{O}(X)$ is hypercontinuous." Altogether, the characterization theorem 3.4 contains a comprehensive collection of such equivalent conditions, and by its analogy with Theorem 2.16 it sheds more light on the role of the pointwise topology in comparison with the Isbell topology.

We start with two preliminary propositions (3.2, 3.3) which are of interest in connection with the pointwise topology.

3.1 DEFINITION [See, e.g., Brown (1963, p. 315).] Let W,X ∈ Top, and denote by |W|,|X| the underlying sets. The initial topology on |W| × |X| with respect to

$$\{f: |W| \times |X| \to Y \mid Y \in \mathrm{Top} \text{ and } \forall x \in X: f(\cdot,x) \in \mathrm{Top}(W,Y) \text{ and } \forall w \in W: f(w,\cdot) \in \mathrm{Top}(X,Y)\}$$

is called the <u>topology of separate continuity</u>; |W| × |X| endowed with this topology is denoted by W ⊗ X.

Equivalently, the topology of separate continuity is the initial topology with respect to all separately continuous functions into $\overline{2}$ [Činčura (1979, p. 432)]. It is also possible to define this topology by finality: it is the final topology with respect to all natural inclusions $W \to |W| \times |X|, w' \mapsto (w',x)$, $X \to |W| \times |X|$, $x' \mapsto (w,x')$ $(x \in X, w \in W)$; cf. Wilker (1970, p. 277).

In the following proposition, we list some well-known fundamental facts about the topology of separate continuity, which will be used in the sequel.

3.2 PROPOSITION (1) The topology of separate continuity is always smaller (i.e., finer) than the usual product topology.

(2) $Top_{pw}(X,\cdot)$ is right adjoint to $\cdot \otimes X$.

(3) Moreover, the natural bijection E: $Top_{pw}(W \otimes X, Y) \to Top_{pw}(W, Top_{pw}(X,Y))$ is a homeomorphism.

By Proposition 3.2(2), $\otimes$ yields a symmetric monoidal closed structure on Top [Eilenberg and Kelly (1966, pp. 475, 535)]. Činčura (1979, 2.16) showed that it is the only one; however, there is a proper class of (non-symmetric) monoidal closed structures on Top [Greve (1980, theorem 3)].

In analogy to Proposition 2.12(1), the pointwise topology on $Top(X,\overline{2})$ can be described by means of the lattice order (Proposition 3.3). This was already observed by Isbell (1975b, p. 329); see also Hoffmann (1979, 2.3(c)).

Recall that for a complete lattice L, the upper topology of L is generated by the complements of the "lower rays" $\downarrow x = \{y \in L \mid y \leq x\}$; we denote it by up(L). In case of $L = Top(X,\overline{2})$, we also use the abbreviation $up(X,\overline{2})$ [instead of $up(Top(X,\overline{2}))$].

3.3 PROPOSITION The upper topology of $Top(X,\overline{2})$ coincides with the pointwise topology.

Recall that a complete lattice L is said to be hypercontinuous [the Compendium (1980, III-3.22)] iff for every point $x \in L$, $x = \sup\{y \mid x \in int_{up(L)} \uparrow y\}$; see also Gierz and Lawson (1981; 6.1,6.4), the Compendium (1980, p. 166f), Erné (1982, theorem 6). Isbell (1975a, p. 53) called a topological space locally finite-bottomed iff each point has a basis of neighborhoods V containing finite sets F "such that no relatively open proper subset of V contains F," i.e., such that every neighborhood of F contains V.

We are now in position to formulate a collection of characterizations for the equality of the pointwise topology and continuous convergence. Many of these equivalences are known [Isbell (1975a, 2.12); Banaschewski (1977, proposition 6) and (1981)]. In our proof, we have tried to make the intrinsic connections between these conditions clearly visible.

3.4 THEOREM Let I be a nonsingleton injective T_0-space, e.g., $I = \overline{2}$. The following conditions are equivalent:

(1) X is locally finite-bottomed.

(2) For every $x \in X$ and every open neighborhood U of x, there exist a finite subset F of U and an open neighborhood V of x such that every open set containing F contains V.

(3) $\mathcal{O}(X)$ is a hypercontinuous lattice.

(4) $\mathcal{O}(X)$ is a continuous lattice and $\sigma(\mathcal{O}(X)) = up(\mathcal{O}(X))$.

(5) $s(\mathcal{O}(X)) = up(\mathcal{O}(X))$.

(6) $Top_{pw}(X,\overline{2}) = Top_c(X,\overline{2})$.

(7) For all $Y \in Top$, $Top_{pw}(X,Y) = Top_c(X,Y)$.

(8) $pw(X,\overline{2})$ is conjoining.

(9) For all $Y \in Top$, $pw(X,Y)$ is conjoining.

(10) For all $W,Y \in Top$, the natural map $E: Top(W \times X, Y) \to Top(W, Top_{pw}(X,Y))$ is a bijection.

(11) For all $W,Y \in Top$, the natural map $E: Top_{pw}(W \times X, Y) \to Top_{pw}(W, Top_{pw}(X,Y))$ is a homeomorphism.

(12) For all $W \in Top$, $W \times X = W \otimes X$.

(13) $\cdot \otimes X$ preserves embeddings.

(14) $Top_{pw}(X,I)$ is an injective (T_0-)space.

(15) $Top_{pw}(X,\cdot)$ preserves Top-injectives.

(16) $Top_{pw}(X,\cdot)$ preserves T_0Top-injectives.

Proof: (1) ⇔ (2). Immediate.

Condition (3) means

(3′) For every $U \in \mathcal{O}(X)$ and every $x \in U$, there exist a finite number of open sets $W_1,\ldots,W_n \in \mathcal{O}(X)$ and $V \in \mathcal{O}(X)$ with $x \in V$ and $U \in (\mathcal{O}(X) - {\downarrow}W_1) \cap \cdots \cap (\mathcal{O}(X) - {\downarrow}W_n) \subset {\uparrow}V$.

(2) ⇒ (3′). Say $F = \{x_1,\ldots,x_n\}$. Choose $W_k = X - \mathrm{cl}\,\{x_k\}$ ($k = 1, \ldots,n$).

(3′) ⇒ (2). Choose $x_k \in U - W_k$ ($k = 1,\ldots,n$) and $F = \{x_1,\ldots,x_n\}$.

(3) ⇔ (4). See the Compendium [1980, III-3.23(1,2)].

(4) ⇔ (5). The Scott topology is the topological modification of the Scott convergence. Apply the Compendium (1980, II-1.8).

(5) ⇔ (6). Propositions 2.12(1) and 3.3.

(6) ⇒ (7). pw(X,Y) and c(X,Y) are initial topologies with respect to $(\mathrm{Top}(X,h)\colon \mathrm{Top}(X,Y) \to \mathrm{Top}_{pw}(X,\bar{2}) = \mathrm{Top}_c(X,\bar{2}) \mid h \in \mathrm{Top}(Y,\bar{2}))$ cf. text preceding Lemma 2.7).

(7) ⇒ (9). Lemma 2.6(1).

(9) ⇒ (8). Trivial.

(8) ⇒ (6). Proposition 2.13(1), Lemma 2.6(1).

(7) ⇒ (12). $1_{|W|\times|X|} \in \mathrm{Top}(W \otimes X, W \otimes X)$ implies $1^*_{|W|\times|X|} \in \mathrm{Top}(W, \mathrm{Top}_{pw}(X, W \otimes X)) = \mathrm{Top}(W,\mathrm{Top}_c(X, W \otimes X))$ by Proposition 3.2(3) and (7) above, and by Proposition 2.3, we conclude $1_{|W|\times|X|} \in \mathrm{Top}(W \times X, W \otimes X)$. Apply Proposition 3.2(1).

(12) ⇒ (11). Proposition 3.2(3).

(11) ⇒ (10) ⇒ (9). Trivial.

(4) ⇒ (15). By Proposition 3.3, $pw(X,\bar{2}) = up(X,\bar{2}) = \sigma(X,\bar{2})$, hence $\mathrm{Top}_{pw}(X,\cdot) = \mathrm{Top}_{is}(X,\cdot)$. Apply Theorem 2.16(2,8).

(15) ⇒ (16). Theorem 2.7(3).

(16) ⇒ (14). Obvious.

(14) ⇒ (4). As in the proof of Theorem 2.16, (10) ⇒ (2), we conclude that $\mathrm{Top}_{pw}(X,\bar{2})$ is an injective T_0-space. Hence $\mathrm{Top}(X,\bar{2})$ is a continuous lattice in the specialization order induced by $pw(X,\bar{2})$, and $pw(X,\bar{2})$ is the corresponding Scott topology. But the specialization order of $\mathrm{Top}_{pw}(X,\bar{2})$ is the pointwise order. Now apply Proposition 3.3.

(13) ⇔ (15). Top is a category in the sense of Banaschewski (1981, remark 3); cf. Banaschewski (1981, example 7). Apply Banaschewski (1981, lemma). ∎

In analogy to Theorem 2.16, the class $\{\bar{2}\}$ in (6) and (8) of Theorem 3.4 may be replaced by an arbitrary initially dense subclass $\underline{D}$ of Top. In the same way, it suffices to consider $Y \in \underline{D}$ in conditions (10) and (11).

3.5 REMARK (1) The equivalence of (1) and (14) in Theorem 3.4 is contained in Isbell (1975a, 2.12). Conditions (2), (8)-(10), (12), and (16) are the conditions (1) and (3)-(7) of Banaschewski (1977, proposition 6). We are indebted to R.-E. Hoffmann for pointing out to us that condition (2) in Banaschewski (1977, proposition 6) is equivalent to the other conditions, i.e., is not affected by the mistake appearing in Banaschewski (1977, corollary 2 of proposition 3). Indeed, for every space X, the space $\mathrm{Top}_{pw}(X,\bar{2})$

is essentially complete by Hoffmann (1979, theorem 2.2) and, consequently, it cannot have an injective hull unless it is itself injective.

(2) Spaces X with the property that $\cdot \otimes X$ preserves embeddings are also called flat. For more information about flatness, see Banaschewski (1981).

(3) The connections between the pointwise topology on function spaces Top(X,Y) and the hypercontinuity of $\mathcal{O}(X)$ are also considered in a recent preprint of Lawson (1982). We are grateful to K. H. Hofmann for this information. Lawson (1982) contains – for T_0-spaces – large parts of Theorem 3.4 and, in addition, the following interesting characterization: The sobrification of X in its specialization order is a quasicontinuous poset equipped with its Scott topology. However, Lawson does not consider connections with continuous convergence.

(4) Hofmann (1977, 1.11) reformulates Isbell's (1975b, 2.3) fundamental theorem on function spaces and continuous lattices. Hofmann's remark that $\sigma(X,Y) = pw(X,Y)$ whenever X is core compact and Y an injective T_0-space is not correct. Under these assumptions, $\sigma(X,Y)$ coincides with c(X,Y) [Proposition 2.10, Theorem 2.16(1,4)]; hence the asserted equality holds (if and) only if X is locally finite-bottomed [Theorem 3.4(1,6)]. [By the way, the assertion was already disproved by Isbell (1975a, 2.12).] The reason for the above error may probably be found in Scott (1972); cf. the introduction of this section.

Consider the equivalence of (12) and (3) in Theorem 3.4. In case of complete lattices in their Scott topologies, "hypercontinuous" in (3) may be weakened to "continuous." As in the Compendium (1980), denote by ΣL the complete lattice L in its Scott topology.

3.6 PROPOSITION [cf. the Compendium (1980, II-4.11)] For a complete lattice L_1 and a nonsingleton complete lattice L_2, the following are equivalent:

(1) $\Sigma L \otimes \Sigma L_1 = \Sigma L \times \Sigma L_1$ for every complete lattice L.

(2) $\sigma(L_1)$ is a continuous lattice.

(3) $\mathrm{Top}(\Sigma L \otimes \Sigma L_1, \Sigma L_2) = \mathrm{Top}(\Sigma L \times \Sigma L_1, \Sigma L_2)$ for every complete lattice L.

(4) $\Sigma(L \times L_1) = \Sigma L \times \Sigma L_1$ for every complete lattice L.

Proof: (2) and (4) are equivalent by the Compendium (1980, II-4.11). Since $\Sigma(L \times L_1) = \Sigma L \otimes \Sigma L_1$ by the Compendium (1980, II-2.9), we have (1) $\Leftrightarrow$

(4). By Proposition 2.14, $\{\Sigma L_2\}$ is initially dense in Top; hence both function sets in (3) are initial sources in Top. Thus (3) implies (1). The converse implication is obvious. ∎

When Scott (1972, 2.6) proved the important fact that a function of several variables between complete lattices is Scott continuous iff it is separately Scott continuous, he thought that he had shown the equivalence of usual continuity and separate continuity of such functions. The latter is disproved by Proposition 3.6(2,3).

3.7 EXAMPLES (1) Every A(lexandroff)-discrete space is locally finite-bottomed because each point has a smallest neighborhood.

(2) A T_1-space is locally finite-bottomed iff it is discrete [Isbell (1975a, p. 41); Banaschewski (1977, corollary of proposition 6)]. Consequently, an R_0-space is locally finite-bottomed iff its topology is generated by a partition of the underlying set.

(3) Every injective space X is locally finite-bottomed, since $\mathcal{O}(X)$ is a completely distributive and thus a hypercontinuous lattice.

The following diagram contains the main connections among the function space topologies considered in this paper.

3.8 DIAGRAM

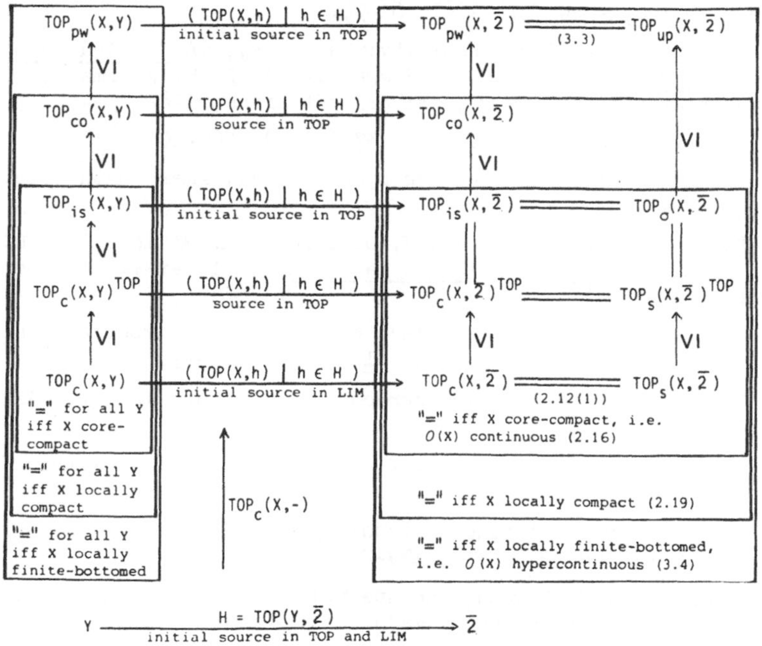

(Note that, obviously, $Top_\sigma(X,\overline{2}) \cong (\mathcal{O}(X),\sigma(\mathcal{O}(X)))$.)

REFERENCES

Arens, R. and J. Dugundji, Topologies for function spaces, Pacific J. Math. 1 (1951), 5-31.

Banaschewski, B., Essential extensions of T_0-spaces, Gen. Topol. Appl. 7 (1977), 233-246.

Banaschewski, B., A lemma on flatness, Algebra Universalis 12 (1981), 154-159.

Brown, R., Ten topologies for X x Y, Quart. J. Math. Oxford 2nd Ser. 14 (1963), 303-319.

Choquet, G., Convergences, Ann. Univ. Grenoble Sect. Sci. Math. Phys. (N.S.) 23 (1948), 57-112.

Činčura, J., Tensor products in the category of topological spaces, Commentationes Math. Univ. Carolinae 20 (1979), 431-446.

Cook, C. H., and H. R. Fischer, On equicontinuity and continuous convergence, Math. Ann. 159 (1965), 94-104.

Day, B. J., and G. M. Kelly, On topological quotient maps preserved by pullbacks or products, Proc. Camb. Phil. Soc. 67 (1970), 553-558.

Dugundji, J., Topology, Allyn and Bacon, Boston 1966.

Eilenberg, S., and G. M. Kelly, Closed categories, in Proceedings of the Conference on Categorical Algebra (La Jolla 1965) (S. Eilenberg, D. K. Harrison, S. MacLane, and H. Röhrl, eds.), Springer-Verlag, Berlin-Heidelberg-New York 1966, 421-562.

Erné, M., Scott convergence and Scott topology in partially ordered sets II, in Continuous Lattices (Proc. Conf. Bremen 1979) (B. Banaschewski and R.-E. Hoffmann, eds.), Lecture Notes Math. 871, Springer-Verlag, Berlin-Heidelberg-New York 1981, 61-96.

Erné, M., Convergence and distributivity: a survey, in Continuous Lattices and Related Topics (Proc. Conf. Bremen 1981) (R.-E. Hoffmann, ed.), Mathematik-Arbeitspapiere 27, Universität Bremen 1982, 39-50.

Fischer, H. R., Limesräume, Math. Ann. 137 (1959), 269-303.

Fox, R. H., On topologies for function spaces, Bull. Amer. Math. Soc. 51 (1945), 429-432.

Gierz, G., K. H. Hofmann, K. Keimel, J. D. Lawson, M. Mislove, and D. S. Scott, A Compendium of Continuous Lattices, Springer-Verlag, Berlin-Heidelberg-New York 1980.

Gierz, G., and J. D. Lawson, Generalized continuous and hypercontinuous lattices, Rocky Mountain J. Math. 11 (1981), 271-296.

Greve, G., How many monoidal closed structures are there in TOP?, Arch. Math. 34 (1980), 538-539.

Hoffmann, R.-E., Essentially complete T_0-spaces, Manuscripta Math. 27 (1979), 401-432.

Hofmann, K. H., Continuous lattices, topology, and topological algebra, Topology Proceedings 2 (1977), 179-212.

Hofmann, K. H., and J. D. Lawson, The spectral theory of distributive continuous lattices, Trans. Amer. Math. Soc. 246 (1978), 285-310.

Isbell, J. R., Meet-continuous lattices, Symposia Mathematica 16 (Convegni del Marzo 1973 e del Gennaio 1974, Roma: INDAM), Academic Press, London-New York 1975, 41-54.

Isbell, J. R., Function spaces and adjoints, Math. Scand. 36 (1975), 317-339.

Keimel, K., and G. Gierz, Halbstetige Funktionen und stetige Verbände, in Continuous Lattices and Related Topics (Proc. Conf. Bremen 1981) (R.-E. Hoffmann, ed.), Mathematik-Arbeitspapiere 27, Universität Bremen 1982, 59-67.

Kent, D. C., Convergence functions and their related topologies, Fund. Math. 54 (1964), 125-133.

Kowalsky, H.-J., Limesräume und Komplettierung, Math. Nachr. 12 (1954), 301-340.

Lambrinos, P. T., Locally bounded spaces, Proc. Edinburgh Math. Soc. 2nd Ser. 19 (1974-75), 321-325.

Lambrinos, P. T., The bounded-open topology on function spaces, Manuscripta Math. 36 (1981), 47-66.

Lambrinos, P. T., and B. Papadopoulos, The (strong) Isbell topology and (weakly) continuous lattices, this volume, chap. 11).

Lawson, J. D., T_0-spaces and pointwise convergence, Preprint 1982.

Machado, A., Espaces d'Antoine et pseudo-topologies, Cahiers Top. Géom. Diff. 14-3 (1973), 309-327.

Schwarz, F., Cartesian closedness, exponentiality, and final hulls in pseudotopological spaces, Quaest. Math. 5 (1982), 289-304.

Schwarz, F., Exponential objects in categories of (pre)topological spaces and their natural function spaces, C.R. Math. Rep. Acad. Sci. Canada 4 (1982), 321-326.

Schwarz, F., Topological continuous convergence, Institut für Mathematik 142, Universität Hannover 1982.

Schwarz, F., Powers and exponential objects in initially structured categories and applications to categories of limit spaces, Quaest. Math. 6 (1983), 227-254.

Scott, D., Continuous lattices, in Toposes, Algebraic Geometry and Logic (Proc. Conf. Halifax 1971) (F. W. Lawvere, ed.), Lecture Notes Math. 274, Springer-Verlag, Berlin-Heidelberg-New York 1972, 97-136.

Ward, A. S., Problem, in Proceedings of the International Symposium on Topology and Its Applications (Herceg-Novi 1968) (Đ. R. Kurepa, ed.), Beograd 1969, 352.

Węck, S., Scott convergence and Scott topology in partially ordered sets I, in Continuous Lattices (Proc. Conf. Bremen 1979) (B. Banaschewski and R.-E. Hoffmann, eds.), Lecture Notes Math. 871, Springer-Verlag, Berlin-Heidelberg-New York 1981, 372-383.

Wilker, P., Adjoint product and hom functors in general topology, Pacific J. Math. 34 (1970), 269-283.

Wyler, O., Injective spaces and essential extensions in TOP, Gen. Topol. Appl. 7 (1977), 247-249.

16

Projectiveness with Regard to a Right Adjoint Functor

MANUELA SOBRAL
Universidade de Coimbra
Coimbra, Portugal

0. INTRODUCTION

Let U: $\underline{A} \to \underline{K}$ be a functor with left adjoint F, $\underset{\sim}{T} = \langle T, \eta, \mu \rangle$ the monad it induces in $\underline{K}$ and Φ: $\underline{A} \to \underline{K}^{\underset{\sim}{T}}$ the comparison functor. In 1,1.16(a,b), R.-E. Hoffmann raised the following questions:

1. Under what conditions will the restricted comparison functor $\overline{\Phi}$: Proj U $\to$ Proj $U^{\underset{\sim}{T}}$ be an equivalence of categories?

2. Assuming that $\overline{\Phi}$: Proj U $\to$ Proj $U^{\underset{\sim}{T}}$ is an equivalence of categories when is it possible to find a category $\underline{C}$ and a functor V: $\underline{C} \to \underline{K}$, inducing the same monad $\underset{\sim}{T}$ in $\underline{K}$, such that:

(a) $\underline{A}$ is a full reflective subcategory of $\underline{C}$.

(b) U = V$\cdot$E, E being the embedding of $\underline{A}$ in $\underline{C}$.

(c) EF is left adjoint to V.

(d) The comparison functor Φ': $\underline{C} \to \underline{K}^{\underset{\sim}{T}}$ has a full and faithful left adjoint and $\Phi' \cdot E = \Phi$.

In the following we review the answer to question 1 obtained in [4] and delineate an answer to question 2.

1. THE RESTRICTED COMPARISON FUNCTOR

Recall that, for a right adjoint functor U: $\underline{A} \to \underline{K}$, belonging to an adjunction $\langle F, U; \eta, \epsilon \rangle$: $\underline{K} \longrightarrow \underline{A}$, an object P of $\underline{A}$ is said to be U-<u>projective</u> if and only if P is a retract in $\underline{A}$ of some FX, with X $\in$ Obj $\underline{K}$.

1.1 LEMMA [1] Let U: $\underline{A} \to \underline{K}$ be part of an adjunction $\langle F, U; \eta, \epsilon \rangle$: $\underline{K} \longrightarrow \underline{A}$ and P be an object of $\underline{A}$. Then P is U-projective if and only if ϵ_P is a split epi.

We shall denote by Proj U the full subcategory of $\underline{A}$ with objects all U-projectives.

1.2 PROPOSITION Let U: $\underline{A} \to \underline{K}$ be a right adjoint functor and $\underset{\sim}{P}$ be the class of all $\underline{A}$-morphisms whose images under U are split epis. Then (Obj Proj U, $\underset{\sim}{P}$) is a projective structure of $\underline{A}$ in the sense of Maranda [2, p. 100], that is,

(i) $\underline{A}(P,r)$ is a surjection for $P \in$ Proj U and $r \in \underset{\sim}{P}$.
(ii) If $\underline{A}(P,f)$ is a surjection for every $P \in$ Proj U, then $f \in \underset{\sim}{P}$.
(iii) If $\underline{A}(A,r)$ is a surjection for every $r \in \underset{\sim}{P}$, then $A \in$ Proj U.
(iv) For every $\underline{A}$-object A there exists a morphism $r: P \to A$ with $r \in \underset{\sim}{P}$ and $P \in$ Proj U.

The preceding proposition collects results that are contained in [1, 1.3 and 1.5 (b)].

In the following, we shall consider the adjunction $\langle F, U; \eta, \epsilon \rangle$: $\underline{K} \rightharpoonup \underline{A}$, $\underset{\sim}{T} = \langle T,\eta,\mu \rangle$ the monad it induces in $\underline{K}$, and Φ: $\underline{A} \to \underline{K}^{\underset{\sim}{T}}$ the comparison functor, $\underline{K}^{\underset{\sim}{T}}$ being the Eilenberg-Moore category of $\underset{\sim}{T}$-algebras in $\underline{K}$. Furthermore, $\langle F^{\underset{\sim}{T}}, U^{\underset{\sim}{T}}; \eta^{\underset{\sim}{T}}, \epsilon^{\underset{\sim}{T}} \rangle$: $\underline{K} \rightharpoonup \underline{K}^{\underset{\sim}{T}}$ denotes the associated canonical adjunction.

If A is U-projective, then $\Phi A = (UA, U\epsilon_A)$ is $U^{\underset{\sim}{T}}$-projective, because $\epsilon^{\underset{\sim}{T}}_{\Phi A} = \Phi\epsilon_A$ is a split epi in $\underline{K}^{\underset{\sim}{T}}$ since ϵ_A is a split epi in $\underline{A}$. Then Φ can be restricted and corestricted to a functor $\bar{\Phi}$: Proj U $\to$ Proj $U^{\underset{\sim}{T}}$.

1.3 LEMMA [4] The restricted comparison functor $\bar{\Phi}$ is full and faithful.

We shall now provide an answer to question 1.

1.4 THEOREM [4] The functor $\bar{\Phi}$: Proj U $\to$ Proj $U^{\underset{\sim}{T}}$ is an equivalence if and only if for all $(X,\xi) \in$ Proj $U^{\underset{\sim}{T}}$ the pair $(\epsilon_{FX}, F\xi)$ has a coequalizer in $\underline{A}$.

It is natural to ask when is $\bar{\Phi}$ an isomorphism of categories. Recall that a functor U: A $\to \underline{K}$ lifts isomorphisms uniquely if for every $\underline{A}$-object A and every isomorphism f: UA $\to$ X in $\underline{K}$ there is a unique isomorphism g: A $\to$ B in $\underline{A}$ with Ug = f (in particular, UB = X; there may be other morphisms h: A $\to$ C in $\underline{A}$ with Uh = f, but these must fail to be isomorphisms).

1.5 PROPOSITION Let $\bar{\Phi}$: Proj U $\to$ Proj $U^{\underset{\sim}{T}}$ be an equivalence of categories. Then $\bar{\Phi}$ is an isomorphism if and only if the restriction U′: Proj U $\to \underline{K}$ of U to Proj U lifts isomorphisms uniquely.

Proof: Since $U^{\underset{\sim}{T}} \cdot \Phi = U$ and $U^{\underset{\sim}{T}}$ lifts isomorphisms uniquely, then U′ lifts isomorphisms uniquely if and only if the same holds for $\bar{\Phi}$. And an equivalence of categories is an isomorphism if and only if it lifts isomorphisms uniquely. ∎

1.6 EXAMPLE In [1, sec. 2], R.-E. Hoffmann determines the Ω-projectives for Ω: Sober → Poset, the functor "specialization order" from the category of sober spaces and continuous maps to the category of partially ordered sets and isotone maps. This functor has a left adjoint, and so we obtain a monad $\underset{\sim}{T}$ in Poset whose algebras are the d-posets [1, 2.6,2.7].

In this case, $\overline{\Phi}$: Proj $\Omega \to$ Proj $\Omega^{\underset{\sim}{T}}$ is an isomorphism by Proposition 1.5.

2. EXISTENCE OF ($\underline{C}$,$\underline{V}$)

In the second question in Sec. 0, it is a consequence of the assumptions that Φ has a left adjoint.

2.1 PROPOSITION If Φ has a left adjoint, then $\overline{\Phi}$: Proj U → Proj $U^{\underset{\sim}{T}}$ is an equivalence of categories.

Proof: Let Φ be part of an adjunction $\langle L, \Phi; \alpha, \beta\rangle$: $\underline{K}^{\underset{\sim}{T}} \to \underline{A}$. Since $U^{\underset{\sim}{T}} \Phi = U$, there is a natural isomorphism θ: $LF^{\underset{\sim}{T}} \to F$ such that

$$U\,\theta_X \cdot U^{\underset{\sim}{T}}\,\alpha_{F^{\underset{\sim}{T}}X} \cdot \eta^{\underset{\sim}{T}}_X = \eta_X$$

for each $\underline{K}$-object X. Then, since $\eta^{\underset{\sim}{T}} = \eta$ and $F^{\underset{\sim}{T}} = \Phi\, F$, we have that

$$\Phi\,\theta_X \cdot \alpha_{\Phi FX} = 1_{\Phi FX}$$

and so $\theta_X = \beta_{FX}$. Thus β_{FX} is an isomorphism for each $\underline{K}$-object X.

For (X,ξ) an object in $\underline{K}^{\underset{\sim}{T}}$, the pair $(\Phi\epsilon_{FX},\Phi F\xi)$ has coequalizer ξ in $\underline{K}^{\underset{\sim}{T}}$. Hence, $L\xi$ is the coequalizer in $\underline{A}$ of the pair $(L\,\Phi\,\epsilon_{FX},\ L\,\Phi\,F\xi)$ since L, being a left adjoint, preserves colimits. Thus $L\,\xi\cdot\beta^{-1}_{FX}$ is the coequalizer in $\underline{A}$ of $(\epsilon_{FX},F\xi)$.

We have proved that $\underline{A}$ has coequalizers of all pairs $(\epsilon_{FX},F\xi)$ for $(X,\xi) \in \mathrm{Obj}\ \underline{K}^{\underset{\sim}{T}}$, and so, by Theorem 1.4, we conclude that $\overline{\Phi}$ is an equivakence of categories. ■

Note that, if Φ has a left adjoint L, then the unit of the adjunction is, for each (X,ξ) in $\underline{K}^{\underset{\sim}{T}}$, the unique $\underline{K}^{\underset{\sim}{T}}$-morphism $\alpha_{(X,\xi)}$ such that $\alpha_{(X,\xi)} \cdot \xi = \Phi e$, with e: FX → L(X,$\xi$) the coequalizer of $(\epsilon_{FX},F\xi)$ in $\underline{A}$.

2.2 THEOREM Suppose that U is a faithful right adjoint functor and that the comparison functor belongs to an adjunction $\langle L, \Phi; \alpha, \beta\rangle$: $\underline{K}^{\underset{\sim}{T}} \to \underline{A}$. Then there exists a category $\underline{C}$ and a functor V: $\underline{C} \to \underline{K}$, inducing the same monad $\underset{\sim}{T}$ in $\underline{K}$, such that:

(i) $\underline{A}$ is a full isomorphism-closed reflective subcategory of $\underline{C}$.

(ii) U = V·E, E: $\underline{A} \to \underline{C}$ being the embedding.

(iii) EF is left adjoint to V.

(iv) The comparison functor Φ': $\underline{C} \to \underline{K}^{\tilde{T}}$ has a full and faithful left adjoint and $\Phi' \cdot E = \Phi$.

Sketch of proof: We construct a category $\underline{C}$ with

$$\text{Obj } \underline{C} = \text{Obj } \underline{A} \cup \{(X,\xi) \in \text{Obj } \underline{K}^{\tilde{T}}: \alpha_{(X,\xi)} \text{ is not an isomorphism}\}$$

$$\underline{C}(A,A') = \underline{A}(A,A') \qquad \underline{C}((X,\xi), (Y,\theta)) = \underline{K}^{\tilde{T}} ((X,\xi), (Y,\theta))$$

$$\underline{C}((X,\xi),A) = \underline{K}^{\tilde{T}} ((X,\xi), \Phi A)$$

and $\underline{C}(A,(X,\xi))$ is the set of all $\underline{K}^{\tilde{T}}$-morphisms f: $\Phi A \to (X,\xi)$ such that $\alpha_{(X,\xi)} \cdot f = \Phi\bar{f}$ for some $\underline{A}$-morphism $\bar{f}$: $A \to L(X,\xi)$.

The composition $g \circ f$ in $\underline{C}$ is composition in $\underline{A}$ (denoted by $g{\cdot}f$) if g and f are $\underline{A}$-morphisms and is composition in $\underline{K}^{\tilde{T}}$ (also denoted by $g{\cdot}f$) if they are $\underline{K}^{\tilde{T}}$-morphisms, except when $f \in \underline{C}(A,(X,\xi))$ and $g \in \underline{C}((X,\xi),A')$. In this case, we have that $g \circ f$ is the composition in $\underline{A}$ of the morphisms $\bar{g}$ and $\bar{f}$ with $\Phi\,\bar{g} \cdot \alpha_{(X,\xi)} = g$ and $\Phi\,\bar{f} = \alpha_{(X,\xi)} \cdot f$, i.e., $g \circ f = \bar{g}{\cdot}\bar{f}$.

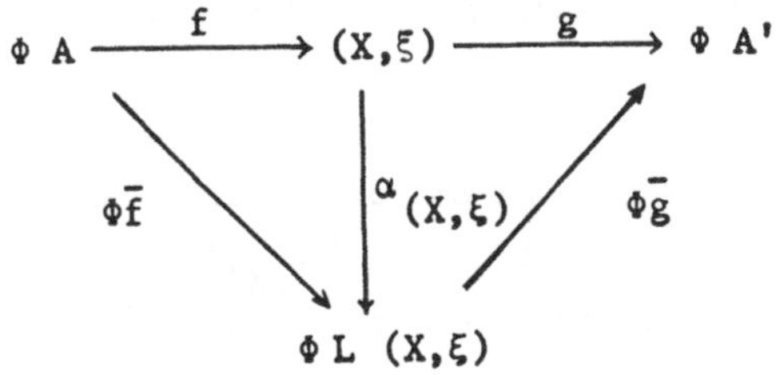

There exist morphisms $\bar{g}$ and $\bar{f}$ satisfying the equalities above because $\alpha_{(X,\xi)}$ is universal from (X,ξ) to Φ, and by definition of $\underline{C}(A,(X,\xi))$, respectively. The composition is well defined because U, and so Φ, is faithful.

If f: $A \to A'$ and g: $A' \to (X,\xi)$, then $g \circ f = g \cdot \Phi f$, and if f: $(X,\xi) \to A$ and g: $A \to A'$, then $g \circ f = \Phi\, g{\cdot}f$.

With this data we obtain a category $\underline{C}$.

(i) The functor E: $\underline{A} \to \underline{C}$ defined by $E(A) = A$ and $E(f) = f$ is a full reflective embedding with reflection maps

$$r_A = 1_A \qquad \text{for } A \in \text{Obj } \underline{A}$$

and

$$r_{(X,\xi)} = \alpha_{(X,\xi)} \in \underline{C}(X,\xi),L(X,\xi))$$

for the remaining objects of $\underline{C}$. Furthermore, it is clear that if there is an isomorphism $f: A \to (X,\xi)$ in $\underline{C}$, then $\alpha_{(X,\xi)}$ is an isomorphism, and so we have a contradiction since, in this case, (X,ξ) is not a $\underline{C}$-object.

(ii) Let us define $V: \underline{C} \to \underline{A}$ by $VE = U$ and $V(X,\xi) = U^{\underset{\sim}{T}}(X,\xi)$, and $Vf = U^{\underset{\sim}{T}}f$ for all objects and morphisms belonging to $\underline{C} \cap \underline{K}^{\underset{\sim}{T}}$. It is easy to verify that V is a functor. For example, if $f: A \to (X,\xi)$ and $g: (X,\xi) \to A'$,

$$V(g \circ f) = V(\overline{g}\cdot\overline{f}) = U\overline{g} \cdot U\overline{f} = U^{\underset{\sim}{T}}(\Phi\overline{g} \cdot \Phi\overline{f}) = U^{\underset{\sim}{T}}(g\cdot f) = Vg \cdot Vf$$

(iii) EF is left adjoint to V, η being the unit of the adjunction. Indeed, for $f: X \to V(Y,\theta)$, there exists a unique $\underline{K}^{\underset{\sim}{T}}$ morphism $g: F^{\underset{\sim}{T}}X \to (Y,\theta)$ such that $U^{\underset{\sim}{T}}g \cdot \eta_X = f$. Since $\Phi FX = F^{\underset{\sim}{T}}X$ is not a $\underline{C}$-object, because $\alpha_{\Phi FX}$ is an isomorphism, it remains to prove that $g \in \underline{C}(FX,(Y,\theta))$, i.e., that there exists an $\underline{A}$-morphism $\overline{g}$: $FX \to L(Y,\theta)$ such that $\alpha_{(Y,\theta)} \cdot g = \Phi\overline{g}$. Let e be the coequalizer of $(\epsilon_{FY}, F\theta)$ in $\underline{A}$. Then we have

$$\begin{aligned}
\alpha_{(Y,\theta)} \cdot g &= \alpha_{(Y,\theta)} \cdot g \cdot \Phi\epsilon_{FX} \cdot \Phi F\eta_X && (\text{because } \epsilon_{FX} \cdot F\eta_X = 1_{FX})\\
&= \alpha_{(Y,\theta)} \cdot \theta \cdot \Phi Fg \cdot \Phi F\eta_X && (\text{since } g \text{ is a } \underset{\sim}{T}\text{-morphism})\\
&= \Phi\, e \cdot \Phi Fg \cdot \Phi F\eta_X && (\text{by definition of } \alpha_{(Y,\theta)})\\
&= \Phi(e \cdot Ff)
\end{aligned}$$

and so $g \in \underline{C}(FX,(Y,\theta))$.

Since $\alpha_{(X,\xi)} \cdot \xi = \Phi e$ with $e = \mathrm{coeq}(\epsilon_{FX}, F\xi)$, then $\xi \in \underline{C}(FX,(X,\xi))$ for all $(X,\xi) \in \mathrm{Obj}\, \underline{K}^{\underset{\sim}{T}}$ for which $\alpha_{(X,\xi)}$ is not an isomorphism. In this case, ξ is the unique $\underline{C}$-morphism such that $V\xi \cdot \eta_{V(X,\xi)} = 1_{V(X,\xi)}$. Then the counit of the adjunction $\langle EF, V; \eta, \epsilon'\rangle: \underline{K} \to \underline{C}$ is $\epsilon'_A = \epsilon_A$ for all $\underline{A}$-objects, and $\epsilon'_{(X,\xi)} = \xi$ for the remaining objects of $\underline{C}$.

(iv) The functor $\Phi': \underline{C} \to \underline{K}^{\underset{\sim}{T}}$ satisfies $\Phi' \cdot E = \Phi$, and $\Phi'(X,\xi) = (V(X,\xi), V\epsilon'_{(X,\xi)}) = (X,\xi)$.

We now define

$$\alpha'_{(X,\xi)} = \begin{cases} \alpha_{(X,\xi)} & \text{if } \alpha_{(X,\xi)} \text{ is an isomorphism} \\ 1_{(X,\xi)} & \text{otherwise} \end{cases}$$

In this way we obtain a universal arrow from (X,ξ) to Φ' for each $(X,\xi) \in \mathrm{Obj}\, \underline{K}^{\underline{T}}$. Indeed, if $\alpha_{(X,\xi)}$ is an isomorphism and $f\colon (X,\xi) \to \Phi'(Y,\theta) = (Y,\theta)$, then $g = f \cdot \alpha^{-1}_{(X,\xi)} \in \underline{C}(L(X,\xi),(Y,\theta))$, because $\alpha_{(Y,\theta)} \cdot g = \Phi Lf$, and is the unique $\underline{C}$-morphism such that $\Phi' g \cdot \alpha_{(X,\xi)} = f$. The other case is trivially verified.

There is a unique way to define a functor $L'\colon \underline{K}^{\underline{T}} \to \underline{C}$ for which α' is a natural transformation from $\mathrm{Id}_{\underline{K}^{\underline{T}}}$ to $\Phi' L'$. The functor so obtained is a full and faithful left adjoint to Φ'. ∎

2.3 COROLLARY With the data of Theorem 2.2, if L is faithful, then $\underline{A}$ is a full isomorphism-closed bireflective subcategory of $\underline{C}$.

Proof: If L is faithful, then $\alpha_{(X,\xi)}$ is a monomorphism in $\underline{K}^{\underline{T}}$ for all objects of this category.

If $(X,\xi) \in \mathrm{Obj}\, \underline{C}$, then $\alpha_{(X,\xi)} \in \underline{C}((X,\xi),L(X,\xi))$ is also a monomorphism in $\underline{C}$. Then $\underline{A}$ is a full monoreflective subcategory of $\underline{C}$ and so it is bireflective in $\underline{C}$ [3, 16.5.4 (C)]. ∎

Example 1.6 satisfies the conditions of Corollary 2.3. Thus the category Sober can be bireflectively embedded in a category $\underline{C}$ satisfying the conditions (2)-(4) of question 2. I still have not been able to retrieve the category of d-spaces in this abstract categorical setting. (This problem is implicit in [1, 1.8].

REFERENCES

1. Hoffmann, R.-E., Projective sober spaces, in Proceedings of the Conference on Topological and Categorical Aspects of Continuous Lattices, LNM 871 (Springer, Berlin, 1981), 124-158.

2. Maranda, J. M., Injective structures, Trans. Amer. Math. Soc. 110 (1964), 98-135.

3. Schubert, H., Categories, Springer, Berlin-Heidelberg-New York, 1972.

4. Sobral, M., Restricting the comparison functor of an adjunction to projective objects, Quaestiones Math. 6 (1983), 303-312.

17

Lattices and Semilattices: A Convex Point of View

MARCEL VAN DE VEL

Free University of Amsterdam
Amsterdam, The Netherlands

1. INTRODUCTION

A convexity on a set X can be described as an algebraic lattice of subsets of X, including $\emptyset$ and X. In fact, every algebraic lattice is isomorphic to a convexity on a suitably chosen set [7]. In his 1974 dissertation [11], Jamison introduced a compact zero-dimensional topology on a convexity, which he called the inclusion-exclusion topology. This is exactly the topology that nowadays can be obtained from the fundamental theorem for algebraic lattices [7]. This topology proved useful in obtaining a variety of results in convexity theory [11,13,34,39,42].

On the other hand, if a set is equipped with both a topology and a convexity, then the opposite lattice of convex closed sets is a particularly interesting kind of object in the theory of continuous lattices: see Lawson [19], Tiller [31].

Connections between convexity and (semi-)lattice theory also exist in a different direction. Lattices and semilattices themselves can be convexified in several ways, and yield a rich class of examples in convexity to which general results apply. An early example is a paper of Ellis [5] where a general type of separation property was obtained, from which both the Kakutani theorem on vector spaces and the Stone theorem on prime ideals in a distributive lattice could be derived. Some other examples of a topological flavor will be presented below. Investigations of convexity in lattices or semilattices were carried out in recent years by, e.g., Doignan [4] (Helly numbers in crystallographic lattices), Jamison [12] (order generation and the Krein-Milman theorem), Varlet [44] (Prenowitz convexity in distributive lattices), Jamison [14] (Tietze's local convexity theorem in chainwise connected semilattices), and Isbell [10] (median-stable sets).

Our purpose is to give a survey of some more recent developments in the area of convexity [39,40,41], closely related to lattice and semilattice theory. Some new results have been included.

In Section 2 we describe how a convex structure induces a family of base-point orders on the underlying set. Under mild conditions, these orders become semilattice orders. Compare Muenzenberger-Smithson [24] for trees, and Ovchinnikov [25] for subsets of distributive lattices. In particular, every convex interval becomes a lattice. These base-point orders have been computed for a semilattice, convexified with the order-convex subsemilattices.

After discussing some conditional completeness concepts, we construct an intrinsic topology. Examples are: the Lawson topology [7] on a complete continuous semilattice (convexified as above); the interval topology [2] on a complete and distributive lattice (where the order-convex sublattices are taken convex); and the weak finite topology [16] of a vector space.

In general, no criteria are known for the base-point orders to be continuous, but some answers are given in special cases.

In Section 3, some properties of the intrinsic topology are discussed for a class of Helly number 2 convexities, which form a natural generalization of distributive lattices. The main result is the equivalence of the following conditions:

(i) The intrinsic topology is Hausdorff.
(ii) The resulting topological convexity has a Hahn-Banach type separation property.
(iii) All convex intervals are completely distributive lattices.
(iv) All base-point orders are continuous semilattice orders.

The proofs (given in [39]) rely on the analogous results that are known for completely distributive lattices.

Section 4 is concerned with dimension theory. Two results on equality of various dimension functions are given. The first one (from [40]) deals with Helly number 2 convexities, and it applies in particular to completely distributive lattices. The second one is new, and it applies to the class of chainwise connected Lawson semilattices that are (conditionally) join continuous. In finite dimensions, join continuity is shown to imply continuity of opposite semilattices by a method involving a geometric notion of inside of a convex set. The above result is a topological version of a result in Lawson [20]. As a consequence, all base-point orders are continuous.

Finally, Section 5 deals with continuous selections. Two applications of a general selection theorem of [41] are quoted: one concerning the approximation of lsc functions by continuous functions, and one concerning a lattice-theoretic formulation of the classical Urysohn and Tietze extension theorems in topology [6].

The above-described results and interactions between convexity and (semi)lattice theory lead to various questions in both theories. In each section some sample problems are formulated.

Let us recall that a partially ordered set (poset) is continuous if it is up complete, the points that are way below ($\ll$) a given point x form an up-directed set, and x is the supremum of this set [7]. For our present purposes, it seems better to weaken up-completeness to conditional up-completeness in the sense that each updirected set with an upper bound has a supremum. A lattice is continuous if it is complete and continuous as a poset.

2. INDUCED ORDERINGS AND INTRINSIC TOPOLOGY

2.1 SOME ADDITIONAL CONCEPTS IN CONVEXITY

A set with convexity (or a convex structure) will be denoted with a set symbol X only. Then "the" convexity of X is denoted by $\mathcal{C}(X)$. With the obvious definition of a convex hull operator h, every set of type h(F), $F \subset X$ finite, will be called a polytope. Note that these sets are exactly the compact elements of the algebraic lattice $\mathcal{C}(X)$. By an interval is meant the hull of a two-point set. A half space of X is a convex set [that is, a member of $\mathcal{C}(X)$] with a convex complement.

The reader may be guided throughout by the following two examples:

(1) A semilattice with the convexity of all order-convex subsemilattices.

(2) A lattice with the convexity of all order-convex sublattices.

For other types of examples, the reader may consult [11,14,15,26,37,41]. A function $X \to Y$ between convex structures is called convexity preserving (CP) if it converts convex sets of Y into convex sets of X. For instance, a semilattice homomorphism is CP, and a lattice homomorphism $L_1 \to L_2$ or $L_1 \to L_2^{op}$ is a CP function.

A convex structure X is said to have the separation property

S_1, if every singleton is convex.

S_2, if every two distinct points are in complementary half spaces.

S_3, if each convex set is an intersection of half spaces.

S_4, if every two disjoint convex sets are included in complementary half spaces.

NOTE: It is assumed throughout that all convex structures are S_1.

By using the compactness of $\mathcal{C}(X)$ in its natural topology, Jamison [11] proved that a convex structure X is S_4 if every two disjoint <u>polytopes</u> are in complementary half spaces.

2.2 APPLICATION A semilattice has the separation property S_4 [41].

Proof: For $C \subset S$, put

$$\geq (C) = \{x: x \geq c \text{ for some } c \in C\}$$
$$\leq (C) = \{x: x \leq c \text{ for some } c \in C\}$$

If A, B are convex sets in S with $A \cap B = \emptyset$, then one can show that either $\leq (A) \cap B = \emptyset$ or $\leq (B) \cap A = \emptyset$. Let F,G be finite sets with disjoint hulls. Then, for instance, $\leq (h(F))$ is disjoint from h(G) and, consequently,

$$\geq (h(G)) \cap h(F) = \emptyset$$

Now

$$\geq h(G) = \{x: x \geq \inf G\}$$

which is easily seen to be a half space of S. Then apply the above-quoted result. ∎

For lattices, the situation is somewhat different:

2.3 THEOREM A lattice is S_4 iff it is distributive.

The proof is based on the classical Stone theorem [8]. See Varlet [44] or the author [39]. ∎

2.4 INDUCED PARTIAL ORDERINGS Let X be a convex structure with the separation property S_3, and let $b \in X$ be fixed. According to [32], the following determines a partial order on X:

(1) $u \leq_b v$ iff $h\{b,u\} \subset h\{b,v\}$ [iff $u \in h\{b,v\}$].

$\leq_b$ is called a <u>base-point order</u> with base-point b. It is known that

(2) If $u \leq_b w$, then $u \leq_b v \leq_b w$ iff $v \in h\{u,w\}$.

(3) If $c \in X$, then $\leq_b$ and $\leq_c$ are mutually inverse on $h\{b,c\}$.

2.5 EXAMPLE Let S be a semilattice, let $b \in S$ be a base point, and let $a \in S$.

(1) If $a \nleq b$, then

$$a \leq_b x \text{ iff } a \leq x \quad \text{and} \quad a \wedge b = x \wedge b$$

The set

$$A = \{x\colon a \leq_b x\}$$

is convex in S and the orderings $\leq_b$ and $\leq$ coincide on A.
(2) If $a < b$, then

$$a \leq_b x \text{ iff } x \leq a$$

The set

$$A = \{x\colon a \leq_b x\}$$

is convex in S, and the orderings $\leq_b$ and $\geq$ coincide on A.

Proof of (1): $a \leq_b x$ means $a \in h\{b,x\}$, or, explicitly:

$$b \wedge x \leq a \qquad a \leq b \quad \text{or} \quad a \leq x$$

The first of these alternatives ($a \leq b$) can be dropped. Then $b \wedge x \leq a$ and $a \leq x$ imply that $a \wedge b = x \wedge b$.

If x is such that

$$a \wedge b = x \wedge b \qquad a \leq x$$

then clearly $a \in h\{b,x\}$, or $a \leq_b x$.

Convexity of the set $A = \geq_b (\{a\})$ holds in any S_4 (in fact, S_3) convexity Suppose $x \leq_b y$ are in A. From $a \leq_b x \leq_b y$ and 2.4(2) we find that $x \in h\{a,y\}$, that is,

$$a \wedge y \leq x \qquad x \leq a \quad \text{or} \quad x \leq y$$

Each of these alternatives gives us $x \leq y$.

If $x \leq y$ are in A, then

$$b \wedge y = b \wedge a \leq a \leq x \leq y$$

showing that $x \in h\{b,y\}$ and hence that $x \leq_b y$.

Proof of (2): Left to the reader. ∎

We note that if $b < a$, then (1) applies, and clearly

$$a \leq_b x \text{ iff } a \leq x$$

[the second condition in (1) is automatically fulfilled]. To visualize $\leq_b$, one should distinguish between three types of areas:

Type I: $\{x: x \geq b\}$

Type II: $\{ x: b \wedge x = c, x \not\leq b\}$ (where c is constant)

Type III: $\{x: x \leq b\}$

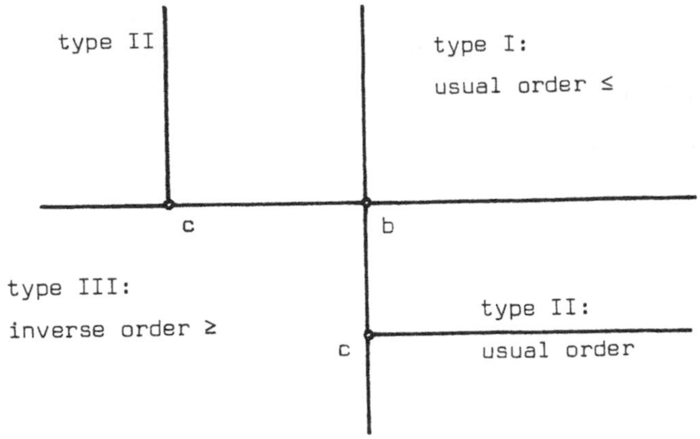

2.6 COMPLETENESS First, let $(Y,\leq)$ be a poset, $C \subset Y$. There C is down-(up-)complete in Y provided that for each down-(up-)directed set $A \subset C$ that has a lower (upper) bound in Y, inf A (sup A) exists and is in C (as in [7], a down- or up-directed set is assumed nonempty). Note that these are notions of conditional and relative completeness.

Let X now be an S_3 convex structure. A set $C \subset X$ is down-(up-)complete in X provided that for each $b \in X$, C is down-(up-)complete in $(X,\leq_b)$. We note that the prefixes down and up are somewhat misleading here: what is "down" for one base point may be "up" from another point of view. But these terms describe very well what is meant.

In [39] an example is given of a down-complete (in itself) convex structure that is not up-complete, but such examples seem rather pathological because the two notions are quite close to each other.

2.7 THEOREM Let X be an S_3 convex structure in which each interval (see 2.1) is up-complete. Then a subset of X is up-complete in X iff it is down-complete in X.

Proof: Let $C \subset X$, and assume first that C is up-complete in X. Fix a base point $b \in X$, and let $A \subset C$ be down-directed for $\leq_b$. We may assume that all $x \in A$ satisfy $x \leq_b a$ for some fixed $a \in A$. By 2.4(3), A becomes an up-directed set for $\leq_a$ that is bounded from above by b. Hence the "a-sup" of A exists and is in C. Also, a-sup A $\leq_a$ b, whence a-sup A $\in$ h{b,a}. (Up-completeness of intervals in X is not needed here.) By 2.4(3) again,

$$\text{a-sup } A = \text{b-inf } A$$

establishing down-completeness of C in X.

Assume next that C is down-complete in X. Fix b $\in$ X and let A $\subset$ C be a b-updirected set with an upper bound d. Note that then

$$A \subset h\{b,d\}$$

Upcompleteness of intervals in X implies that b-sup A exists and is in h{b,d}. Then it follows from 2.4(3) that

$$\text{b-sup } A = \text{d-inf } A$$

C being down-complete in X, the latter point is in C again. It follows that C is up-complete in $(X,\leq_b)$. ∎

Let us again take a look at a semilattice S with a base point b. If D $\subset$ S is a b-updirected (downdirected) set, then clearly D must remain eventually within a set of type I, II, or III, and D becomes an ordinary up- or downdirected set in S. Convex intervals in S are special semilattices with a bottom element and with one or two maximal elements. Hence, if D (as above) is in such an interval, then D has lower and upper bounds. Consequently, the intervals of S are up-complete in S iff these intervals are complete as a semilattice.

2.8 THEOREM Let X be an S_4 convex structure in which each interval is up-complete. Then each base-point order in X is a semilattice order.

Proof: Fix a base point b $\in$ X, and let $a_1, a_2 \in X$. Suppose u,v are both in

$$A = \{x\colon x \leq_b a_1,\ x \leq_b a_2\}$$

If $h\{u,a_1\}$ and $h\{v,a_2\}$ are disjoint, then choose a half space H of X with

$$u,a_1 \in H \qquad v,a_2 \notin H$$

We have, for instance, b $\in$ H, and hence that

$$v \in h\{b,a_1\} \subset H$$

a contradiction. So consider a point w in $h\{u,a_1\} \cap h\{v,a_2\}$. By 2.4(2) we find

$$u \le_b w \le_b a_1 \qquad v \le_b w \le_b a_2$$

Hence $w \in A$ is a b-upper bound for u and v, and A is b-updirected. Also, A is bounded from above (by a_1 and by a_2). By the up-completeness of intervals in X, we conclude that sup A exists and is in

$$h\{b,a_1\} \cap h\{b,a_2\} = A \qquad \blacksquare$$

We note the following particular case: if S is a semilattice in which the intervals are semilattice complete, then each base-point order on S leads to a semilattice structure with the chosen base point as the "bottom" element.

It is natural to ask under what circumstances each of these semilattice structures is continuous. Keeping in mind the above description of base-point order in a semilattice, it can be seen that this is the case iff the original S is continuous and for each $b \in S$ the semilattice $[0,b]^{op}$ (0: bottom element of S) is continuous. A sufficient condition for the latter to be true will be given in Section 4 below.

A similar question could be raised for general S_4-convexities in which intervals are up-complete. For a moment, let us consider a vector space X with its natural convexity, which is S_4 (Kakutani), and in which intervals (line segments between two points) are up-complete. If $b \in X$ is taken as a base point, then $u \le_b v$ iff u,v,b are collinear, and $u \le v$ in the usual ordering of a real half line (where $b = 0$). Such semilattice orders are always continuous.

The following result is from [39].

2.9 THEOREM Let X be a Hausdorff topological space with an additional S_3-convexity such that each interval is compact and each base-point order has a closed graph. Then all intervals are up-complete in X. Consequently, down- and up-completeness are equivalent concepts for a subset of X. Also, every closed subset of X is (down-)complete in X.

2.10 INTRINSIC TOPOLOGY Let X be an S_3 convex structure. A subset of X which is both down- and up-complete in X will be called <u>complete</u> in X. Then the <u>intrinsic topology</u> of X is the one that has as a subbase of closed sets the family of all convex sets that are complete in X.

We note that this is a "weak" topology, that is, a topology generated by closed sets that are also convex (compare with locally convex linear spaces where the "weak" topology is defined as the one induced by the continuous linear functions onto $\mathbb{R}$: such a topology has a subbase as described above).

2.11 EXAMPLES [39] (1) Let S be a complete continuous semilattice. Then the intrinsic topology of S is exactly the Lawson topology.

(2) Let L be a complete and distributive lattice. Then the intrinsic topology of L is the interval topology.

(3) Let X be a vector space. Then the intrinsic topology of X is the weak finite topology.

The finite topology of a vector space is the one in which a set is closed iff it meets every finite dimensional linear subspace in a closed set of the euclidean topology [16]. The corresponding weak topology is the one generated by the closed convex sets [32]. If $C \subset X$ is a linearly compact convex set, then all base-point orders in C are complete continuous. For a fixed base point b, the corresponding Lawson topology makes C into a bunch of lines held together in b. This topology reflects the choice of a base point so heavily that it becomes unacceptable for other base points.

The above construction of an intrinsic topology was devised in [39], but it has been studied explicitly for a restricted class of convex structures only (see Section 3). Before proceeding with these results, let us mention the main general problems.

2.12 QUESTIONS Let X be an S_3 or S_4 convex structure in which intervals are up-complete.

(1) Are intervals (polytopes) compact in the intrinsic topology?

(2) Under what circumstances is the intrinsic topology Hausdorff?

(3) When is each base-point order a continuous (semilattice) order?

(4) Note by 2.4(3) and Theorem 2.8 that each interval is a (complete) lattice in the order of an endpoint. What is the meaning of distributivity for the convex structure?

These and other problems have been solved satisfactorily for the following class of convexities.

3. THE HELLY NUMBER 2 CASE

3.1 HELLY NUMBER; BINARY CONVEXITY A convex structure X is said to have a Helly number $\leq n$ (where $n < \infty$), in symbols, $h(X) \leq n$, provided that for each finite set $F \subset X$ with n + 1 or more points,

$$\bigcap_{x \in F} h(F \setminus \{x\}) \neq \emptyset$$

Equivalently [26]. $\underline{h}(X) \le n$ iff every finite family of convex sets meeting n by n has a nonempty intersection. Note:

$$\underline{h}(X) = \begin{cases} 0 & \text{iff } X = \emptyset \\ 1 & \text{iff } X \text{ is a singleton} \end{cases}$$

A convex structure with a Helly number ≤ 2 is called binary.

The lattice convexity described in 2.1 is easily seen to be binary. A quite different example is the following one: let X be a tree (a connected, locally connected Hausdorff space in which every two points can be separated by a third point). The convex sets of X are the connected subsets. See [32] for details, and see [21,37] for other examples. For binary convexities, it is known [39] that the separation properties S_2, S_3, and S_4 are equivalent. Compare with the statement that "compact Hausdorff spaces are normal." The point of view that binary S_2 convexities are "the compact Hausdorff spaces of convexity" has been defended in [22] with additional facts provided by [37].

For binary S_3 convexities, there is no difference at all between the notions of down- and up-completeness, and the induced orderings are always semilattice orders: see [39].

Let us first link up binary convexities with distributive lattices.

3.2 THEOREM [39] (1) If X is a binary S_3 convex structure, and if $u,v \in X$, then $h\{u,v\}$ is a distributive lattice under $\le_u$, with bottom u and top v.

(2) The transitions

binary S_3 interval $\rightleftarrows$ distributive lattice with top and bottom

are mutually inverse.

(3) If L_1, L_2 are distributive lattices, then the CP functions (see 2.1) $L_1 \to L_2$ are exactly the lattice homomorphisms $L_1 \to L_2$ or $L_1 \to L_2^{op}$.

(4) A distributive lattice with top and bottom is lattice complete iff it is complete as a convex structure.

If X is merely an S_4 convex structure in which intervals are up-complete, then by Theorem 2.8 and by 2.4(3) each interval $h\{u,v\}$ is a lattice under $\le_u$ with top v and bottom u. In contrast to binary convexities, this lattice structure does not reflect all properties of the convexity on $h\{u,v\}$. As an example, consider the following (distributive) lattice.

top:

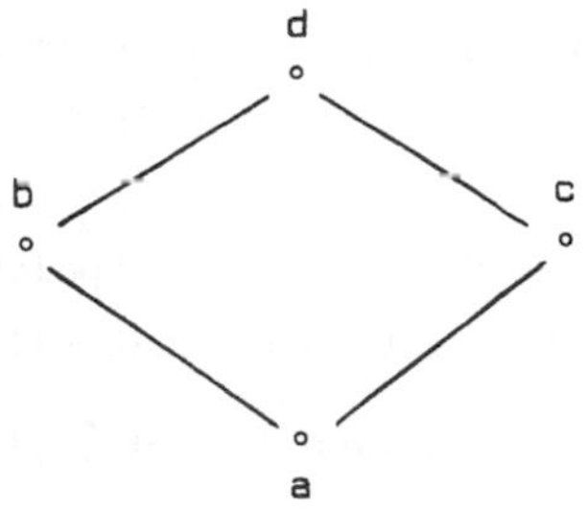

bottom:

Let $\mathcal{C}$ be a convexity on $\{a,b,c,d\}$ from which this lattice structure arises. Then certain sets (such as $\{a,b\}$, $\{a,c\}$, $\{b,d\}$, and $\{c,d\}$) must be convex, and other sets (like $\{a,b,d\}$) will not be convex. But one cannot decide whether or not $\{b,c\}$ is convex. If $\mathcal{C}$ is required to be binary, however, then it can be recovered from the lattice data.

At this stage, the reader may wonder how extensive the class of binary S_3 convex structures is in comparison to the class of distributive lattices. Here is the answer.

3.3 THEOREM [39] Up to isomorphism, the binary S_3 convex structures are exactly the median-stable subsets of distributive lattices with the relative convexity.

Recall [2] that the median operator of a distributive lattice L is the function

$$m: L^3 \to L$$
$$m(x,y,z) = (x \wedge y) \vee (y \wedge z) \vee (z \wedge x)$$
$$= (x \vee y) \wedge (y \vee z) \wedge (z \vee x)$$

A subset X of L is median stable provided $m(X^3) \subset X$. In [23], the term triple-convex is used. We note that the (restricted) operator m on X corresponds with the convexity of X as follows: $m(x,y,z)$ is the unique point in the set

$$h\{x,y\} \cap h\{y,z\} \cap h\{z,x\}$$

We also note that the distributive lattice in Theorem 3.3 can be taken equal to a power of $\{0,1\}$. Hence this result gives no adequate information concerning the intrinsic topology.

The median algebras studied by Isbell [10] form a slightly more general class of structures when compared with binary S_3 convex structures. It follows from Theorem 3.3 that the latter are in fact algebraic objects (in contrast to their definition). A topological version of Theorem 3.3 (involving Tychonov cubes) was already obtained in [23].

Before turning to the properties of the intrinsic topology, we recall some results and concepts from the theory of topological convex structures (sets with a topology and a convexity such that all polytopes are closed).

3.4 TOPOLOGICAL SEPARATION A function $f: X \to [0,1]$ separates the sets $A,B \subset X$ if there exists $t < t'$ with

$$A \subset f^{-1}[0,t],\ B \subset f^{-1}[t',1] \qquad \text{(or vice versa)}$$

A topological convex structure X is said to be

Semiregular if every convex closed set C and every $x \in X\setminus C$ can be separated by a continuous CP function $X \to [0,1]$.

Regular if every convex closed set C and every polytope $P \subset X\setminus C$ can be separated by a continuous CP function $X \to [0,1]$.

Normal if every two disjoint convex closed sets can be separated by a continuous CP function $X \to [0,1]$.

It is known that semiregular $\Rightarrow S_3$ and regular $\Rightarrow S_4$ [32]. We note that normality is a very strong requirement, which is rarely fulfilled on noncompact spaces. For instance, a topological vector space of dimension > 1 is never normal as a convex structure, but it is regular iff it is locally convex [32].

In [17] it is shown that on a complete continuous semilattice, points can be separated by continuous semilattice homomorphisms (hence CP maps) into [0,1]. With more effort, it can be shown that the convexity of a complete continuous semilattice is normal in the Lawson topology. Note that continuous semilattices are locally convex in their topology, and vice versa.

3.5 THEOREM [39]. Let X be a (conditionally) complete binary S_3 convex structure. Then all polytopes of X are compact and closed in the intrinsic topology, and the following are equivalent:

(1) The intrinsic topology of X is Hausdorff.

(2) The convexity of X is normal in the intrinsic topology; in particular, each base-point order has a closed graph.

(3) Every interval in X is completely distributive.

Moreover, if X is equipped with any Hausdorff topology making polytopes compact and making the convexity at least semiregular, then the corresponding weak topology is the intrinsic topology.

If X is replaced by a distributive lattice with top and bottom elements, the equivalence of (1), (2), and (3) corresponds to well-known results in [30]. In fact, these results are used very essentially in the proof of Theorem 3.5.

Concerning the final part of Theorem 3.5, it was already observed in [37] that a binary convexity admits at most one weak topology making polytopes compact and making the convexity at least semiregular. This result led us to attempts to find an "intrinsic" topology.

A distributive lattice L is completely distributive iff both L and L^{op} are continuous lattices [7]. The following conclusion is readily obtained.

3.6 COROLLARY Let X be a complete and binary S_3 convex structure. Then the intrinsic topology of X is Hausdorff iff each base-point order is continuous.*

As an application of base-point techniques, the following result is obtained in [39].

3.7 THEOREM Let X,Y be normal binary convex structures with compact polytopes, and let f: X → Y be a CP function onto. Then f is continuous in the weak topologies of X and Y iff its fibers are closed.

The restriction to surjective functions is essential.

4. RESULTS INVOLVING DIMENSION

A number of results have recently been obtained for topological convex structures that involve dimension. In these results, the "geometry" of a convex structure is captured in a special dimension function "cind," the convex small inductive dimension [32,33,34,35,38,40]. Its definition is repeated below.

To link up these results with established topological concepts, it has been necessary to relate cind with one or more of the usual topological dimension functions. Such a program has been carried out in [33,41,22]. On the other hand, the most common dimension function in the theory of (semi-)lattices is the so-called cohomological dimension.

4.1 CONVEX DIMENSION Let X be a topological convex structure. The <u>convex dimension</u> of X, cind X, is defined by the following two rules:

---

*Recall that only conditional up-completeness is required; cf. the introduction.

(1) cind $X = -1$ iff $X = \emptyset$.

(2) cind $X \le n + 1$ (where $n < \infty$) iff for each convex closed set C and for each $x \notin C$ there exist convex closed sets A, B with

$$A \cup B = X \quad C \subset A\setminus B \quad x \in B\setminus A \quad \text{cind } (A \cap B) \le n$$

Here, $A \cap B$ is equipped with the "relative" convexity from X. Note that $A \cap B$ is a topological separator between C and x.

For general results on cind, see [33]. This dimension function has proved most successful for semiregular convexities with connected convex sets, and with a closure-stability property: the closure of each convex set is convex.

In a compact Hausdorff topological semilattice (convexified as in 2.1), connectedness of convex sets is equivalent to chainwise connectedness, and a continuous semilattice is even a normal convex structure in its Lawson topology. We are left with identifying closure stability:

4.2 THEOREM Let S be a complete continuous semilattice such that each subsemilattice of type

$$[0,b] = \{x: 0 \le x \le b\}$$

is join continuous (conditional join continuity). Then S is closure stable.

Proof: S has small semilattices in the Lawson topology (which is considered throughout). In fact, it is easy to see that S is even locally convex, in the sense that each point has a base of convex neighborhoods. The following is an elementary fact in convexity theory, and its proof is left to the reader:

Let X be locally convex and S_4. Then X is closure stable iff the closure of each half space in X is convex. (Warning: The closure of a half space need not be a half space. Even "nice" semilattices can fail to have that property.) We continue with our proof. Let $H \subset S$ be a half space. Then H is a subsemilattice, and hence H^- is a subsemilattice. We check that H^- is also order convex.

First case: $0 \in H$ (0 = the bottom element). Let $x \le y$ be in H^-. Let y_i be a net in H converging to y. As $\wedge$ is continuous, we find from $x \le z \le y$ that $z \wedge y_i$ converges to $z \wedge y = z$. Now $z \wedge y_i \le y_i$, whence

$$z \wedge y_i \in h\{0,y_i\} \subset H$$

This shows that $z \in H^-$.

Second case: $0 \notin H$. Let $x,y \in H^-$ and $x \le z \le y$. We may assume that $x = \inf H^-$. Let $x_i \to x$ and $y_j \to y$, where $x_i, y_j \in H$. Then $z_j = z \wedge y_j \to z \wedge y = z$. Fix j for a moment. H being a semilattice with lower bound x, we may assume that the net x_i is downdirected and that $x_i \le y_j$ for all i. Note that x_i, z_j are in $[0,y_j]$, and by join continuity,

$$(*) \quad z_j = z_j \vee x = z_j \vee (\inf_i x_i) = \inf_i (z_j \vee x_i)$$

Hence $z_j \vee x_i \to z_j$. Now $X \backslash H$ is convex and contains 0. If $z_j \vee x_i \notin H$, then $0 \le x_i \le z_j \vee x_i$ implies $x_i \notin H$, a contradiction. We conclude that $z_j \vee x_i \in H$, and hence that $z_j \in H^-$. As $z_j \to z$, we finally obtain $z \in H^-$. ∎

We note that conditional join continuity is not necessary for closure stability. As an example, let S be the convex hyperspace of the closed 2-cell (with its natural convexity). S is a continuous semilattice under the operation

$$C \wedge D = \text{closed convex hull of } C \cup D$$

[7]. In order to see that S is closure stable, one can check the (slightly stronger) condition that the convex closed sets in S form a compact subspace of the entire hyperspace of S. It is easy to see that S is not conditionally join continuous.

On the other hand, a semilattice, described in [41], is not closure stable, although it is continuous, metrizable, chainwise connected, and two-dimensional.

4.3 QUESTION Find necessary and sufficient conditions on a (continuous) semilattice S — to be stated in terms of the semilattice structure — for S to be closure stable.

Throughout, cd will denote cohomological dimension [3]. For background information on the topological dimension functions dim (Lebesgue dimension), ind, and Ind (small, resp., large inductive dimension), see [6].

4.4 THEOREM [40] Let X be a compact space with a normal binary convexity. Then

(1) cind $X \ge n$ iff there is an interval in X with cind $\ge n$ $(n < \infty)$.

(2) cind X = cd X = ind X = dim X = Ind X.

In a sense, (1) reduces the problem of determining the dimension of X to the problem of determining the dimension of the various completely distributive lattices that occur in X as intervals (cf. Theorem 3.5).

The proof of (2) starts with a sequence of inequalities

cd ≤ dim ≤ ind ≤ Ind

[3,6] and consists in showing that cind is both a lower and an upper bound for this sequence. A similar procedure will be used below for semilattices.

The second main theorem of [40] is as follows:

4.5 EMBEDDING THEOREM Let X be a compact space with a normal binary convexity. Then

(1) The decomposition space (space of components) dX of X carries one and only one normal binary convexity making the decomposition map d: X → dX CP (see 2.1).

(2) If X is n-dimensional, then there is an n-dimensional connected quotient q: X → qX of X, together with a unique normal binary convexity on qX making q CP, such that the map

(q,d): X → qX × dX

is an embedding of convex structures.

Here, qX × dX is equipped with the product convexity [11], which is again normal and binary [32,26]. The phenomenon of dimension-raising lattice homomorphisms [9] shows that qX is <u>not</u> a universal connected quotient [40].

If the space X above is a completely distributive lattice, then so are dX and qX, and all CP maps in consideration become lattice homomorphisms (Theorem 3.2). In particular, (q,d) is a lattice embedding. This is a particularly fancy result because the <u>connected</u> and the <u>totally disconnected</u> lattices happen to be the best known.

Among the applications of Theorem 4.5, given in [40], are the following.

4.6 COROLLARIES (1) Up to isomorphism, the completely distributive lattices of dimension ≤ n are exactly the closed sublattices of a generalized n-cube times a Cantor cube.

(2) A finite dimensional completely distributive lattice is metrizable iff it is separable and its decomposition lattice has but countably many local minima.

(3) The breadth $\underset{\sim}{b}(L)$ of a completely distributive lattice L of dimension n satisfies the (sharp) inequalities $n \le \underset{\sim}{b}(L) \le n + \underset{\sim}{b}(dL)$.

Part (1) involves the corresponding result for connected lattices (Baker and Stralka [1]), and (2) involves a result quoted in Hofmann, Mislove, and Stralka [9] concerning totally disconnected lattices. An extension of (2) has been obtained in [42] for arbitrary compact subsets of a normal binary convex structure. The result (3) involves the equality of breadth and dimension in the connected case (Lawson [18]).

4.7 THEOREM Let X be a compact space with a semiregular, closure-stable convexity of connected convex sets. Then,

$$\text{cind } X = \text{cd } X = \dim X = \text{ind } X = \text{Ind } X$$

Proof: If "closure-stable" is replaced by "uniformizable" (see below), then Theorem 4.7 (without cohomological dimension) is a result of [22]. Our present proof gives a sharpening of this result, and it combines ideas from [22] with [40]. We therefore omit most details.

We start with the inequalities

$$\text{cd} \le \dim \le \text{ind} \le \text{Ind}$$

(1) cind $X \le$ cd: Suppose $n \le$ cind X, where $n < \infty$. According to [33], there is a continuous CP function

$$f: X \to [0,1]^n$$

which is onto (the n-cube is convexified as a lattice). The fibers of f are convex and hence [41] cohomologically trivial. By a Vietoris-Begle theorem for Čech cohomology over a field [28], we find for each closed $A \subset [0,1]^n$ that the pairs

$$(X, f^{-1}(A)) \qquad ([0,1]^n, A)$$

have isomorphic cohomology, from which it follws that $n \le$ cd X.

(2) Ind $X \le$ cind X: Similar to [22]. ∎

We have the following consequence for semilattices:

4.8 COROLLARY Let S be a closure-stable and chainwise connected Lawson semilattice. Then the following are true:

(1) cind $S \ge n$ iff there is an interval $[x,y]$ with $x \le y$ and cind $[x,y] \ge n$ $(n < \infty)$.

(2) cind S = cd S = ind S = Ind S = dim S.

Proof: S satisfies the assumptions of Theorem 4.7, giving us (2). As for (1), it follows from a result in [33] that cind $S \ge n$ $(n < \infty)$ iff there is a polytope $P = h(A)$ ($A \subset S$ finite) with cind $P \geq n$. Now

$$P = \bigcup_{x \in A} [\inf A, x]$$

whence, by the sum theorem for cind [33],

$$\text{cind } P = \max_{x \in A} \text{cind } [\inf A, x]$$

establishing (1). ■

By a result of Lawson [18], the breadth of a compact chainwise connected semilattice of dimension n equals n or n + 1, and it equals n if S has a top element. Hence, by Corollary 4.8, the various dimension functions are also equal to the maximum of the breadths of the intervals $[x,y]$, $x \leq y \in S$.

Corollary 4.8(2) was already obtained in [38], but without cohomological dimension, and for a somewhat more restricted class of semilattices.

In view of Theorem 4.4, we are led to ask the following:

4.9 QUESTION Is equality of dimension functions still valid for Lawson semilattices that are not (chainwise) connected, or are not closure stable?

Now that we have linked up convex dimension with other "standard" dimension functions, we can describe certain results on semilattices in somewhat more acceptable terms.

4.10 THEOREM Let S be a chainwise connected and conditionally join-continuous Lawson semilattice of finite dimension. Then, for each $b \in S$, $[0,b]^{op}$ is a continuous semilattice, and hence S has continuous base-point orders.

We note that a semilattice as above also has finite breadth by Lawson [18], and hence that the first part of Theorem 4.10 can also be obtained through another result of Lawson [20]. Our method of proof is very direct.

Proof: Let us recall the following general fact from [32]. Let X be a semiregular, closure-stable, convex structure with connected convex sets. If cind $X < \infty$, then the collection

$$iX = \bigcap \{C: C \subset X \text{ dense and convex}\}$$

is itself dense (and convex) in X. It is called the <u>inside</u> of X.

Take $b \in S$. We show that if $y \in i[x,b]$ (where $x \leq b$), then y is way below x in $[0,b]^{op}$. Indeed, let $D \subset [0,b]$ be a downdirected set with $\inf D \leq x$. By join continuity of $[0,b]$, we find that

$$\inf \{x \vee d : d \in D\} = x$$

where $x \vee d \in [x,b]$. It follows from join continuity again that

$$\cup_{d \in D} [x \vee d, b]$$

is a dense subset of $[x,b]$, which is also convex since the points $x \vee d$ form a downdirected set. y being inside of $[x,b]$, there must be a $d \in D$ with

$$y \in [x \vee d, b]$$

and we find that

$$d \leq x \vee d \leq y$$

Having established our claim, it follows from the density of $i[x,b]$ in $[x,b]$ that the way-below relation of $[0,b]^{op}$ is approximating. ∎

The inside of a binary convex structure has also been used in [43] to obtain an extension to normal binary spaces of a result of Shirley and Stralka [27] concerning continuity of surjective lattice homomorphisms between finite dimensional connected lattices.

Let us give one other application of the inside notion.

4.11 THEOREM Let S be a chainwise connected and closure-stable Lawson semilattice of dimension n, $n < \infty$.

(1) If the upper segment $\geq(x)$ (see 2.2) is n-dimensional, then its interior is nonempty.

(2) If an n-dimensional convex set in S is covered by a family of half spaces of S, then one of them has a nonempty interior.

Proof: (1) Suppose $\operatorname{int} \geq(x) = \emptyset$. As $\geq(x)$ is a half space, it follows that $S \setminus \geq(x)$ is a dense half space, the boundary of which is $\geq(x)$. A boundary of a half space is called a <u>hyperplane</u>. By [33], the (convex) dimension of a hyperplane is smaller than the (finite) dimension of the entire space.

(2) By [33], some point of the n-dimensional set is inside of S. One of the half spaces covers this point, and hence it must have nonempty interior. ∎

Note that the converse of (1) is false and that the set of all $x \in S$ with $\operatorname{int} \geq(x) \neq \emptyset$ equals $\leq (i(S))$, the lower set of $i(S)$.

5. SELECTING IN SEMILATTICES AND IN LATTICES

We present here two applications of a selection theorem to (semi)lattices. First, we describe the main concept used in this theorem.

5.1 UNIFORMIZABLE AND METRIZABLE CONVEXITY Let X be a uniform space with an additional convexity $\mathcal{C}$. Then $\mathcal{C}$ is compatible with the uniformity of X provided that

(1) All polytopes are closed in the uniform topology.
(2) For each uniform entourage U there exists a uniform entourage V with the following property:

$$\forall C \in \mathcal{C}: hV[C] \subset U[C]$$

(Equivalently, the hulls of two V-close sets are U-close.)

A topological convex structure X is uniformizable (metrizable) if there is a (metric) uniformity that induces the X-topology and that is compatible with the X-convexity.

If X is a compact space, then a topological convexity on X is uniformizable iff the hull operator of X is continuous on the hyperspace of closed sets. See [41] for details.

If S is a Lawson semilattice and if $F \subset S$ is closed, then

$$h(F) = \{x\colon \inf F \le x \le u \text{ for some } u \in F\}$$

Continuity of S implies that inf is continuous on the hyperspace of closed sets. It follows that S is uniformizable as a convex structure iff the interval operator

$$[.,.]\colon \text{graph} \le \ \to \text{hyperspace of } S$$
$$(x,y) \to [x,y]$$

is continuous [41].

Uniformizable convex structures are always closure stable [41], and hence the above condition on a semilattice is somewhat stronger than closure stability. Also, if S is even a topological lattice, then the interval operator is automatically continuous.

5.2 SELECTION THEOREM [41] Let X be a separable space equipped with a metrizable S_4 convexity, having connected convex sets and compact polytopes. Let Y be a normal topological space and let F: Y → X be a lower semicontinuous multifunction with F(y) compact and convex for each $y \in Y$. Then F has a continuous selection.

5.3 COROLLARY Let S be a chainwise connected Lawson semilattice of countable weight with a continuous interval operator. Let Y be a compact Hausdorff space. As in [7], [Y,S] denotes the (continuous) semilattice of all lsc functions Y → S.

(1) If $f_0 \ll f_1$ are in $[Y,S]$, then there is a continuous $f: Y \to S$ with $f_0 \le f \ll f_1$.

(2) Each $f \in [Y,S]$ is the supremum of an updirected family of continuous functions $Y \to S$.

Proof: Statement (1) is well known for S a closed interval in $\mathbb{R}$ [7]. For a proof of (1), one has to use a characterization of $\ll$ in $[Y,S]$ to construct two finitely-valued step functions between f_0 and f_1. These functions bound a lower semicontinuous multifunction. Theorem 5.2 then yields a continuous f with $f_0 \le f \le f_1$. To obtain the desired result, the interpolation property of $\ll$ is used: consider $f_1' \in [Y,S]$ with $f_0 \ll f_1' \ll f_1$, and apply the above procedure to f_0, f_1'. Statement (2) follows directly from (1). ∎

5.4 COROLLARY Let L be a compact, connected, and metrizable topological lattice with small inf-semilattices. If Y is a normal topological space, and if $f_0, f_1: Y \to L$ are such that

$$f_0 \le f_1 \qquad f_0 \text{ is usc} \qquad f_1 \text{ is lsc}$$

and if $f: A \to L$ is a continuous function on a closed set $A \subset Y$ with

$$f_0|A \le f \le f_1|A$$

then there is a continuous extension f' of f over Y such that $f_0 \le f' \le f_1$.

Proof: Convexify L as an inf-semilattice. Note that connectedness of L implies chainwise connectedness. Then construct a multivalued function

$$F(y) = [f_0(y), f_1(y)] \subset L \qquad y \in Y$$

This F is lower semicontinuous, and f is a partial selection for F on A. As a consequence of Theorem 5.2, such partial selections can always be extended to full selections. ∎

The above f_0, f_1 can be taken as the characteristic function of an open or closed subset of Y. Then Corollary 5.4 can be seen as a formulation of the classical Urysohn and Tietze theorems in terms of lattices.

As another application of the selection theorem 5.2, a result has been obtained in [41] on approximating upper semicontinuous multifunctions by continuous multifunctions in a Fréchet vector space. This result quite resembles Corollary 5.3(1) above. The proof involves the hyperspace of compact convex sets, which is a noncompact semilattice (in fact, it is a conditionally complete semilattice in the terminology of Section 2).

This prevented us from using the theory of complete continuous semilattices, their natural topology, and their natural convexity. Instead, we considered a somewhat coarser convexity that does not relate directly to the semilattice structure, but is known from other studies [35,36].

5.5 QUESTION Is there a satisfactory theory (topologically and also convexly) of semilattices with complete continuous polytopes? (Convex hyperspaces tend to have this property.)

REFERENCES

1. K. A. Baker and A. R. Stralka, Compact distributive lattices of finite breadth, Pacific J. Math. 34 (1970), 311-320.
2. G. Birkhoff, Lattice Theory, Amer. Math. Soc. Coll. Publ., 3d edition, Providence, R.I., 1967.
3. H. Cohen, A cohomological definition of dimension for compact Hausdorff spaces, Duke Math. J. 21 (1954), 209-224.
4. J. P. Doignon, Convexity in crystallographic lattices, J. Geometry 3 (1973), 71-85.
5. J. W. Ellis, A general set-separation theorem, Duke Math. J. 19(1) (1952), 417-421.
6. R. Engelking, General Topology, PWN- Polish Scientific Publishers, Warszawa, 1977.
7. G. Gierz et al., A Compendium of Continuous Lattices, Springer-Verlag, Berlin, 1980.
8. G. Grätzer, Lattice Theory, W. H. Freeman and Co., San Francisco, 1971.
9. K. H. Hofmann, M. Mislove, and A. R. Stralka, Dimension raising maps in topological algebra, Math. Z. 135 (1973), 1-36.
10. J. R. Isbell, Median algebra, Trans. Amer. Math. Soc. 260 (1980), 319-362.
11. R. E. Jamison, A General Theory of Convexity, Dissertation, Univ. of Washington, Seattle, 1974.
12. R. E. Jamison, Extreme points, order alignments, and semilattices, unpublished manuscript, 1975.
13. R. E. Jamison, Some intersection and generation properties of convex sets, Comp. Math. 35(2) (1977), 147-161.
14. R. E. Jamison, Tietze's convexity theorem for semilattices, Semigroup Forum 15 (1978), 357-373.
15. R. E. Jamison, Convexity and related combinatorial geometry, in Proc. 2nd Univ. of Oklahoma Conf. (D. C. Kay and M. Breen, eds.), Dekker, New York, 1982, 113-150.
16. S. Kakutani and V. Klee, The finite topology of a linear space, Arch. Math. 14 (1963), 55-58.
17. J. D. Lawson, Topological semilattices with small semilattices, J. London Math. Soc. 1 (1969), 719-724.

18. J. D. Lawson, The relation of breadth and codimension in topological semilattices II, Duke Math. J. 38 (1971), 555-559.

19. J. D. Lawson, Embeddings of compact convex sets and locally compact cones, Pacific J. Math. 66 (1976), 443-453.

20. J. D. Lawson, Algebraic conditions leading to continuous lattices, Proc. Amer. Math. Soc. 78 (1980), 477-481.

21. J. van Mill and M. van de Vel, Path connectedness, contractibility, and LC properties of superextensions, Bull. Acad. Polon. Sci. 26(3) (1978), 261-269.

22. J. van Mill and M. van de Vel, Equality of the Lebesgue and the inductive dimension functions for compact spaces with a uniform convexity, Colloq, Math., to appear.

23. J. van Mill and E. Wattel, An external characterization of spaces which admit binary normal subbases, Amer. J. Math. 100(5) (1978), 987-994.

24. T. B. Muenzenberger and R. E. Smithson, Semilattice structure on dendritic spaces, in Proc. 1977 Topology Conf. (LSU, Baton Rouge, La. 1977) I, Top. Proc. 2 (1977), 243-260 (1978).

25. S. V. Ovchinnikov, Convexity in subsets of lattices, Stochastika 4(2) (1980), 129-140.

26. G. Sierksma, Carathéodory and Helly numbers of convex-product-structures, Pacific J. Math. 61(1) (1975), 275-282.

27. E. D. Shirley and A. R. Stralka, Homomorphisms on connected topological lattices, Duke Math. J. (1971), 483-490.

28. E. Spanier, Algebraic Topology, McGraw-Hill, New York, 1966.

29. A. R. Stralka, Locally convex topological lattices, Trans. Amer. Math. Soc. 151 (1970), 629-640.

30. D. P. Strauss, Topological lattices, Proc. London Math. Soc. (3)18 (1968), 217-230.

31. J. A. Tiller, Continuous Lattices and Convexity Theory, Dissertation, McMaster Univ., Hamilton, Ontario, 1980.

32. M. van de Vel, Pseudo-boundaries and pseudo-interiors for topological convexities, Diss. Math. 210 (1983), 1-72.

33. M. van de Vel, Finite dimensional convex structures I: General results, Top. Appl. 14 (1982), 201-225.

34. M. van de Vel, On the rank of a topological convexity, Fund. Math. 119 (1983), 17-48.

35. M. van de Vel, Dimension of convex hyperspaces, Fund. Math. 122 (1984), 11-31.

36. M. van de Vel, Two-dimensional convexities are join-hull commutative, Top. Appl. 16 (1983), 181-206.

37. M. van de Vel, Matching binary convexities, Top. Appl. 16 (1983), 207-235.

38. M. van de Vel, Convex structures and dimension, in General Topology and Its Relations to Modern Analysis and Algebra V (J. Novák, ed.), Proc. 5th Prague Topol. Symp. 1981, Helderman Verlag, Berlin, 1982, 648-652.

39. M. van de Vel, Binary convexities and distributive lattices, Proc. London Math. Soc. (3) 48, 1-33.

40. M. van de Vel, Dimension of binary convex structures, Proc. London Math. Soc. (3) 48, 34-54.

41. M. van de Vel, A selection theorem for topological convex structures, to appear.

42. M. van de Vel, Metrizability of finite dimensional spaces with a binary convexity, Canadian J. Math., to appear.

43. M. van de Vel, Continuity of convexity preserving functions, J. London Math. Soc., to appear.

44. J. C. Varlet, Remarks on distributive lattices, Bull. Acad. Polon. Sci. 23(11) (1975), 1143-1147.

The Continuous Lattices Bibliography

RUDOLF-E. HOFFMANN
Universität Bremen
Bremen
Federal Republic of Germany

KARL H. HOFMANN
Technische Hochschule Darmstadt
Darmstadt
Federal Republic of Germany

DANA S. SCOTT
Carnegie-Mellon University
Pittsburgh, Pennsylvania

The following bibliography is both an extension and an update of the bibliography of A Compendium of Continuous Lattices by G. Gierz, K. H. Hofmann, K. Keimel, J. D. Lawson, M. Mislove, and D. S. Scott (Springer-Verlag, Berlin, 1980). Whereas the bibliography of that book emphasized the connections of the subject with general lattice theory and the study of topological lattices, the present bibliography reflects more broadly the intimate relationships with various branches of general topology: early interest in the filter space and certain other extension spaces (Banaschewski, Flachsmeyer), the study of Hausdorff compactifications by means of binary relations ("proximity type structures"), and some of the work on locally quasicompact spaces done from the point of view of functional analysis after Fell's important discovery of what is now called the Fell compactification, etc. Also, a few classics on function space topologies have been included. On the other hand, the discovery of the correspondence between continuous posets and completely distributive complete lattices has necessitated the inclusion of some additional references from general lattice theory. Further, a few selected references have been included with the understanding that their relationship to continuous lattice theory warrants exploration. Above all, the literature which appeared after the publication of the Compendium had to be gathered.

AARNES, J. F., EFFROS, E. G. and NIELSEN, O. A.

1970 *Locally compact spaces and two classes of C^*-algebras.* **Pacific Journal of Mathematics**, vol. 34 (1970), pp. 1–16. [MR 42:6626 (erratum 42, p. 1825)]

ABBOTT, J. C.

1969 **Sets, Lattices and Boolean Algebras.** Allyn and Bacon, 1969, xiii +282 pp. [MR 39:4052]

ADÁMEK, J.

1974 *Free algebras and automata realizations in the language of categories.* **Commentationes Mathematicae Universitatis Carolinae**, vol. 15 (1974), pp. 589–602. [MR 50:4696]

1977 *Realization theory for automata in categories.* **Journal of Pure and Applied Algebra**, vol. 9 (1977), pp. 281–296. [MR 57:404]

1982 *Construction of free continuous algebras.* **Algebra Universalis**, vol. 14 (1982), pp. 140–166. [MR 83a:08016]

ADÁMEK, J. and KOUBEK, V.

1979 *Least fixed point of a functor.* **Journal of Computer and System Sciences**, vol. 19 (1979), pp. 163–178. [MR 82d:18004]

ADÁMEK, J., NELSON, E. and REITERMAN, J.

1981 *Tree constructions of free continuous algebras.* **Journal of Computer and System Sciences**, vol. 24 (1981), pp. 114–146. [MR 84e:06008]

ADAMS, M. E.

1975 *The poset of prime ideals of a distributive lattice.* **Algebra Universalis**, vol. 5 (1975), pp. 141–142. [MR 52:2986]

ALBERT, M. H.

1984 *Iteratively algebraic posets have the a.c.c.* **Semigroup Forum**, vol. 30 (1984), pp. 371-373.

ALEXANDROV, P. S.

1937 *Diskrete Räume.* **Matematicheskij Sbornik**, vol. 2 (1937), pp. 501–519.

1939 *Bikompakte Erweiterungen topologischer Räume.* **Matematicheskij Sbornik**, vol. 5 (1939), pp. 403–423. [MR 1-318]

ALEXANDROV, P. S. and PONOMAREV, V. I.

1958 *(On bi-) compact extensions of topological spaces (Russian).* **Doklady Akademii Nauk SSSR**, vol. 121 (1958), pp. 575–578. [MR 20:4254]

1960 *(On-bi) compact extensions of topological spaces (Russian).* **Vestnik Moskovskogo Universiteta, Ser. Math. Mech. Astron. Fiz. Chim.**, vol. 14 (1960), pp. 93–108. [MR 22:9956]

ALÒ, R. A. and FRINK, O.

1966 *Topologies of lattice products.* **Canadian Journal of Mathematics**, vol. 18 (1966), pp. 1004–1014. [MR 35:98]

ANDERSON, L. W.

1958 *Topological lattices and n-cells.* **Duke Mathematical Journal**, vol. 25 (1958), pp. 205–208. [MR 20:2402]

1959 *One-dimensional topological lattices.* **Proceedings of the American Mathematical Society**, vol. 10 (1959), pp. 715–720. [MR 21:6401]

1959 *On the breadth and co-dimension of a topological lattice.* **Pacific Journal of Mathematics**, vol. 9 (1959), pp. 327–333. [MR 21:4206]

1959 *On the distributivity and simple connectivity of plane topological lattices.* **Transactions of the American Mathematical Society**, vol. 91 (1959), pp. 102–112. [MR 21:1365]

1961 *Locally compact topological lattices.* **American Mathematical Society Proceedings of Symposia in Pure Mathematics**, vol. 2 (1961), pp. 195–197. [MR 23:A1356]

1962 *The existence of continuous lattice homomorphisms.* **The Journal of the London Mathematical Society**, vol. 37 (1962), pp. 60–62. [MR 24:A3106]

ANDERSON, L. W. and WARD, L. E., JR.

1961 *A structure theorem for topological semilattices.* **Proceedings of the Glasgow Mathematical Association**, vol. 5 (1961), pp. 1–3. [MR 26:5546]

1961 *One-dimensional topological semilattices.* **Illinois Journal of Mathematics**, vol. 5 (1961), pp. 182–186. [MR 22:11069]

ANDIMA, S. J. and LARSON, R. E.

1975 *The lattice of topologies: a survey.* **The Rocky Mountain Journal of Mathematics**, vol. 5 (1975), pp. 177–198. [MR 52:9143]

ANDIMA, S. J. and THRON, W. J.

1978 *Order-induced topological properties.* **Pacific Journal of Mathematics**, vol. 75 (1978), pp. 297–318. [MR 80a:54065]

ARENS, R. F.

1946 *A topology for spaces of transformations.* **Annals of Mathematics**, vol. 47 (1946), pp. 480–495. [MR 8-165]

ARTIN, M., GROTHENDIECK, A. and VERDIER, J.

1972 **Théorie des topos et cohomologie étale des schémas (SGA 4). Lecture Notes in Mathematics**, Springer-Verlag, 1972, xix+525 pp, vol. 269. [MR 50:7130]

ATSUMI, K.

1966 *On complete lattices having the Hausdorff interval topology.* **Proceedings of the American Mathematical Society**, vol. 17 (1966), pp. 197–199. [MR 32:4674]

ATTARDI, G.

1974 *Caratterizzazione dei reticoli continui per la teoria della computazione di Dana Scott.* **Calcolo**, vol. 11 (1974), pp. 33–46. [MR 51:84]

AUMANN, G.

1955 *Bemerkung über Galois-Verbindungen.* **Bayerische Akademie der Wissenschaften. Mathematisch-Naturwissenschaftliche Klasse. Sitzungsberichte**, 1955, pp. 281–284. [MR 17-1180]

AUSLANDER, L. and MOORE, C. C.

1966 **Unitary Representations of Solvable Lie Groups. Memoirs of the American Mathematical Society**, American Mathematical Society, 1966, 199 pp, vol. 62. [MR 34:7723]

AUSTIN, C. W.

1963 *Duality theorems for some commutative semigroups.* **Transactions of the American Mathematical Society**, vol. 109 (1963), pp. 245–256. [MR 27:3737]

AVANN, S. P.

1964 *Dependence of finiteness conditions in distributive lattices.* **Mathematische Zeitschrift**, vol. 85 (1964), pp. 245–256. [MR 30:37]

BAARTZ, A. P.

1967 *The measure algebra of a locally compact semigroup.* **Pacific Journal of Mathematics**, vol. 21 (1967), pp. 199–214. [MR 35:4678]

BAKER, J. W., PYM, J. S. and VASUDEVA, H. L.

1983 *Multipliers for some measure algebras on compact semilattices.* In: **Recent Developments in the Algebraic, Analytical, and Topological Theory of Semigroups**, Oberwolfach, 1981, edited by K. H. Hofmann, H. Jürgensen and H. J. Weinert. **Lecture Notes in Mathematics**, vol. 998, Springer-Verlag, 1983, pp. 8–30.

BAKER, K. A. and STRALKA, A. R.

1970 *Compact distributive lattices of finite breadth.* **Pacific Journal of Mathematics**, vol. 34 (1970), pp. 311–320. [MR 44:129]

BALBES, R.

1971 *On the partially ordered set of prime ideals of a distributive lattice, I.* **Canadian Journal of Mathematics**, vol. 23 (1971), pp. 866–874. [MR 45:128]

1978 *On the partially ordered set of prime ideals of a distributive lattice, II.* **Bulletin de l'Académie Polonaise des Sciences, Série des Sciences Mathématiques**, vol. 26 (1978), pp. 771–773. [MR 80a:06001]

1980 *Catalytic distributive lattices.* **Algebra Universalis**, vol. 11 (1980), pp. 334–340. [MR 82c:06021]

BALBES, R. and DWINGER, PH.

1974 **Distributive Lattices**. University of Missouri Press, 1974, xiii + 294 pp. [MR 51:10185]

BALBES, R. and HORN, A.

1970 *Injective and projective Heyting algebras.* **Transactions of the American Mathematical Society**, vol. 148 (1970), pp. 549–559. [MR 41:1607]

BALL, R. N.

1984 *Distributive Cauchy lattices.* **Algebra Universalis**, vol. 18 (1984), pp. 134-174.

BANASCHEWSKI, B.

1953 ***Untersuchungen über Filterräume***. Doctoral Dissertation, Universität Hamburg, 1953.

1956 *Hüllensysteme und Erweiterung von Quasi-Ordnungen.* **Zeitschrift für Mathematische Logik und Grundlagen der Mathematik**, vol. 2 (1956), pp. 117–130. [MR 18-551]

1964 *Extensions of topological spaces.* **Canadian Mathematical Bulletin**, vol. 7 (1964), pp. 1–22. [MR 28:4501]

1969 *Frames and compactifications.* In: **Contributions to Extension Theory of Topological Structures**, Berlin, 1967, edited by J. Flachsmeyer, H. Poppe and F. Terpe. Deutscher Verlag der Wissenschaften, 1969, pp. 29–33.

1970 *Injectivity and essential extensions in equational classes of algebras.* In: **Proceedings, Conference on Universal Algebra**, Queen's University, Kingston, Ontario, 1969, edited by G. H. Wenzel. Queen's University, 1970, pp. 131–147. [MR 41:3354]

1971 *Projective covers in categories of topological spaces and topological algebras.* In: **General Topology and its Relations to Modern Analysis and Algebra**, Kanpur, 1968, edited by S. P. Franklin, Z. Frolik and V. Koutnik. Academia, 1971, pp. 63–91. [MR 44:1616]

1973 *The filter space of a lattice: its rôle in general topology.* In: **Proceedings of the University of Houston Lattice Theory Conference**, Houston, Texas, March 22-24, 1973, edited by S. Fajtlowicz and K. Kaiser. University of Houston, 1973, pp. 147–155. [MR 55:1306]

1977 *Essential extensions of T_0-spaces.* **General Topology and its Applications**, vol. 7 (1977), pp. 233–246. [MR 56:16557]

1978 *Hulls, kernels, and continuous lattices.* **Houston Journal of Mathematics**, vol. 4 (1978), pp. 517–525. [MR 80j:06009]

1980 *The duality of distributive continuous lattices.* **Canadian Journal of Mathematics**, vol. 32 (1980), pp. 385–394. [MR 81m:54072]

1981 *A lemma on flatness.* **Algebra Universalis**, vol. 12 (1981), pp. 154–159. [MR 83f:18018]

1981 *Coherent frames.* In: **Continuous Lattices**, Proceedings of the Conference on Topological and Categorical Aspects of Continuous Lattices (Workshop IV), University of Bremen, Germany, November 9-11, 1979, edited by B. Banaschewski and R.-E. Hoffmann. **Lecture Notes in Mathematics**, vol. 871, Springer-Verlag, 1981, pp. 1–11.

1981 *The duality of distributive σ-continuous lattices.* In: **Continuous Lattices**, Proceedings of the Conference on Topological and Categorical Aspects of Continuous Lattices (Workshop IV), University of Bremen, Germany, November 9-11, 1979, edited by B. Banaschewski and R.-E. Hoffmann. **Lecture Notes in Mathematics**, vol. 871, Springer-Verlag, 1981, pp. 12–19.

1983 *The power of the ultrafilter theorem.* **The Journal of the London Mathematical Society**, vol. 27 (1983), pp. 193–202.

1985 *On the topologies of injective spaces.* In: **Continuous Lattices and Their Applications**, edited by R.-E. Hoffmann and K. H. Hofmann. Marcel Dekker, Inc., 1985, pp. 1–8.

BANASCHEWSKI, B. and BRUNS, G.

1967 *Categorical characterization of the MacNeille completion.* **Archiv der Mathematik**, vol. 18 (1967), pp. 369–377. [MR 36:5036]

BANASCHEWSKI, B. and HOFFMANN, R.-E. (eds.)

1981 **Continuous Lattices.** Proceedings of the Conference on Topological and Categorical Aspects of Continuous Lattices (Workshop IV), University of Bremen, Germany, November 9-11, 1979, **Lecture Notes in Mathematics**, vol. 871, Springer-Verlag, 1981, x+413 pp. [MR 83d:06001]

BANASCHEWSKI, B. and MULVEY, C. J.

1980 *Stone-Čech compactification of locales, I.* **Houston Journal of Mathematics**, vol. 6 (1980), pp. 301–312. [MR 82b:06010]

BANASCHEWSKI, B. and NELSON, E.

1982 *Completions of partially ordered sets.* **SIAM Journal on Computing**, vol. 11 (1982), pp. 521–528. [MR 84k:06001]

BANDELT, H.-J.

1980 *The tensor product of continuous lattices.* **Mathematische Zeitschrift**, vol. 172 (1980), pp. 89–96. [MR 81j:06014]

1980 *Regularity and complete distributivity.* **Semigroup Forum**, vol. 19 (1980), pp. 123–126. [MR 81d:20058]

1981 *Complemented continuous lattices.* **Archiv der Mathematik**, vol. 36 (1981), pp. 474–475. [MR 82m:06006]

1982 *M-distributive lattices.* **Archiv der Mathematik**, vol. 39 (1982), pp. 436–442.

1982 *Tight residuated mappings and d-extensions.* In: **Universal Algebra**, edited by B. Csákány, E. Fried and E. T. Schmidt. North-Holland Publishing Company, 1982, pp. 61–72. [MR 83f:06007]

1983 *Coproducts of bounded (α, β)-distributive lattices.* **Algebra Universalis**, vol. 17 (1983), pp. 92–100. [MR 84i:06011]

BANDELT, H.-J. and ERNÉ, M.

1983 *The category of Z-continuous posets.* **Journal of Pure and Applied Algebra**, vol. 30 (1983), pp. 219–226.

1984 *Representations and embeddings of M-distributive lattices.* **Houston Journal of Mathematics**, vol. 10 (1984), pp. 315–324.

BARENDREGT, H.

1981 **The Lambda-calculus: Its Syntax and Semantics.** North-Holland Publishing Company, 1981, xiv+615 pp. [MR 83b:03016]

BATBEDAT, A.

1978 *Des schémas en demi-groupes commutatifs.* **Semigroup Forum**, vol. 16 (1978), pp. 473–481. [MR 80f:20070b]

1980 *Le spectre plein d'un ensemble ordonné.* **Revue Roumaine de Mathématiques Pures et Appliquées**, vol. 25 (1980), pp. 1167–1177. [MR 826:18007]

BAUER, H.

1982 *Ein Verfeinerungssatz für Produkte geordneter topogischer Räume.* In: **Continuous Lattices and Related Topics**, Proceedings of the Conference on Topological and Categorical Aspects of Continuous Lattices (Workshop V), Universität Bremen, Germany, May 8–10, 1981, edited by R.-E. Hoffmann. **Mathematik-Arbeitspapiere, Universität Bremen**, vol. 27, 1982, pp. 1–38.

1983 *Compact semilattices.* Manuscript, Technische Hochschule Darmstadt, 1983, 18 pp.

BAUER, H., KEIMEL, K. and KÖHLER, P.

1981 *Verfeinerungs - und Kürzungssätze für Produkte geordneter topologischer Räume und für Funktionen (-halb-) verbände.* In: **Continuous Lattices**, Proceedings of the Conference on Topological and Categorical Aspects of Continuous Lattices (Workshop IV), University of Bremen, Germany, November 9-11, 1979, edited by B. Banaschewski and R.-E. Hoffmann. **Lecture Notes in Mathematics**, vol. 871, Springer-Verlag, 1981, pp. 20–44.

BEER, G.

1982 *Upper semicontinuous functions and the Stone approximation theorem.* **Journal of Approximation Theory**, vol. 34 (1982), pp. 1–11. [MR 83h:26005]

BENADO, M.

1951 *La notion de normalité et les théorèmes de décomposition de l'algèbre (Romanian, Russian and French summaries).* **Académie de la République Populaire Roumaine, Stud. Cerc. Mat.**, vol. 1 (1951), pp. 282–317. [MR 16-212]

BILLERA, L. J.

1971 *Topologies for 2^X. Set-valued functions and their graphs.* **Transactions of the American Mathematical Society**, vol. 155 (1971), pp. 137–147. [MR 42:8462 (erratum 43, p. 1697)]

BIRKHOFF, G.

1937 *Moore-Smith convergence in general topology.* **Annals of Mathematics**, vol. 38 (1937), pp. 39–56.

1967 **Lattice Theory**. American Mathematical Society, 1967, vi+418 pp. [MR 37:2638]

BIRKHOFF, G. and FRINK, O.

1948 *Representations of lattices by sets.* **Transactions of the American Mathematical Society**, vol. 64 (1948), pp. 299-316. [MR 10-279]

BLAIR, R. L.

1955 *Stone's topology for a binary relation.* **Duke Mathematical Journal**, vol. 22 (1955), pp. 271–280. [MR 16-1138]

BLOOM, S.

1976 *Varieties of ordered algebras.* **Journal of Computer and System Sciences**, vol. 13 (1976), pp. 200–212. [MR 55:239]

BLYTH, T. S. and JANOWITZ, M. F.

1972 **Residuation Theory**. Pergamon Press, Oxford, 1972, ix+379 pp. [MR 53:226]

BOBOC, N., BUCUR, GH. and CORNEA, A.

1979 *Natural topologies on H-cones. Weak completeness.* **Revue Roumaine de Mathématiques Pures et Appliquées**, vol. 24 (1979), pp. 1013–1026. [MR 80i:31009]

1981 **Order and Convexity in Potential Theory: H-Cones. Lecture Notes in Mathematics**, Springer-Verlag, 1981, iv+286 pp, vol. 853. [MR 82i:31011]

BORREGO, J. T.

1970 *Continuity of the operation of a semilattice.* **Colloquium Mathematicum**, vol. 21 (1970), pp. 49–52. [MR 41:113]

BOSBACH, B.

1982 *A representation theorem for completely join-distributive algebraic lattices.* **Periodica Mathematica Hungarica**, vol. 13 (1982), pp. 113–118.

BOWEN, W. G.

1981 ***Lattice theory and topology***. Doctoral Dissertation, Oxford, 1981.

BOWMAN, T. T.

1974 *Analogue of Pontryagin character theory for topological semigroups.* **Proceedings of the American Mathematical Society**, vol. 46 (1974), pp. 95–105. [MR 50:527]

BRACHO, F.

1983 ***Continuously generated fixed points***. Doctoral Dissertation, Oxford University, 1983, 223 pp.

BROWN, D. R.

1965 *Topological semilattices on the 2-cell.* **Pacific Journal of Mathematics**, vol. 15 (1965), pp. 35–46. [MR 31:725]

BROWN, D. R. and STEPP, J. W.

1983 *Inner points in certain holoidal algebras.* In: **Recent Developments in the Algebraic, Analytical, and Topological Theory of Semigroups**, Oberwolfach, 1981, edited by K. H. Hofmann, H. Jürgensen and H. J. Weinert. **Lecture Notes in Mathematics**, vol. 998, Springer-Verlag, 1983, pp. 31–49.

BROWN, D. R. and STRALKA, A. R.

1973 *Problems on compact semilattices.* **Semigroup Forum**, vol. 6 (1973), pp. 265–270. [MR 51:8326]

1977 *Compact totally instable zero-dimensional semilattices.* **General Topology and its Applications**, vol. 7 (1977), pp. 151–159. [MR 55:3144]

BRUNS, G.

1961 *Distributivität und subdirekte Zerlegbarkeit vollständiger Verbände.* **Archiv der Mathematik**, vol. 12 (1961), pp. 61–66. [MR 23:A1561]

1962 *Darstellungen und Erweiterungen geordneter Mengen. I.* **Journal für die Reine und Angewandte Mathematik**, vol. 209 (1962), pp. 167–200. [MR 26:1270a]

1962 *Darstellungen und Erweiterungen geordneter Mengen. II.* **Journal für die Reine und Angewandte Mathematik**, vol. 210 (1962), pp. 1–23. [MR 26:1270b]

1967 *A lemma on directed sets and chains.* **Archiv der Mathematik**, vol. 18 (1967), pp. 561–563. [MR 36:3683]

BÜCHI, J. R.

1952 *Representation of complete lattices by sets.* **Portugaliae Mathematica**, vol. 11 (1952), pp. 151–167. [MR 14:940]

BULMAN-FLEMING, S., FLEISCHER, I. and KEIMEL, K.

1979 *The semilattices with distinguished endomorphisms which are equationally compact.* **Proceedings of the American Mathematical Society**, vol. 73 (1979), pp. 7–10. [MR 80d:08008]

BUNCE, J. W.

1973 *Irreducible representations of the C^*-algebra generated by a quasinormal operator.* **Transactions of the American Mathematical Society**, vol. 183 (1973), pp. 487–494. [MR 48:4805]

CARRUTH, J. H.

1968 *A note on partially ordered compacta.* **Pacific Journal of Mathematics**, vol. 24 (1968), pp. 229–231. [MR 36:5902]

CARSTENS, A. M.

1969 *The lattice of pretopologies on an arbitrary set S.* **Pacific Journal of Mathematics**, vol. 29 (1969), pp. 67–71. [MR 39:4797]

1969 *The lattice of pseudotopologies on S.* **Notices of the American Mathematical Society**, vol. 16 (1969), p. 150.

CHAN, W. M., HIGGS, D. A. and LENIHAN, B.

1974 *Compactness as a type of injectivity.* **General Topology and its Applications**, vol. 4 (1974), pp. 103–107. [MR 49:9809]

CHIMENTI, F. A.

1973 *Tychonoff's theorem for hyperspaces.* **Proceedings of the American Mathematical Society**, vol. 37 (1973), pp. 281–286. [MR 46:6262]

CHOE, T. H.

1969 *The breadth and dimension of a topological lattice.* **Proceedings of the American Mathematical Society**, vol. 23 (1969), pp. 82–84. [MR 40:2011]

1969 *Intrinsic topologies in a topological lattice.* **Pacific Journal of Mathematics**, vol. 28 (1969), pp. 49–52. [MR 39:1365]

1969 *Notes on locally compact connected topological lattices.* **Canadian Journal of Mathematics**, vol. 21 (1969), pp. 1533–1536. [MR 40:5508]

1969 *On compact topological lattices of finite dimension.* **Transactions of the American Mathematical Society**, vol. 140 (1969), pp. 223–237. [MR 39:1366]

1969 *Remarks on topological lattices.* **Kyungpook Mathematical Journal**, vol. 9 (1969), pp. 59–62. [MR 41:7640]

1971 *Locally compact lattices with small lattices.* **Michigan Mathematical Journal**, vol. 18 (1971), pp. 81–85.

1973 *Injective compact distributive lattices.* **Proceedings of the American Mathematical Society**, vol. 37 (1973), pp. 241–245. [MR 46:8923]

1973 *Projective compact distributive topological lattices.* **Proceedings of the American Mathematical Society**, vol. 39 (1973), pp. 606–608. [MR 47:5843]

1979 *Partially ordered topological spaces.* **Anais da Academia Brasileira de Ciências**, vol. 51 (1979), pp. 53–63. [MR 80k:54058]

CHOE, T. H. and HONG, Y. H.

1976 *Extension of completely regular ordered spaces.* **Pacific Journal of Mathematics**, vol. 66 (1976), pp. 37–48. [MR 56:1274]

CHOE, T. H. and PARK, Y. S.

1979 *Embedding ordered topological spaces into topological semilattices.* **Semigroup Forum**, vol. 17 (1979), pp. 189–199. [MR 80e:54041]

1979 *Wallman's type order compactification.* **Pacific Journal of Mathematics**, vol. 82 (1979), pp. 339–347. [MR 80m:54034]

ČINČURA, J.

1979 *Tensor products in the category of topological spaces.* **Commentationes Mathematicae Universitatis Carolinae**, vol. 20 (1979), pp. 431–446. [MR 81d: 18015]

CLARK, C. E. and EBERHART, C.

1968 *A characterization of compact connected planar lattices.* **Pacific Journal of Mathematics**, vol. 24 (1968), pp. 233–240. [MR 38:5178]

CLINKENBEARD, D.

1976 **Lattices of congruences on compact topological lattices.** Doctoral Dissertation, University of California at Riverside, 1976, 54 pp.

1979 *Simple compact topological lattices.* **Algebra Universalis**, vol. 9 (1979), pp. 322–328. [MR 80k:06005]

1981 *The lattice of closed congruences on a topological lattice.* **Transactions of the American Mathematical Society**, vol. 263 (1981), pp. 457–467. [MR 82d:06005]

COHEN, H.

1954 *A cohomological definition of dimension for locally compact Hausdorff spaces.* **Duke Mathematical Journal**, vol. 21 (1954), pp. 209–224.

COHEN, M. and RUBIN, M.

1982 *Lattices of continuous monotonic functions.* **Proceedings of the American Mathematical Society**, vol. 86 (1982), pp. 685–691. [MR 84k:54031]

COURCELLE, B. and RAOULT, J. C.

1980 *Completions of ordered magmas.* **Fundamenta Informaticae**, vol. 3 (1980), pp. 105–115. [MR 82b:68041]

COUSOT, P. and COUSOT, R.

1979 *Constructive versions of Tarski's fixed point theorems.* **Pacific Journal of Mathematics**, vol. 82 (1979), pp. 43–57. [MR 82d:06004]

CRAWLEY, P. and DILWORTH, R. P.

1973 **Algebraic Theory of Lattices**. Prentice-Hall, Inc., 1973, vi+201 pp.

CRAWLEY, W.

1976 *A note on epimorphisms of compact Lawson semilattices.* **Semigroup Forum**, vol. 13 (1976), pp. 92–94. [MR 54:5380]

1977 *Amalgamation of compact Lawson semilattices.* Manuscript, Shippensburg State College, 1977.

1978 *Epimorphisms and H-coextensions.* **Archiv der Mathematik**, vol. 30 (1978), pp. 449–457. [MR 80g:18009]

1980 *Epimorphisms and semilattices of semigroups.* **Semigroup Forum**, vol. 21 (1980), pp. 95–112. [MR 81k:20084]

CROWN, G. D.

1970 *Projectives and injectives in the category of complete lattices with residuated mappings.* **Mathematische Annalen**, vol. 187 (1970), pp. 295–299. [MR 42:5868]

CSÁSZÁR, A.

1963 **Grundlagen der allgemeinen Topologie**. Akadémiai Kiadó, 1963, 367 pp. [MR 26:6917]

CUNNINGHAM, F. and ROY, N. M.

1974 *Extreme functionals on an upper semicontinuous function space.* **Proceedings of the American Mathematical Society**, vol. 42 (1974), pp. 461–465. [MR 48:6921]

CURTIS, D. and SCHORI, R.

1974 *2^X and $C(X)$ are homeomorphic to the Hilbert cube.* **Bulletin of the American Mathematical Society**, vol. 80 (1974), pp. 927–931. [MR 50:5719]

DAVEY, B. A.

1973 *A note on representable posets.* **Algebra Universalis**, vol. 3 (1973), pp. 345–347. [MR 50:2006]

1979 *On the lattice of subvarieties.* **Houston Journal of Mathematics**, vol. 5 (1979), pp. 183–192. [MR 80j:08007]

DAVEY, B. A. and PRIESTLEY, H. A.

1984 *Lattices of lattice homomorphisms.* Manuscript, La Trobe University, Bundoora, Australia, 1984, 74 pp.

DAVEY, B. A. and RIVAL, I.

1982 *Exponents of lattice-ordered algebras.* **Algebra Universalis**, vol. 14 (1982), pp. 87–98. [MR 83m: 06007]

DAVIES, E. B.

1968 *The existence of characters on topological lattices.* **The Journal of the London Mathematical Society**, vol. 43 (1968), pp. 217–220. [MR 37:112]

DAVIS, A. C.

1955 *A characterization of complete lattices.* **Pacific Journal of Mathematics**, vol. 5 (1955), pp. 311–319. [MR 17:574]

DAY, A.

1975 *Filter monads, continuous lattices and closure systems.* **Canadian Journal of Mathematics**, vol. 27 (1975), pp. 50–59. [MR 51:3258]

DAY, B. J.

1972 *A reflection theorem for closed categories.* **Journal of Pure and Applied Algebra**, vol. 2 (1972), pp. 1–11. [MR 45:5187]

DAY, B. J. and KELLY, G. M.

1970 *On topological quotient maps preserved by pullbacks or products.* **Proceedings of the Cambridge Philosophical Society**, vol. 67 (1970), pp. 553–558. [MR 40:8024]

DE BAKKER, J. W.

1976 *Least fixed points revisited.* **Theoretical Computer Science**, vol. 2 (1976), pp. 155–181. [MR 53:12055]

DE JONGH, D. H. J. and TROELSTRA, A. S.

1966 *On the connection of partially ordered sets with some pseudo-Boolean algebras.* **Indagationes Mathematicae**, vol. 28 (1966), pp. 317–329. [MR 33:5537]

DELAROCHE, C.

1969 *Extensions localement quasi compactes d'un espace localement quasi compact par un autre.* **Comptes Rendus de l' Académie des Sciences, Paris, Série A-B**, vol. 269 (1969), pp. A953-A955. [MR 41:6165]

1969 *Spectres des extensions de C^*-algèbres.* **Comptes Rendus de l' Académie des Sciences, Paris, Série A-B**, vol. 269 (1969), pp. A1003-A1005. [MR 41:6166]

1970 *Limites et valeurs d'adhérence d'un filtre sur le spectre d'une C^*-algèbre.* **Comptes Rendus de l' Académie des Sciences, Paris, Série A-B**, vol. 271 (1970), pp. A434-A437. [MR 44:4531]

1972 **Extensions des C*-algébres.** Société Mathématique de France, 1972, 142 pp. [MR 58:30290]

DERDERIAN, J. C.

1967 *Residuated mappings.* **Pacific Journal of Mathematics**, vol. 20 (1967), pp. 35–43. [MR 35:1511]

DILWORTH, R. P. and CRAWLEY, P.

1960 *Decomposition theory for lattices without chain conditions.* **Transactions of the American Mathematical Society**, vol. 96 (1960), pp. 1–22. [MR 22:9461]

DIXMIER, J.

1968 *Sur les espaces localement quasi-compacts.* **Canadian Journal of Mathematics**, vol. 20 (1968), pp. 1093–1100. [MR 38:5171]

DOBBERTIN, H.

1982 *On Vaught's criterion for isomorphisms of countable Boolean algebras.* **Algebra Universalis**, vol. 15 (1982), pp. 95–114. [MR 83m:06017]

1983 ***Verfeinerungsmonoide, Vaught Monoide und Boolesche Algebren.*** Doctoral Dissertation, Universität Hannover, 1983.

DOBBERTIN, H. and ERNÉ, M.

1981 ***Intrinsic topologies on lattices.*** Technical Report, no. 123, Universität Hannover, 1981.

DOBBERTIN, H., ERNÉ, M. and KENT, D. C.

1982 ***A note on order convergence in complete lattices.*** Technical Report, no. 140, Universität Hannover, 1982, 13 pp.

DOBBS, D. E.

1982 *Posets admitting a unique order-compatible topology.* **Discrete Mathematics**, vol. 41 (1982), pp. 235–240. [MR 84a:06007]

DOLCHER, M.

1955 *Topologia delle famiglie di filtri.* **Rendiconti del Seminario Matematico dell'Università di Padova**, vol. 24 (1955), pp. 443–473. [MR 17-397]

DOWKER, C. H. and PAPERT, D.

1966 *Quotient frames and subspaces.* **Proceedings of the London Mathematical Society**, vol. 16 (1966), pp. 275–296. [MR 34:2510]

DOWNING, J. S.

1977 *Absolute retracts and extensors for non-normal spaces.* **General Topology and its Applications**, vol. 7 (1977), pp. 275–281. [MR 57:7514]

DRAKE, D. and THRON, W. J.

1965 *On the representation of an abstract lattice as the family of closed subsets of a topological space.* **Transactions of the American Mathematical Society**, vol. 120 (1965), pp. 57–71. [MR 32:6390]

DUBREIL-JACOTIN, M. L., LESIEUR, L. and CROISOT, R.

1953 **Leçons sur la théorie des treillis**. Gauthier-Villars, 1953, viii+385 pp. [MR 15-279]

DUFFUS, D. and RIVAL, I.

1978 *A logarithmic property for exponents of partially ordered sets.* **Canadian Journal of Mathematics**, vol. 30 (1978), pp. 797–807. [MR 58:16432]

DWINGER, PH.

1979 *Classes of completely distributive complete lattices.* **Indagationes Mathematicae**, vol. 41 (1979), pp. 411–42. [MR 81m:06029]

1981 *Structure of completely distributive complete lattices.* **Indagationes Mathematicae**, vol. 43 (1981), pp. 361–373.

1982 *Characterization of the complete homomorphic images of a completely distributive complete lattice, I.* **Indagationes Mathematicae**, vol. 44 (1982), pp. 403–414.

1983 *Characterization of the complete homomorphic images of a completely distributive complete lattice, II.* **Indagationes Mathematicae**, vol. 45 (1983), pp. 43–49.

DYER, E. and SHIELDS, A.

1959 *Connectivity of topological lattices.* **Pacific Journal of Mathematics**, vol. 9 (1959), pp. 443–448. [MR 21:4205]

EDMONDSON, D. E.

1956 *A non-modular compact connected topological lattice.* **Proceedings of the American Mathematical Society**, vol. 7 (1956), pp. 1157–1158. [MR 18:461]

1969 *A modular topological lattice.* **Pacific Journal of Mathematics**, vol. 29 (1969), pp. 271–277. [MR 39:4062]

1969 *Modularity in topological lattices.* **Proceedings of the American Mathematical Society**, vol. 21 (1969), pp. 81–82. [MR 39:2132]

EFFROS, E. G.

1965 *Convergence of closed subsets in a topological space.* **Proceedings of the American Mathematical Society**, vol. 16 (1965), pp. 929–931. [MR 31:6208]

EGLI, H.

1973 ***An analysis of Scott's λ-calculus models.*** Technical Report, no. TR 73-191, Cornell University, 1973.

EGLI, H. and CONSTABLE, R. L.

1976 *Computability concepts for programming language semantics.* **Theoretical Computer Science**, vol. 2 (1976), pp. 133–145. [MR 54:4166]

EHRESMANN, CH.

1957 *Gattungen von lokalen Strukturen.* **Jahresbericht der Deutschen Mathematiker-Vereinigung**, vol. 60 (1957), pp. 59–77. [MR 20:2392]

EILENBERG, S. and KELLY, G. M.

1966 *Closed categories.* In: **Proceedings of the Conference on Categorical Algebra at La Jolla**, La Jolla, Calif., June 7-12, 1965, edited by S. Eilenberg, D. K. Harrison, S. Mac Lane and H. Röhrl. Springer-Verlag, 1966, pp. 421–562. [MR 37:1432]

EILENBERG, S. and MOORE, J. C.

1965 *Adjoint functors and triples.* **Illinois Journal of Mathematics**, vol. 9 (1965), pp. 381–398. [MR 32:2455]

ELLIOTT, G. A.

1977 *Some C*-algebras with outer derivations III.* **Annals of Mathematics**, vol. 106 (1977), pp. 121–143. [MR 56:6403]

ERNÉ, M.

1979 ***Verallgemeinerungen der Verbandstheorie , I. Halbgeordnete Mengen und das Prinzip der Vervollständigungs-Invarianz.*** Technical Report, no. 109, Universität Hannover, 1979.

1979 ***Verallgemeinerungen der Verbandstheorie, II. m-Ideale in halbgeordneten Mengen und Hüllenräumen.*** Habilitationsschrift, Universität Hannover, 1979, 132 pp.

1980 *Order-topological lattices.* **Glasgow Mathematical Journal**, vol. 21 (1980), pp. 57–68. [MR 82f:06010]

1980 *Topologies on products of partially ordered sets, I. Interval topologies.* **Algebra Universalis**, vol. 11 (1980), pp. 295–311. [MR 82m:54028a]

1980 *Topologies on products of partially ordered sets, II. Ideal topologies.* **Algebra Universalis**, vol. 11 (1980), pp. 312–319. [MR 82m:54028b]

1980 *Separation axioms for interval topologies.* **Proceedings of the American Mathematical Society**, vol. 79 (1980), pp. 185–190. [MR 81d:54012]

1981 *Completion-invariant extension of the concept of continuous lattices.* In: **Continuous Lattices**, Proceedings of the Conference on Topological and Categorical Aspects of Continuous Lattices (Workshop IV), University of Bremen, Germany, November 9-11, 1979, edited by B. Banaschewski and R.-E. Hoffmann. **Lecture Notes in Mathematics**, vol. 871, Springer-Verlag, 1981, pp. 45–60.

1981 ***Homomorphisms of M-generated and M-distributive posets.*** Technical Report, no. 125, Universität Hannover, 1981.

1981 *Scott convergence and Scott topology in partially ordered sets, II.* In: **Continuous Lattices**, Proceedings of the Conference on Topological and Categorical Aspects of Continuous Lattices (Workshop IV), University of Bremen, Germany, November 9-11, 1979, edited by B. Banaschewski and R.-E. Hoffmann. **Lecture Notes in Mathematics**, vol. 871, Springer-Verlag, 1981, pp. 61–96.

1981 *Topologies on products of partially ordered sets, III. Order convergence and order topology.* **Algebra Universalis**, vol. 13 (1981), pp. 1–23. [MR 82m:54028c]

1982 *Convergence and distributivity: a survey.* In: **Continuous Lattices and Related Topics**, Proceedings of the Conference on Topological and Categorical Aspects of Continuous Lattices (Workshop V), Universität Bremen, Germany, May 8–10, 1981, edited by R.-E. Hoffmann. **Mathematik-Arbeitspapiere, Universität Bremen**, vol. 27, 1982, pp. 39–50.

1982 *Distributivgesetze und Dedekind'sche Schnitte.* **Abhandlungen der Braunschweigischen Wissenschaftlichen Gesellschaft**, vol. 33 (1982), pp. 117–145. [MR 84i:06002]

1982 **Einführung in die Ordnungstheorie.** Bibliographisches Institut, Mannheim, 1982, v+296 pp. [MR 84b:06001]

1983 *Adjunctions and standard constructions for partially ordered sets.* In: **Contributions to General Algebra II**, Teubner, Stuttgart, 1983, edited by G. Eigenthaler, K. Kaiser, W. B. Müller and W. Nöbauer. 1983, pp. 78–106.

1983 ***Compact generation in partially ordered sets.*** Technical Report, no. 161, Universität Hannover, 1983, 18 pp.

1983 *On the existence of decompositions in lattices.* **Algebra Universalis**, vol. 16 (1983), pp. 338–343. [MR 84f:06009]

198? **Order, Topology and Closure**. Springer-Verlag, 198?. (To appear).

ERNÉ, M. and GATZKE, H.

1985 *Convergence and continuity in partially ordered sets and semilattices.* In: **Continuous Lattices and Their Applications**, edited by R.-E. Hoffmann and K. H. Hofmann. Marcel Dekker, Inc., 1985, pp. 9–40.

ERNÉ, M. and WECK, S.

1980 *Order convergence in lattices.* **The Rocky Mountain Journal of Mathematics**, vol. 10 (1980), pp. 805–818. [MR 83f:46009]

ERNÉ, M. and WILKE, G.

1983 *Standard completions for quasiordered sets.* **Semigroup Forum**, vol. 27 (1983), pp. 351–376.

ERŠOV, JU. L.

1972 *Computable functionals of finite types. (Russian).* **Algebra i Logika**, vol. 11 (1972), pp. 367–437, 492.

1972 *Continuous lattices and A-spaces. (Russian).* **Doklady Akademii Nauk SSSR**, vol. 207 (1972), pp. 523–526. [MR 51:2892]

1973 *Continuous lattices and A-spaces. (English translation).* **Soviet Mathematics Doklady**, vol. 13 (1973), pp. 1551–1555. [MR 51:2892]

1973 *The theory of A-spaces. (Russian).* **Algebra i Logika**, vol. 12 (1973), pp. 369–416, 492. [MR 54:7236]

1974 *Computable functionals of finite types. (English translation).* **Algebra and Logic**, vol. 11 (1974), pp. 203–242. [MR 50:12688]

1975 *The theory of A-spaces. (English translation).* **Algebra and Logic**, vol. 12 (1975), pp. 209-232. [MR 54:7236]

1976 *Model C of partial continuous functions.* In: **Logic Colloquium '76**, Oxford, 1976, edited by R. O. Gandy and J. M. E. Hyland. **Studies in Logic and Foundations of Mathematics**, vol. 87, North-Holland Publishing Company, 1976, pp. 455–467. [MR 58:21541]

EVANS, E.

1980 *Complete co-atomic lattices with enough primes.* **Semigroup Forum**, vol. 21 (1980), pp. 113–126. [MR 83b:06005]

EVERETT, C. J.

1944 *Closure operators and Galois theory in lattices.* **Transactions of the American Mathematical Society**, vol. 55 (1944), pp. 514–525. [MR 6-36]

FELL, J. M. G.

1961 *The structure of algebras of operator fields.* **Acta Mathematica** , vol. 106 (1961), pp. 233–280. [MR 29:1547]

1962 *A Hausdorff topology for the closed subsets of a locally compact non-Hausdorff space.* **Proceedings of the American Mathematical Society**, vol. 13 (1962), pp. 472–476. [MR 25:2573]

FLACHSMEYER, J.

1961 *Zur Spektralentwicklung topologischer Räume.* **Mathematische Annalen**, vol. 144 (1961), pp. 253–274. [MR 26:735]

1964 *Verschiedene Topologisierungen im Raum der abgeschlossenen Mengen.* **Mathematische Nachrichten**, vol. 26 (1964), pp. 321–337. [MR 30:4233]

1965 *Einige topologische Fragen in der Theorie der Booles. .ı Algebren.* **Archiv der Mathematik**, vol. 16 (1965), pp. 25–33. [MR 32:65]

1979 *On the Busemann metrization of the hyperspace.* **Mathematische Nachrichten**, vol. 89 (1979), pp. 51–56. [MR 80i:51022]

FLEISCHER, I.

1976 *Even every join-extension solves a universal problem.* **Journal of the Australian Mathematical Society (Series A)**, vol. 21 (1976), pp. 220–223. [MR 53:216]

FLETCHER, P. and LINDGREN, W. F.

1982 **Quasi-uniform Spaces. Lecture Notes in Pure and Applied Mathematics**, Marcel Dekker, Inc., New York, 1982, viii+216 pp, vol. 77. [MR 84h:54026]

FLOYD, E. E.

1955 *Boolean algebras with pathological order topologies.* **Pacific Journal of Mathematics**, vol. 5 (1955), pp. 687–689. [MR 17-450]

FOURMAN, M. P. and SCOTT, D. S.

1979 *Sheaves and logic.* In: **Applications of Sheaves**, Proceedings of the Research Symposium on Applications of Sheaf Theory to Logic, Algebra, and Analysis, Durham, U.K., July 9-21, 1978, edited by M. P. Fourman, C. J. Mulvey and D. S. Scott. **Lecture Notes in Mathematics**, vol. 753, Springer-Verlag, 1979, pp. 302–401. [MR 82d:03061]

FOX, R. H.

1945 *On topologies for function spaces.* **Bulletin of the American Mathematical Society**, vol. 51 (1945), pp. 429–432. [MR 6-278]

FREUDENTHAL, H.

1942 *Neuaufbau der Endentheorie.* **Annals of Mathematics**, vol. 43 (1942), pp. 261–279. [MR 3-315]

FREYD, P.

1964 **Abelian Categories**. Harper and Row, 1964, xi+164 pp. [MR 29:3517]

FRIEDBERG, M.

1972 *Metrizable approximations of semigroups.* **Colloquium Mathematicum**, vol. 25 (1972), pp. 63–69 and 164. [MR 46:3678]

FRINK, O.

1942 *Topology in lattices.* **Transactions of the American Mathematical Society**, vol. 51 (1942), pp. 569–582. [MR 3-313]

1954 *Ideals in partially ordered sets.* **American Mathematical Monthly**, vol. 61 (1954), pp. 223–234. [MR 15-848]

FUCHSSTEINER, B.

1974 *Lattices and Choquet's theorem.* **Journal of Functional Analysis**, vol. 17 (1974), pp. 377-387. [MR 51:13630]

1981 *Generalized Hewitt-Nachbin spaces arising in state-space completions.* In: **Functional Analysis**, Rio de Janeiro, 1978, edited by S. Machado. **Lecture Notes in Mathematics**, vol. 843, Springer-Verlag, 1981, pp. 296–318. [MR 82i:46037]

GASKILL, H. S.

1973 *Classes of semilattices associated with an equational class of lattices.* **Canadian Journal of Mathematics**, vol. 25 (1973), pp. 361–365. [MR 47:6566]

GEISSINGER, L. and GRAVES, W.

1972 *The category of complete algebraic lattices.* **Journal of Combinatorial Theory (A)**, vol. 13 (1972), pp. 332–338. [MR 46:5193]

GEORGESCU, G. and LUNGULESCU, B.

1969 *Sur les propriétés topologiques des structures ordonnées.* **Revue Roumaine de Mathématiques Pures et Appliquées**, vol. 14 (1969), pp. 1453–1456. [MR 41:1614]

GIERZ, G.

1982 *Bündel und stetige Verbände.* In: **Continuous Lattices and Related Topics**, Proceedings of the Conference on Topological and Categorical Aspects of Continuous Lattices (Workshop V), Universität Bremen, Germany, May 8–10, 1981, edited by R.-E. Hoffmann. **Mathematik-Arbeitspapiere, Universität Bremen**, vol. 27, 1982, pp. 51–58.

1982 **Bundles of Topological Vector Spaces and Their Duality. Lecture Notes in Mathematics**, Springer-Verlag, 1982, iv+296 pp, vol. 955. [MR 84c:46076]

1982 *Colimits of continuous lattices.* **Journal of Pure and Applied Algebra**, vol. 23 (1982), pp. 137–144. [MR 83b:06019]

GIERZ, G. and HOFMANN, K. H.

1978 *On a lattice-theoretical characterization of compact semilattices.* Manuscript, Tulane University, 1978.

GIERZ, G., HOFMANN, K. H., KEIMEL, K., LAWSON, J. D., MISLOVE, M. W. and SCOTT, D. S.

1980 **A Compendium of Continuous Lattices**. Springer-Verlag, New York, 1980, xx+371 pp. [MR 82h:06005]

GIERZ, G. and KEIMEL, K.

1976 *Topologische Darstellung von Verbänden.* **Mathematische Zeitschrift**, vol. 150 (1976), pp. 83–99. [MR 55:2694]

1977 *A lemma on primes appearing in algebra and analysis.* **Houston Journal of Mathematics**, vol. 3 (1977), pp. 207–224. [MR 57:193]

1981 *Continuous ideal completions and compactifications.* In: **Continuous Lattices**, Proceedings of the Conference on Topological and Categorical Aspects of Continuous Lattices (Workshop IV), University of Bremen, Germany, November 9-11, 1979, edited by B. Banaschewski and R.-E. Hoffmann. **Lecture Notes in Mathematics**, vol. 871, Springer-Verlag, 1981, pp. 97–124.

GIERZ, G. and LAWSON, J. D.

1981 *Generalized continuous and hypercontinuous lattices.* **The Rocky Mountain Journal of Mathematics**, vol. 11 (1981), pp. 271–296. [MR 82h:54069]

GIERZ, G., LAWSON, J. D. and STRALKA, A. R.

1983 *Metrizability conditions for completely distributive lattices.* **Canadian Mathematical Bulletin**, vol. 26 (1983), pp. 446–453. [MR 85c:06013]

1983 *Quasicontinuous posets.* **Houston Journal of Mathematics**, vol. 9 (1983), pp. 191–208. [MR 85b:06009]

GIERZ, G. and STRALKA, A. R.

1982 *The injective hull of a distributive lattice and congruence extension.* Manuscript, University of California at Riverside, 1982, 16 pp.

1982 *Natural topologies on distributive lattices.* Manuscript, University of California at Riverside, 1982.

1982 *Reductive lattices and essential embeddings.* Manuscript, University of California at Riverside, 1982, 18 pp.

1985 *Natural topologies, essential extensions, reductive lattices and congruence extension.* In: **Continuous Lattices and Their Applications**, edited by R.-E. Hoffmann and K. H. Hofmann. Marcel Dekker, Inc., 1985, pp. 41–55.

GILES, R.

1977 *Continuous lattices in the foundations of physics.* Manuscript, Queen's University, Kingston, Ont., 1977.

1979 *The concept of a proposition in classical and quantum physics.* **Studia Logica**, vol. 38 (1979), pp. 337–353. [MR 81f:81005]

GINGRAS, A. R.

1976 *Convergence lattices.* **The Rocky Mountain Journal of Mathematics**, vol. 6 (1976), pp. 85–104. [MR 52:10532]

1976 *Order convergence and order ideals.* In: **Proceedings of the Conference on Convergence Spaces**, University of Nevada, Reno, 1976, edited by D. C. Kent and E. F. Wagner. 1976, pp. 45–59. [MR 55:228]

1978 *Complete distributivity and order convergence.* Manuscript, 1978.

GIULI, E.

1971 *Una caratterizzazione degli spazi ad aperti localmente compatti.* **Atti della Accademia Nazionale dei Lincei. Rendiconti. Classe di Scienze Fisiche, Mathematiche e Naturali. Serie VIII**, vol. 50 (1971), pp. 24–28. [MR 45:7667]

GLEASON, A. M.

1958 *Projective topological spaces.* **Illinois Journal of Mathematics**, vol. 2 (1958), pp. 482–489. [MR 22:12509]

GLIMM, J.

1962 *Families of induced representations.* **Pacific Journal of Mathematics**, vol. 12 (1962), pp. 885–911. [MR 26:3819]

GOGUEN, J. A., THATCHER, J. W., WAGNER, E. G. and WRIGHT, J. B.

1977 *Initial algebra semantics and continuous algebras.* **Journal of the Association for Computing Machinery**, vol. 24 (1977), pp. 68–95.

GOODYKOONTZ, J. T., JR.

1977 *C(X) is not necessarily a retract of 2^X.* **Proceedings of the American Mathematical Society**, vol. 67 (1977), pp. 177–178. [MR 56:16580]

GORBACHËV, N. V.

1981 *Essential extensions of C^*-algebras for a C-dual C^*-algebra (Russian. English summary).* **Vestnik Moskovskogo Universiteta, Ser. I. Mat. Mekh.**, 1981, pp. 27–30, 115. [MR 84h:46073]

GRÄTZER, G.

1978 **General Lattice Theory**. Birkhäuser, Basel, 1978, xiii+381 pp. [MR 80c:06001]

GROSS, J. I.

1967 *A third definition of local compactness.* **American Mathematical Monthly**, vol. 74 (1967), pp. 1120–1122.

GUESSARIAN, I.

1979 *On continuous completions.* In: **Theoretical Computer Science: 4th GI Conference**, Aachen, March 26-28, 1979, edited by K. Weihrauch. **Lecture Notes in Computer Science**, vol. 67, Springer-Verlag, 1979, pp. 142–152. [MR 83e:06010]

HALMOS, P. R.

1963 **Lectures on Boolean Algebras**. D. Van Nostrand Company, Inc., 1963, v+147 pp. [MR 29:4713]

1974 **Lectures on Boolean Algebras**. Springer-Verlag, 1974, v+147 pp. (Reprint of [Halmos 1963]).

HARVEY, J. M.

1983 *Categorical characterization of uniform hyperspaces.* **Mathematical Proceedings of the Cambridge Philosophical Society**, vol. 94 (1983), pp. 229–233.

HERMES, H.

1967 **Einführung in die Verbandstheorie**. Springer-Verlag, Berlin, 1967, xii+209 pp. [MR 36:3686]

HIGGS, D. A.

1971 *Lattices isomorphic to their ideal lattices.* **Algebra Universalis**, vol. 1 (1971), pp. 71–72. [MR 45:123]

HOCHSTER, M.

1969 *Prime ideal structure in commutative rings.* **Transactions of the American Mathematical Society**, vol. 142 (1969), pp. 43–60. [MR 40:4257]

HOFFMANN, R.-E.

1975 *Charakterisierung nüchterner Räume.* **Manuscripta Mathematica**, vol. 15 (1975), pp. 185–191. [MR 51:11405]

1976 *Topological functors admitting generalized Cauchy-completions.* In: **Categorical Topology**, Mannheim, 1975, edited by E. Binz and H. Herrlich. **Lecture Notes in Mathematics**, vol. 540, Springer-Verlag, 1976, pp. 286–344. [MR 54:30994]

1977 *Irreducible filters and sober spaces.* **Manuscripta Mathematica**, vol. 22 (1977), pp. 365–380. [MR 57:4107]

1979 *Continuous posets and adjoint sequences.* **Semigroup Forum**, vol. 18 (1979), pp. 173–188. [MR 80h:18002]

1979 *Essentially complete T_0-spaces.* **Manuscripta Mathematica**, vol. 27 (1979), pp. 401–432. [MR 81c:54060]

1979 *On the sobrification remainder ${}^{S}X$-X.* **Pacific Journal of Mathematics**, vol. 83 (1979), pp. 145–156. [MR 81b:54024]

1979 *Sobrification of partially ordered sets.* **Semigroup Forum**, vol. 17 (1979), pp. 123–138. [MR 81c:54013]

1979 *Topological spaces admitting a "dual".* In: **Categorical Topology**, Proceedings of the International Conference, Berlin, August 27th to September 2nd, 1978, Berlin, August 27th to September 2nd, 1978, edited by H. Herrlich and G. Preuss. **Lecture Notes in Mathematics**, vol. 719, Springer-Verlag, 1979, pp. 157–166. [MR 80j:54001]

1981 *Essential extensions of T_1-spaces.* **Canadian Mathematical Bulletin**, vol. 24 (1981), pp. 237–240. [MR 82j:18009]

1981 *Continuous posets, prime spectra of completely distributive complete lattices, and Hausdorff compactifications.* In: **Continuous Lattices**, Proceedings of the Conference on Topological and Categorical Aspects of Continuous Lattices (Workshop IV), University of Bremen, Germany, November 9-11, 1979, edited by B. Banaschewski and R.-E. Hoffmann. **Lecture Notes in Mathematics**, vol. 871, Springer-Verlag, 1981, pp. 159–208. (Corrections in: **Zentralblatt für Mathematik**, vol. 476, 1982, review 06005.).

1981 *The CL-compactification of a continuous poset.* Manuscript, Universität Bremen, 1981, 64 pp.

1981 *Projective sober spaces.* In: **Continuous Lattices**, Proceedings of the Conference on Topological and Categorical Aspects of Continuous Lattices (Workshop IV), University of Bremen, Germany, November 9-11, 1979, edited by B. Banaschewski and R.-E. Hoffmann. **Lecture Notes in Mathematics**, vol. 871, Springer-Verlag, 1981, pp. 125-158. [MR 82b:54017] (Corrections in **Zentralblatt für Mathematik**, vol. 476, 1982, review 06004.).

1982 *The CL-compactification and the injective hull of a continuous poset.* Manuscript, Universität Bremen, 1982, 51 pp.

1982 ***Continuous posets: injective hull and MacNeille completion.*** Technical Report, Universität Bremen, 1982, 36 pp.

1982 *The injective hull and the CL-compactification of a continuous poset.* **Seminarbericht Mathematik der Fernuniversität Hagen**, vol. 16 (1982), pp. 31-92.

1982 *Essentially complete* T_0*-spaces, II. A lattice-theoretic approach.* **Mathematische Zeitschrift**, vol. 179 (1982), pp. 73–90. [MR 83i:54038]

1983 *Duality for distributive compact multiplicative continuous lattices.* Manuscript, Universität Bremen, 1983, 16 pp.

1985 *The Fell compactification revisited.* In: **Continuous Lattices and Their Applications**, edited by R.-E. Hoffmann and K. H. Hofmann. Marcel Dekker, Inc., 1985, pp. 57–116.

1985 *The trace of the weak topology and of the* Γ*-topology of* L^{op} *coincide on the pseudo-meet-prime elements of a continuous lattice* L. In: **Continuous Lattices and Their Applications**, edited by R.-E. Hoffmann and K. H. Hofmann. Marcel Dekker, Inc., 1985, pp. 117–119.

HOFFMANN, R.-E. (ed.)

1982 **Continuous Lattices and Related Topics.** Proceedings of the Conference on Topological and Categorical Aspects of Continuous Lattices (Workshop V), Universität Bremen, Germany, May 8–10, 1981, **Mathematik-Arbeitspapiere, Universität Bremen**, vol. 27, 1982, vii+314 pp.

HOFFMANN, R.-E. and HOFMANN, K. H. (eds.)

1985 **Continuous Lattices and Their Applications.** Proceedings of the Conference on Topological and Categorical Aspects of Continuous Lattices (Workshop VI), University of Bremen, Germany, July 2–3, 1982, **Lecture Notes in Pure and Applied Mathematics**, Marcel Dekker, Inc., 1985, 00+000 pp.

HOFMANN, K. H.

1970 *A general invariant metrization theorem for compact spaces.* **Fundamenta Mathematicae**, vol. 68 (1970), pp. 281–296. [MR 42:2428]

1978 *Continuous lattices, topology and topological algebra.* In: **Topology Proceedings, 2**, Auburn, AL, 1977, edited by W. Kuperberg, G. M. Reed and Ph. Zenor. Auburn University, 1978, pp. 179–212. [MR 80k:06011]

1980 *A note on Baire spaces and continuous lattices.* **Bulletin of the Australian Mathematical Society**, vol. 21 (1980), pp. 265–279. [MR 81k:54049]

1985 *Complete distributivity and the essential hull of a* T_0*-space.* In: **Continuous Lattices and Their Applications**, edited by R.-E. Hoffmann and K. H. Hofmann. Marcel Dekker, Inc., 1985, pp. 121–127.

198? *The order-theoretical aspects of the essential hull of a topological T_0-space.* **Discrete Mathematics**, (198?). (To appear).

HOFMANN, K. H. and KEIMEL, K.

1972 **A General Character Theory for Partially Ordered Sets and Lattices. Memoirs of the American Mathematical Society**, American Mathematical Society, 1972, iv+121 pp, vol. 122. [MR 49:4885]

HOFMANN, K. H. and LAWSON, J. D.

1976 *Irreducibility and generation in continuous lattices.* **Semigroup Forum**, vol. 13 (1976), pp. 307–353. [MR 55:7868]

1978 *The spectral theory of distributive continuous lattices.* **Transactions of the American Mathematical Society**, vol. 246 (1978), pp. 285–310. [MR 80c:54045]

1984 *On the order theoretical foundation of a theory of quasicompactly generated spaces without separation axiom.* **Journal of the Australian Mathematical Society (Series A)**, vol. 36 (1984), pp. 194–212.

HOFMANN, K. H. and MISLOVE, M. W.

1973 *Lawson semilattices do have a Pontryagin duality.* In: **Proceedings of the University of Houston Lattice Theory Conference**, Houston, Texas, March 22-24, 1973, edited by S. Fajtlowicz and K. Kaiser. University of Houston, 1973, pp. 200–205. [MR 53:10970]

1975 *Epics of compact Lawson semilattices are surjective.* **Archiv der Mathematik**, vol. 26 (1975), pp. 337–345. [MR 51:12636]

1976 *Amalgamation in categories with concrete duals.* **Algebra Universalis**, vol. 6 (1976), pp. 327–347. [MR 56:5676]

1977 *The lattice of kernel operators and topological algebra.* **Mathematische Zeitschrift**, vol. 154 (1977), pp. 175–188. [MR 56:2884]

1981 *Local compactness and continuous lattices.* In: **Continuous Lattices**, Proceedings of the Conference on Topological and Categorical Aspects of Continuous Lattices (Workshop IV), University of Bremen, Germany, November 9-11, 1979, edited by B. Banaschewski and R.-E. Hoffmann. **Lecture Notes in Mathematics**, vol. 871, Springer-Verlag, 1981, pp. 209–248.

1984 *Errata for* "**A Compendium of Continuous Lattices**". Manuscript, TH Darmstadt, 1984, 5 pp.

1985 *Free objects in the category of completely distributive lattices.* In: **Continuous Lattices and Their Applications**, edited by R.-E. Hoffmann and K. H. Hofmann. Marcel Dekker, Inc., 1985, pp. 129–150.

HOFMANN, K. H., MISLOVE, M. W. and STRALKA, A. R.

1973 *Dimension raising maps in topological algebra.* **Mathematische Zeitschrift**, vol. 135 (1973), pp. 1–36. [MR 49:3019]

1974 **The Pontryagin Duality of Compact 0-Dimensional Semilattices and its Applications. Lecture Notes in Mathematics**, Springer-Verlag, 1974, xvi+122 pp, vol. 396. [MR 50:7398]

1975 *On the dimensional capacity of semilattices.* **Houston Journal of Mathematics**, vol. 1 (1975), pp. 43–55. [MR 54:452]

HOFMANN, K. H. and MOSTERT, P. S.

1966 **Elements of Compact Semigroups**. Merrill, 1966, xiii+384 pp. [MR 35:285] (See also: J. H. Carruth, K. H. Hofmann, M. W. Mislove, *Errors in **"Elements of Compact Semigroups"***, **Semigroup Forum**, vol. 5, 1972, pp. 285-322 [MR 47:8757]).

HOFMANN, K. H. and STRALKA, A. R.

1973 *Mapping cylinders and compact monoids.* **Mathematische Annalen**, vol. 205 (1973), pp. 219–239. [MR 48:2299]

1973 *Push-outs and strict projective limits of semilattices.* **Semigroup Forum**, vol. 5 (1973), pp. 243–261. [MR 47:5167]

1976 *The algebraic theory of compact Lawson semilattices. Applications of Galois connections to compact semilattices.* **Dissertationes Mathematicae**, vol. 137 (1976), pp. 1–54. [MR 55:213]

HOFMANN, K. H. and THAYER, F. J.

1980 *Approximately finite dimensional C^*-algebras.* **Dissertationes Mathematicae**, vol. 174 (1980), pp. 1–59. [MR 81h:46072]

HOFMANN, K. H. and WATKINS, F.

1981 *The spectrum as a functor.* In: **Continuous Lattices**, Proceedings of the Conference on Topological and Categorical Aspects of Continuous Lattices (Workshop IV), University of Bremen, Germany, November 9-11, 1979, edited by B. Banaschewski and R.-E. Hoffmann. **Lecture Notes in Mathematics**, vol. 871, Springer-Verlag, 1981, pp. 249–263.

HONG, S. S.

1975 *Extensive subcategories of the category of T_0-spaces.* **Canadian Journal of Mathematics**, vol. 27 (1975), pp. 311–318. [MR 51:11406]

1980 *0-dimensional compact ordered spaces.* **Kyungpook Mathematical Journal**, vol. 20 (1980), pp. 159–167. [MR 82j:54065]

HORN, A. and KIMURA, N.

1971 *The category of semilattices.* **Algebra Universalis**, vol. 1 (1971), pp. 26–38. [MR 47:6568]

HOSONO, CH. and SATO, M.

1977 *The retracts in $P\omega$ do not form a continuous lattice – a solution to Scott's problem.* **Theoretical Computer Science**, vol. 4 (1977), pp. 137–142. [MR 58:8442]

HYLAND, J. M. E.

1979 *Continuity in spatial toposes.* In: **Applications of Sheaves**, edited by M. P. Fourman, C. J. Mulvey and D. S. Scott. **Lecture Notes in Mathematics**, vol. 753, Springer-Verlag, 1979, pp. 442–465. [MR 83g: 18009]

1981 *Function spaces in the category of locales.* In: **Continuous Lattices**, Proceedings of the Conference on Topological and Categorical Aspects of Continuous Lattices (Workshop IV), University of Bremen, Germany, November 9-11, 1979, edited by B. Banaschewski and R.-E. Hoffmann. **Lecture Notes in Mathematics**, vol. 871, Springer-Verlag, 1981, pp. 264–281.

INSEL, A. J.

1963 *A compact topology for a lattice.* **Proceedings of the American Mathematical Society**, vol. 14 (1963), pp. 382–385. [MR 26:6940]

1964 *A relationship between the complete topology and the order topology of a lattice.* **Proceedings of the American Mathematical Society**, vol. 15 (1964), pp. 847–850. [MR 29:4712]

ISBELL, J. R.

1972 *Atomless parts of spaces.* **Symposia Mathematica**, vol. 31 (1972), pp. 5–32. [MR 50:11184]

1975 *Function spaces and adjoints.* **Symposia Mathematica**, vol. 36 (1975), pp. 317–339. [MR 53:9134]

1975 *Meet-continuous lattices.* **Symposia Mathematica**, vol. 16 (1975), pp. 41–54.

1981 *Product spaces in locales.* **Proceedings of the American Mathematical Society**, vol. 81 (1981), pp. 116–118. [MR 82c:54006]

1982 *Direct limits of meet-continuous lattices.* **Journal of Pure and Applied Algebra**, vol. 23 (1982), pp. 33–35. [MR 83a:18015]

1982 *Completion of a construction of Johnstone.* **Proceedings of the American Mathematical Society**, vol. 85 (1982), pp. 333–334. [MR 83i:06011]

1983 *A frame with no admissible topology.* **Mathematical Proceedings of the Cambridge Philosophical Society**, vol. 94 (1983), pp. 447–448.

1985 *Discontinuity of meets and joins.* In: **Continuous Lattices and Their Applications**, edited by R.-E. Hoffmann and K. H. Hofmann. Marcel Dekker, Inc., 1985, pp. 151–154.

JAKUBIK, J.

1968 *Higher degrees of distributivity in lattices and lattice-ordered groups.* **Czechoslovak Mathematical Journal**, vol. 18 (1968), pp. 356–376. [MR 37:1283]

JAMISON, R. E.

1974 ***A general theory of convexity.*** Doctoral Dissertation, University of Washington, 1974, 127 pp.

1981 *The space of maximal convex sets.* **Fundamenta Mathematicae**, vol. 111 (1981), pp. 45–59. [MR 82h:52003]

JANSEN, S. L.

1975 *Représentation d'un ensemble ordonné dans un sous-treillis d'un treillis complet et complètement distributif.* **Comptes Rendus de l' Académie des Sciences, Paris, Série A-B**, vol. 281 (1975), pp. A127-A128. [MR 52:10514]

1978 *A subdirect representation of partially ordered sets.* **The Journal of the London Mathematical Society**, vol. 17 (1978), pp. 195–202. [MR 58:5423]

JARZEMBSKI, G.

1982 *Free ω-complete algebras.* **Algebra Universalis**, vol. 14 (1982), pp. 231–234. [MR 83a:08017]

JOHNSTONE, P. T.

1980 *The Gleason cover of a topos, I.* **Journal of Pure and Applied Algebra**, vol. 19 (1980), pp. 171–192. [MR 82a:18002]

1981 *Injective toposes.* In: **Continuous Lattices**, Proceedings of the Conference on Topological and Categorical Aspects of Continuous Lattices (Workshop IV), University of Bremen, Germany, November 9-11, 1979, edited by B. Banaschewski and R.-E. Hoffmann. **Lecture Notes in Mathematics**, vol. 871, Springer-Verlag, 1981, pp. 284–297.

1981 *The Gleason cover of a topos, II.* **Journal of Pure and Applied Algebra**, vol. 22 (1981), pp. 229–247. [MR 83a:18011]

1981 *Scott is not always sober.* In: **Continuous Lattices**, Proceedings of the Conference on Topological and Categorical Aspects of Continuous Lattices (Workshop IV), University of Bremen, Germany, November 9-11, 1979, edited by B. Banaschewski and R.-E. Hoffmann. **Lecture Notes in Mathematics**, vol. 871, Springer-Verlag, 1981, pp. 282–283.

1982 **Stone Spaces.** Cambridge University Press, 1982, xxi+370 pp.

1982 *The Vietoris monad on the category of locales.* In: **Continuous Lattices and Related Topics**, Proceedings of the Conference on Topological and Categorical Aspects of Continuous Lattices (Workshop V), Universität Bremen, Germany, May 8–10, 1981, edited by R.-E. Hoffmann. **Mathematik-Arbeitspapiere, Universität Bremen**, vol. 27, 1982, pp. 162–179.

1983 *The point of pointless topology.* **Bulletin of the American Mathematical Society**, vol. 8 (1983), pp. 41–53. [MR 84f:01043]

1985 *Vietoris locales and localic semilattices.* In: **Continuous Lattices and Their Applications**, edited by R.-E. Hoffmann and K. H. Hofmann. Marcel Dekker, Inc., 1985, pp. 155–180.

JOHNSTONE, P. T. and JOYAL, A.

1982 *Continuous categories and exponentiable toposes.* **Journal of Pure and Applied Algebra**, vol. 25 (1982), pp. 255–296. [MR 83k:18005]

JONES, L. W., JR.

1980 ***Freeness and Continuity in Semilattices.*** Doctoral Dissertation, Tulane University, 1980, vi+146 pp.

JÓNSSON, B.

1967 *Algebras whose congruence lattices are distributive.* **Symposia Mathematica**, vol. 21 (1967), pp. 110–121. [MR 38:5689]

JUNG, A.

1983 ***Stetige Verbände und ein Approximationssatz für oberhalbstetige Funktionen.*** Diplomarbeit (Master's Thesis), Technische Hochschule Darmstadt, 1983, ii+40 pp.

KAHN, G.

1978 *Concepts fondamentaux de la théorie des modèles.* Manuscript, 1978.

KAMARA, M.

1978 *Treillis continus et treillis complètement distributifs.* **Semigroup Forum**, vol. 16 (1978), pp. 387–388. [MR 58:21881]

KANDA, A.

1982 *A categorization of effective retract calculus.* In: **Continuous Lattices and Related Topics**, Proceedings of the Conference on Topological and Categorical Aspects of Continuous Lattices (Workshop V), Universität Bremen, Germany, May 8–10, 1981, edited by R.-E. Hoffmann. **Mathematik-Arbeitspapiere, Universität Bremen**, vol. 27, 1982, pp. 180-203.

KAPPOS, D. A. and PAPANGELOU, F.

1966 *Remarks on the extension of continuous lattices.* **Mathematische Annalen**, vol. 166 (1966), pp. 277–283. [MR 33:7279]

KEIMEL, K.

1972 *A unified theory of minimal prime ideals.* **Acta Mathematica Academiae Scientiarum Hungaricae**, vol. 23 (1972), pp. 51–69. [MR 47:6586]

KEIMEL, K. and GIERZ, G.

1982 *Halbstetige Funktionen und stetige Verbände.* In: **Continuous Lattices and Related Topics**, Proceedings of the Conference on Topological and Categorical Aspects of Continuous Lattices (Workshop V), Universität Bremen, Germany, May 8–10, 1981, edited by R.-E. Hoffmann. **Mathematik-Arbeitspapiere, Universität Bremen**, vol. 27, 1982, pp. 59–67.

KELLEY, J. L.

1955 **General Topology**. D. Van Nostrand Company, Inc., 1955, xiv+298 pp. [MR 16-1136]

1975 **General Topology**. Springer-Verlag, 1975, xiv+298 pp. [MR 51:6681] (Reprint of [Kelley 1955]).

KENT, D. C.

1966 *On the order topology in a lattice.* **Illinois Journal of Mathematics**, vol. 10 (1966), pp. 90–96. [MR 32:6425]

1967 *The interval topology and order convergence as dual convergence structures.* **American Mathematical Monthly**, vol. 74 (1967), pp. 426–427, 1231. [MR 36:1369 (E)]

KERSTAN, J.

1960 *Tensorielle Erweiterungen distributiver Verbände.* **Mathematische Nachrichten**, vol. 22 (1960), pp. 1–20. [MR 25:3874]

KNIGHT, C. J., MORAN, W. and PYM, J. S.

1970 *The topologies of separate continuity, I.* **Proceedings of the Cambridge Mathematical Society**, vol. 68 (1970), pp. 663–671. [MR 42:2422]

1972 *The topologies of separate continuity, II.* **Proceedings of the Cambridge Mathematical Society**, vol. 71 (1972), pp. 307–319. [MR 45:9282]

KOCH, R. J.

1959 *Arcs in partially ordered spaces.* **Pacific Journal of Mathematics**, vol. 9 (1959), pp. 723–728. [MR 21:7269]

1965 *Connected chains in quasi-ordered spaces.* **Fundamenta Mathematicae**, vol. 56 (1965), pp. 245–249. [MR 30:5263]

KOCH, R. J. and KRULE, I. S.

1960 *Weak cutpoint ordering in hereditarily unicoherent continua.* **Proceedings of the American Mathematical Society**, vol. 11 (1960), pp. 679–681. [MR 22:11356]

KOLIBIAR, M.

1962 *Bemerkungen über Intervalltopologie in halbgeordneten Mengen.* In: **General Topology and its Relations to modern Analysis and Algebra**, Prague, 1961, edited by J. Novák. Academic Press, 1962, pp. 252–253. [MR 26:4940]

1972 *Distributive sublattices of a lattice.* **Proceedings of the American Mathematical Society**, vol. 34 (1972), pp. 359–364. [MR 45:3262]

KOWALSKY, H. J.

1960 *Verbandstheoretische Kennzeichnung topologischer Räume.* **Mathematische Nachrichten**, vol. 21 (1960), pp. 297–318. [MR 22:9952]

KRISHNAN, V. S.

1978 *Categories of preordered spaces.* **Notices of the American Mathematical Society**, vol. 25 (1978), pp. A-668,A-669.

KRULE, I. S.

1957 *Structs on the 1-sphere.* **Duke Mathematical Journal**, vol. 24 (1957), pp. 405–413. [MR 19-669]

KUCERA, T. G. and SANDS, B.

1978 *Lattices of lattice homomorphisms.* **Algebra Universalis**, vol. 8 (1978), pp. 180–190. [MR 57:3019]

KURATOWSKI, K.

1965 *Sur la topologie exponentielle des algèbres Brouweriennes.* In: **Simposio di Topologia**, Messina, 1964, edited by W. Sierpinski. Edizioni Oderisi, 1965, pp. 1–4.

LAMBEK, J.

1968 *A fixpoint theorem for complete categories.* **Mathematische Zeitschrift**, vol. 103 (1968), pp. 151–161. [MR 37:270]

LAMBRINOS, P. TH.

1973 *A topological notion of boundedness.* **Manuscripta Mathematica**, vol. 10 (1973), pp. 289–296. [MR 48:5003]

1974 *Locally bounded spaces.* **Proceedings of the Edinburgh Mathematical Society**, vol. 19 (1974), pp. 321–325. [MR 52:15373]

1980 *Boundedly generated topological spaces.* **Manuscripta Mathematica**, vol. 31 (1980), pp. 425–438. [MR 83m:54024]

1981 *The bounded-open topology on function spaces.* **Manuscripta Mathematica**, vol. 36 (1981), pp. 47–66. [MR 83m:54025]

1981 *On products of R-quotient maps, k-spaces and k_R-spaces.* (Abstract, no. 81T-54-501), **Abstracts of Papers Presented to the American Mathematical Society**, vol. 2 (1981), p. 555.

1981 *On the compact-open and the e-compact-open topology.* (Abstract, no. 786-54-68), **Abstracts of Papers Presented to the American Mathematical Society**, vol. 2 (1981), p. 356.

1981 *Weakly continuous lattices and the strong Scott-topology on function spaces.* (Abstract, no. 788-54-92), **Abstracts of Papers Presented to the American Mathematical Society**, vol. 2 (1981), p. 445.

1982 *On the Scott function space topology.* (Abstract, no. 796-54-282), **Abstracts of Papers Presented to the American Mathematical Society**, vol. 3 (1982), p. 369.

1983 *On the exponential law for function spaces.* (Abstract, no. 801-54-292), **Abstracts of Papers Presented to the American Mathematical Society**, vol. 4 (1983), p. 101.

1983 *The category of boundedly generated spaces is not cartesian closed.* Manuscript, University of Thrace, 1983.

1985 *On the exponential law for function spaces equipped with the compact open topology.* In: **Continuous Lattices and Their Applications**, edited by R.-E. Hoffmann and K. H. Hofmann. Marcel Dekker, Inc., 1985, pp. 181–190.

LAMBRINOS, P. TH. and PAPADOPOULOS, B.

1985 *The (strong) Isbell topology and (weakly) continuous lattices.* In: **Continuous Lattices and Their Applications**, edited by R.-E. Hoffmann and K. H. Hofmann. Marcel Dekker, Inc., 1985, pp. 191–211.

LAMBROU, M. S.

1978 *Semisimple completely distributive lattices are Boolean algebras.* **Proceedings of the American Mathematical Society**, vol. 68 (1978), pp. 217–219. [MR 57:3030]

LAU, A. Y. W.

1972 *Concerning costability of compact semigroups.* **Duke Mathematical Journal**, vol. 39 (1972), pp. 657–664. [MR 47:1999]

1972 *Small semilattices.* **Semigroup Forum**, vol. 4 (1972), pp. 150–155. [MR 45:1808]

1973 *Costability in SEM and TSL.* **Semigroup Forum**, vol. 5 (1973), pp. 370–372. [MR 48:8679]

1973 *Coverable semigroups.* **Proceedings of the American Mathematical Society**, vol. 38 (1973), pp. 661–664. [MR 46:9224]

1975 *The boundary of a semilattice on an n-cell.* **Pacific Journal of Mathematics**, vol. 56 (1975), pp. 171–174. [MR 51:10180]

1976 *Existence of n-cells in Peano semilattices.* In: **Topology (Proceedings, Memphis State University)**, Memphis, TN, 1975, edited by S. P. Franklin and B. V. Smith Thomas. **Lecture Notes in Pure and Applied Mathematics**, vol. 24, Marcel Dekker, Inc., 1976, pp. 197–200. [MR 55:3148]

LAURSEN, K. B. and SINCLAIR, A. M.

1975 *Lifting matrix units in C*-algebras, II.* **Mathematica Scandinavica**, vol. 37 (1975), pp. 167–172. [MR 53:1281]

LAWSON, J. D.

1967 ***Vietoris mappings and embeddings of topological semilattices.*** Doctoral Dissertation, University of Tennessee, 1967, 99 pp.

1969 *Topological semilattices with small semilattices.* **The Journal of the London Mathematical Society**, vol. 1 (1969), pp. 719–724. [MR 40:6516]

1970 *Lattices with no interval homomorphisms.* **Pacific Journal of Mathematics**, vol. 32 (1970), pp. 459–465. [MR 51:1019]

1970 *The relation of breadth and codimension in topological semilattices.* **Duke Mathematical Journal**, vol. 37 (1970), pp. 207–212. [MR 41:3333]

1971 *The relation of breadth and codimension in topological semilattices II.* **Duke Mathematical Journal**, vol. 38 (1971), pp. 555–559. [MR 44:125]

1972 *Dimensionally stable semilattices.* **Semigroup Forum**, vol. 5 (1972), pp. 181–185. [MR 47:4874]

1972 *Joint continuity in semitopological semigroups.* **Illinois Journal of Mathematics**, vol. 18 (1972), pp. 275–285. [MR 49:454]

1973 *Intrinsic lattice and semilattice topologies.* In: **Proceedings of the University of Houston Lattice Theory Conference**, Houston, Texas, March 22-24, 1973, edited by S. Fajtlowicz and K. Kaiser. University of Houston, 1973, pp. 206–260. [MR 53:5403]

1973 *Intrinsic topologies in topological lattices and semilattices.* **Pacific Journal of Mathematics**, vol. 44 (1973), pp. 593–602. [MR 47:6580]

1976 *Additional notes on continuity in semitopological semigroups.* **Semigroup Forum**, vol. 12 (1976), pp. 265–280. [MR 53:9175]

1976 *Applications of topological algebra to hyperspace problems.* In: **Topology (Proceedings, Memphis State University)**, Memphis, TN, 1975, edited by S. P. Franklin and B. V. Smith Thomas. **Lecture Notes in Pure and Applied Mathematics**, vol. 24, Marcel Dekker, Inc., 1976, pp. 201–206.

1976 *Embeddings of compact convex sets and locally compact cones.* **Pacific Journal of Mathematics**, vol. 66 (1976), pp. 443–453. [MR 55:13213]

1977 *Compact semilattices which must have a basis of subsemilattices.* **The Journal of the London Mathematical Society**, vol. 16 (1977), pp. 367–371. [MR 57:518]

1979 *The duality of continuous posets.* **Houston Journal of Mathematics**, vol. 5 (1979), pp. 357–394. [MR 81i:06003]

1980 *Algebraic conditions leading to continuous lattices.* **Proceedings of the American Mathematical Society**, vol. 78 (1980), pp. 477–481. [MR 81g:06002]

1982 *T_0-spaces and pointwise convergence.* Manuscript, Louisiana State University, Baton Rouge, 1982.

1982 *Valuations on continuous lattices.* In: **Continuous Lattices and Related Topics**, Proceedings of the Conference on Topological and Categorical Aspects of Continuous Lattices (Workshop V), Universität Bremen, Germany, May 8–10, 1981, edited by R.-E. Hoffmann. **Mathematik-Arbeitspapiere, Universität Bremen**, vol. 27, 1982, pp. 204–225.

1985 *Obtaining the T_0-essential hull.* In: **Continuous Lattices and Their Applications**, edited by R.-E. Hoffmann and K. H. Hofmann. Marcel Dekker, Inc., 1985, pp. 213–217.

LAWSON, J. D., LIUKKONEN, J. R. and MISLOVE, M. W.

1977 *Measure algebras of semilattices with finite breadth.* **Pacific Journal of Mathematics**, vol. 69 (1977), pp. 125–139. [MR 58:12191]

LAWSON, J. D. and WILLIAMS, W.

1970 *Topological semilattices and their underlying spaces.* **Semigroup Forum**, vol. 1 (1970), pp. 209–223. [MR 42:3221]

LEA, J. W., JR.

1972 *An embedding theorem for compact semilattices.* **Proceedings of the American Mathematical Society**, vol. 34 (1972), pp. 325–331. [MR 55:10337]

1973 *The peripherality of irreducible elements of a lattice.* **Pacific Journal of Mathematics**, vol. 45 (1973), pp. 555–560. [MR 47:4876]

1974 *The codimension of the boundary of a lattice ideal.* **Proceedings of the American Mathematical Society**, vol. 43 (1974), pp. 36–38. [MR 51:7971]

1974 *Sublattices generated by chains in modular topological lattices.* **Duke Mathematical Journal**, vol. 41 (1974), pp. 241–246. [MR 55:2689]

1976 *Breadth two topological lattices with connected sets of irreducibles.* **Transactions of the American Mathematical Society**, vol. 219 (1976), pp. 337–345. [MR 53:5401]

1976 *Continuous lattices and compact Lawson semilattices.* **Semigroup Forum**, vol. 13 (1976), pp. 387–388. [MR 55:7864]

1978 *Quasiplanar topological lattices.* **Houston Journal of Mathematics**, vol. 4 (1978), pp. 85–90. [MR 57:12311]

1980 *Continuous modular lattices of breadth two.* **Semigroup Forum**, vol. 19 (1980), pp. 387–388. [MR 81d:06009]

LEA, J. W., JR. and LAU, A. Y. W.

1975 *Codimension of compact M-semilattices.* **Proceedings of the American Mathematical Society**, vol. 52 (1975), pp. 406–408. [MR 51:10524]

LEHMANN, D. J.

1980 *On the algebra of order.* **Journal of Computer and System Sciences**, vol. 21 (1980), pp. 1–23. [MR 82b:68042]

LEUSCHEN, J. E. and SIMS, B. T.

1972 *Stronger forms of connectivity.* **Rendiconti di Circolo Matematico de la Università di Palermo**, vol. 21 (1972), pp. 255–266. [MR 50:3187]

LEVINE, N.

1965 *Strongly connected sets in topology.* **American Mathematical Monthly**, vol. 72 (1965), pp. 1098–1101. [MR 32:3031]

LIBER, S. A.

1977 *L-free compact lattices (Russian).* **Issledovaniya Alg., Saratov**, vol. 5 (1977), pp. 44–52. [MR 58:10645]

1978 *Free compact lattices (Russian).* **Matematicheskii Zametki**, vol. 24 (1978), pp. 621–627, 733. [MR 80c:06014]

1978 *Free compact lattices (English translation).* **Mathematical Notes of the Academy of Sciences of the USSR**, vol. 24 (1978), pp. 832–835. [MR 80c:06014]

LINTON, F. E. J.

1969 *An outline of functorial semantics.* In: **Seminar on Triples and Categorical Homology Theory**, ETH Zürich, 1966/67, edited by B. Eckmann. **Lecture Notes in Mathematics**, vol. 80, Springer-Verlag, 1969, pp. 7–52. [MR 39:5655]

LIUKKONEN, J. R. and MISLOVE, M. W.

1983 *Measure algebras of locally compact semilattices.* In: **Recent Developments in the Algebraic, Analytical, and Topological Theory of Semigroups**, Oberwolfach, 1981, edited by K. H. Hofmann, H. Jürgensen and H. J. Weinert. **Lecture Notes in Mathematics**, vol. 998, Springer-Verlag, 1983, pp. 202–214.

LONGSTAFF, W. E.

1975 *Strongly reflexive lattices.* **The Journal of the London Mathematical Society**, vol. 11 (1975), pp. 491–498. [MR 52:15036]

1976 *Operators of rank one in reflexive algebras.* **Canadian Journal of Mathematics**, vol. 28 (1976), pp. 19–23. [MR 53:1294]

LORRAIN, F.

1969 *Notes on topological spaces with minimum neighborhoods.* **American Mathematical Monthly**, vol. 76 (1969), pp. 616–627. [MR 40:1966]

LOWEN-COLEBUNDERS, E.

1982 *On the uniformization of the hyperspace of closed convergence.* **Mathematische Nachrichten**, vol. 105 (1982), pp. 35–44. [MR 84a:54005]

LOWIG, H. F. J.

1964 *Note on the self-duality of the unrestricted distributive law in complete lattices.* **Israel Journal of Mathematics**, vol. 2 (1964), pp. 170–172. [MR 31:2176]

LYSTAD, G. S. and STRALKA, A. R.

1979 *Semilattices having bialgebraic congruence lattices.* **Pacific Journal of Mathematics**, vol. 85 (1979), pp. 131–143. [MR 81i:06004]

1981 *Lawson semilattices with bialgebraic congruence lattices.* In: **General Topology and Modern Analysis**, University of California, Riverside, 1980, edited by L. F. McAuley and M. M. Rao. Academic Press, 1981, pp. 247–254. [MR 82h:06006]

MAC LANE, S.

1971 **Categories for the Working Mathematician.** Springer-Verlag, 1971, ix+262 pp. [MR 50:7275]

MACNAB, D. S.

1981 *Modal operators on Heyting algebras.* **Algebra Universalis**, vol. 12 (1981), pp. 5–29. [MR 82j:03082]

MACNEILLE, H. M.

1937 *Partially ordered sets.* **Transactions of the American Mathematical Society**, vol. 42 (1937), pp. 416–460.

MANES, E. G.

1976 **Algebraic Theories**. Springer-Verlag, 1976, vii+356 pp. [MR 54:7578]

MANNA, Z.

1974 **Mathematical Theory of Computation**. McGraw-Hill Book Company, New York, 1974, x+448 pp. [MR 53:4601]

MARANDA, J. M.

1964 *Injective structures.* **Transactions of the American Mathematical Society**, vol. 110 (1964), pp. 98–135. [MR 29:1236]

MARKOWSKY, G.

1976 *Chain-complete posets and directed sets with applications.* **Algebra Universalis**, vol. 6 (1976), pp. 53–68. [MR 53:2764]

1977 *Categories of chain-complete posets.* **Theoretical Computer Science**, vol. 4 (1977), pp. 125–135. [MR 56:11859]

1979 *Free completely distributive lattices.* **Proceedings of the American Mathematical Society**, vol. 74 (1979), pp. 227–228. [MR 80c:06017]

1981 *A motivation and generalization of Scott's notion of a continuous lattice.* In: **Continuous Lattices**, Proceedings of the Conference on Topological and Categorical Aspects of Continuous Lattices (Workshop IV), University of Bremen, Germany, November 9-11, 1979, edited by B. Banaschewski and R.-E. Hoffmann. **Lecture Notes in Mathematics**, vol. 871, Springer-Verlag, 1981, pp. 298–307.

1981 *Propaedeutic to chain-complete posets with basis.* In: **Continuous Lattices**, Proceedings of the Conference on Topological and Categorical Aspects of Continuous Lattices (Workshop IV), University of Bremen, Germany, November 9-11, 1979, edited by B. Banaschewski and R.-E. Hoffmann. **Lecture Notes in Mathematics**, vol. 871, Springer-Verlag, 1981, pp. 308–314.

MARKOWSKY, G. and ROSEN, B. K.

1976 *Bases for chain-complete posets.* **IBM Journal of Research and Development**, vol. 20 (1976), pp. 138–147. [MR 52:13523 (EA 54)]

MARTINEZ, J.

1972 *Unique factorization in partially ordered sets.* **Proceedings of the American Mathematical Society**, vol. 33 (1972), pp. 213–220. [MR 45:1806]

MATSUSHIMA, Y.

1960 *Hausdorff interval topology on a partially ordered set.* **Proceedings of the American Mathematical Society**, vol. 11 (1960), pp. 233–235. [MR 22:2567]

MAURICE, M.

1964 **Compact Ordered Spaces**. Mathematisch Centrum, Amsterdam, 1964, 76 pp. [MR 36:3318]

MCWATERS, M. M.

1969 *A note on topological semilattices.* **The Journal of the London Mathematical Society**, vol. 1 (1969), pp. 64–66. [MR 39:4314]

MESEGUER, J.

1980 *Varieties of chain-complete algebras.* **Journal of Pure and Applied Algebra**, vol. 19 (1980), pp. 347–383. [MR 82g:18005]

1983 *Order completion monads.* **Algebra Universalis**, vol. 16 (1983), pp. 63–82. [MR 84i:18005]

MICHAEL, E.

1951 *Topologies on spaces of subsets.* **Transactions of the American Mathematical Society**, vol. 71 (1951), pp. 152–182. [MR 13-54]

1968 *Local compactness and cartesian products of quotient maps and k-spaces.* **Annales de l'Institut Fourier**, vol. 18 (1968), pp. 281–286. [MR 39:6256]

MISLOVE, M. W.

1982 *An introduction to the theory of continuous lattices.* In: **Ordered Sets**, Proceedings of the NATO Advanced Study Institutute, BANFF, Canada, August 28 to September 12, 1981, edited by I. Rival. **Nato Advanced Study Institute, Series C**, vol. 83, D. Reidel Publishing Company, 1982, pp. 379–406. [MR 83k:06012]

MONTEIRO, A. A.

1951 *Les filtres fermés des espaces compacts.* **Gazeta de Matemática**, vol. 12 (1951), pp. 95–96. [MR 13-965]

MOORE, E. H. and SMITH, H. L.

1922 *A general theory of limits.* **American Journal of Mathematics**, vol. 44 (1922), pp. 102–121.

MORALES, P.

1973 *Pointwise compact spaces.* **Canadian Mathematical Bulletin**, vol. 16 (1973), pp. 545–549. [MR 50:3189]

MOWAT, D. G.

1968 *A Galois problem for mappings.* **Bulletin of the American Mathematical Society**, vol. 74 (1968), pp. 1095–1097. [MR 38:1037]

MRÓWKA, S.

1958 *On the convergence of nets of sets.* **Fundamenta Mathematicae**, vol. 45 (1958), pp. 237–246. [MR 20:4820]

1971 *Some comments on the space of subsets.* In: **Set-Valued Mappings, Selections and Topological Properties of 2^X**, New York, 1969, edited by W.M. Fleischman. **Lecture Notes in Mathematics**, vol. 171, Springer-Verlag, 1971, pp. 59–63. [MR 42:5216]

MURDESHWAR, M. G. and NAIMPALLY, S. A.

1966 **Quasi-uniform topological spaces.** Noordhoff, 1966, vi + 73 pp. [MR 35:2267]

NACHBIN, L.

1949 *On a characterization of the lattice of all ideals of a Boolean ring.* **Fundamenta Mathematicae**, vol. 36 (1949), pp. 137–142. [MR 11-712]

1965 **Topology and Order.** D. Van Nostrand Company, Inc., 1965, vi+122 pp. . [MR 36:2125]

NADLER, S. B.

1978 **Hyperspaces of Sets.** Marcel Dekker, Inc., 1978, xvi + 707 pp. [MR 58:18330]

NAIT-ABDALLAH, A.

1982 ***Faisceaux et Sémantique des Programmes (Thèse d'État, University of Paris, 1980).*** Technical Report, no. cs-82-08, University of Waterloo, 1982, 236 pp.

1982 *A universal domain with types: A_∞.* In: **Continuous Lattices and Related Topics**, Proceedings of the Conference on Topological and Categorical Aspects of Continuous Lattices (Workshop V), Universität Bremen, Germany, May 8–10, 1981, edited by R.-E. Hoffmann. **Mathematik-Arbeitspapiere, Universität Bremen**, vol. 27, 1982, pp. 226–259.

1985 *The local approach to programming language theory.* In: **Continuous Lattices and Their Applications**, edited by R.-E. Hoffmann and K. H. Hofmann. Marcel Dekker, Inc., 1985, pp. 219–236.

NAITO, T.

1960 *On a problem of Wolk in interval topologies.* **Proceedings of the American Mathematical Society**, vol. 11 (1960), pp. 156–158. [MR 22:1528]

NEL, L. D.

1972 *Lattices of lower semicontinuous functions and associated topological spaces.* **Pacific Journal of Mathematics**, vol. 40 (1972), pp. 667–673. [MR 46:8165]

1977 *Cartesian-closed coreflective hulls.* **Quaestiones Mathematicae**, vol. 2 (1977), pp. 269–283. [MR 58:843]

NEL, L. D. and WILSON, R. G.

1972 *Epireflections in the category of T_0-spaces.* **Fundamenta Mathematicae**, vol. 75 (1972), pp. 69–74. [MR 46:6285]

NELSON, E.

1981 *Z-continuous algebras.* In: **Continuous Lattices**, Proceedings of the Conference on Topological and Categorical Aspects of Continuous Lattices (Workshop IV), University of Bremen, Germany, November 9-11, 1979, edited by B. Banaschewski and R.-E. Hoffmann. **Lecture Notes in Mathematics**, vol. 871, Springer-Verlag, 1981, pp. 315–334.

NEWMAN, S. E.

1969 *Measure algebras on idempotent semigroups.* **Pacific Journal of Mathematics**, vol. 31 (1969), pp. 161–169. [MR 43:945]

NGUYEN, V. L. and LASSEZ, J. L.

1981 *A dual problem to least fixed points.* **Theoretical Computer Science**, vol. 16 (1981), pp. 211–221. [MR 83a:06005]

NICKEL, K.

1975 *Verbandstheoretische Grundlagen der Intervall-Mathematik.* In: **Interval Mathematics**, Proceedings of the International Symposium, Karlsruhe, West Germany, May 20-24, 1975, edited by K. Nickel. **Lecture Notes in Computer Science**, vol. 29, Springer-Verlag, 1975, pp. 251–262.

NIEFIELD, S. B.

1982 *Exactness and projectivity.* In: **Category Theory**, Gummersbach, 1981, edited by K. H. Kamps, D. Pumplün and W. Tholen. **Lecture Notes in Mathematics**, vol. 962, Springer-Verlag, 1982, pp. 221–227.

1982 *Topologically indexed function spaces and adjoint functors.* **Canadian Mathematical Bulletin**, vol. 25 (1982), pp. 169–178.

1982 *Cartesianness: topological spaces, uniform spaces and affine schemes.* **Journal of Pure and Applied Algebra**, vol. 23 (1982), pp. 147–167. [MR 83b:18012]

NIÑO-SALCEDO, J.

1981 **On continuous posets and their applications.** Doctoral Dissertation, Tulane University, 1981, iii+105 pp.

NOVAK, D.

1982 *On a duality between the concepts "finite" and "directed".* **Houston Journal of Mathematics**, vol. 8 (1982), pp. 545–563. [MR 84i:06007]

1982 *Generalization of continuous posets.* **Transactions of the American Mathematical Society**, vol. 272 (1982), pp. 645–667. [MR 83i:06007]

NUMAKURA, K.

1957 *Theorems on compact totally disconnected semigroups and lattices.* **Proceedings of the American Mathematical Society**, vol. 8 (1957), pp. 623–626. [MR 19-290]

ORE, O.

1944 *Galois connexions.* **Transactions of the American Mathematical Society**, vol. 55 (1944), pp. 493–513. [MR 6-36]

PAPERT, S.

1959 *Which distributive lattices are lattices of closed sets?.* **Proceedings of the Cambridge Philosophical Society**, vol. 55 (1959), pp. 172–176. [MR 21:3354]

PASZTOR, A.

1982 *The epis of Pos(Z).* **Commentationes Mathematicae Universitatis Carolinae**, vol. 23 (1982), pp. 285–299. [MR 83i:06005]

PEDICCHIO, M. C.

1983 *Closed structures on the category of topological spaces determined by systems of filters.* **Bulletin of the Australian Mathematical Society**, vol. 28 (1983), pp. 161–174.

1984 *On the category of topological topologies.* **Cahiers de Topologie et Géométrie Différentielle**, vol. 25 (1984), pp. 3–13.

PENOT, J. P. and THERA, M.

1979 *Semi-continuité des applications et des multiapplications.* **Comptes Rendus de l' Académie des Sciences, Paris, Série A-B**, vol. 288 (1979), pp. A241-A244. [MR 80m:58007]

1982 *Semicontinuous mappings in general topology.* **Archiv der Mathematik**, vol. 38 (1982), pp. 158–166. [MR 83h:54011]

PICKERT, G.

1952 *Bemerkungen über Galois-Verbindungen.* **Archiv der Mathematik**, vol. 3 (1952), pp. 285–289. [MR 14-529]

PLOTKIN, G. D.

1976 *A powerdomain construction: semantics and correctness of programs.* **SIAM Journal on Computing**, vol. 5 (1976), pp. 452–487. [MR 56:4224]

1978 *T^ω as a universal domain.* **Journal of Computer and System Sciences**, vol. 17 (1978), pp. 209–236. [MR 80d:68105]

POPPE, H.

1965 *Stetige Konvergenz und der Satz von Ascoli und Arzelà.* **Mathematische Nachrichten**, vol. 30 (1965), pp. 87–122. [MR 32:6400]

POUZET, M. and RIVAL, I.

1981 *Which ordered sets have a complete linear extension?.* **Canadian Journal of Mathematics**, vol. 33 (1981), pp. 1245–1254. [MR 84d:06012]

PRIESTLEY, H. A.

1970 *Representation of distributive lattices by means of ordered Stone spaces.* **Bulletin of the London Mathematical Society**, vol. 2 (1970), pp. 186–190. [MR 42:153]

1972 *Ordered topological spaces and the representation of distributive lattices.* **Proceedings of the London Mathematical Society**, vol. 24 (1972), pp. 507–530. [MR 46:109]

1975 *The construction of spaces dual to pseudo-complemented distributive lattices.* **The Quarterly Journal of Mathematics (Second Series)**, vol. 26 (1975), pp. 215–228. [MR 52:13548]

1985 *Algebraic lattices as dual spaces of distributive lattices.* In: **Continuous Lattices and Their Applications**, edited by R.-E. Hoffmann and K. H. Hofmann. Marcel Dekker, Inc., 1985, pp. 237–249.

198? *Catalytic distributive lattices and compact zero-dimensional topological lattices.* **Algebra Universalis**, (198?). (To appear).

198? *Ordered sets and duality for distributive lattices.* In: **Proc. Conf. Ens. Ordonnés Appl.**, Lyon, 1982, edited by M. Pouzet and D. Richard. North-Holland Publishing Company, 198?. (To appear).

PROTASOV, I. V.

1978 *On the lattice of subgroups of a pro-finite group (Russian, English summary).* **Akademija Nauk Ukrainskoĭ SSR. Doklady. Serija A. Fiziko-Matematičeskie i Tehničeskie Nauki**, 1978, pp. 975–978, 1053. [MR 80f:20029]

1979 *Topological groups with a compact lattice of closed subgroups (English translation).* **Siberian Mathematical Journal**, vol. 20 (1979), pp. 270–275. [MR 80i:22016]

RANEY, G. N.

1952 *Completely distributive complete lattices.* **Proceedings of the American Mathematical Society**, vol. 3 (1952), pp. 677–680. [MR 14-612]

1953 *A subdirect-union representation for completely distributive complete lattices.* **Proceedings of the American Mathematical Society**, vol. 4 (1953), pp. 518–522. [MR 15-389]

1960 *Tight Galois connections and complete distributivity.* **Transactions of the American Mathematical Society**, vol. 97 (1960), pp. 418–426. [MR 22:10928]

RAUCH, M.

1982 *Stetige Verbände in der axiomatischen Potentialtheorie.* In: **Continuous Lattices and Related Topics**, Proceedings of the Conference on Topological and Categorical Aspects of Continuous Lattices (Workshop V), Universität Bremen, Germany, May 8–10, 1981, edited by R.-E. Hoffmann. **Mathematik-Arbeitspapiere, Universität Bremen**, vol. 27, 1982, pp. 260–308.

REICHMAN, J. Z.

1983 *Semicontinuous real numbers in a topos.* **Journal of Pure and Applied Algebra**, vol. 28 (1983), pp. 81–91. [MR 84j:18002]

RENNIE, B. C.

1951 *Lattices.* **Proceedings of the London Mathematical Society**, vol. 52 (1951), pp. 386–400. [MR 13-7]

1951 **The Theory of Lattices.** Forster and Jagg, 1951, 51 pp. [MR 13-901]

REYNOLDS, J. C.

1972 **Notes on a Lattice-Theoretic Approach to the Theory of Computation.** Syracuse University, 1972, 160 pp.

1975 *On the interpretation of Scott domains.* **Symposia Mathematica**, vol. 15 (1975), pp. 123–135. [MR 54:1704]

RHODES, J.

1973 *Decomposition of semilattices with applications to topological lattices.* **Pacific Journal of Mathematics**, vol. 44 (1973), pp. 299–307. [MR 47:3262]

RICE, M. D.

1979 *Function spaces and cartesian-closed categories.* Manuscript, George Mason University, 1979, 14 pp.

RICHTER, G.

1982 *On the structure of lattices in which every element is a join of join-irreducible elements.* **Periodica Mathematica Hungarica**, vol. 13 (1982), pp. 47–69. [MR 83i:06009]

RIGUET, J.

1948 *Relations binaires, fermetures, correspondances de Galois.* **Bulletin de la Société Mathématique de France**, vol. 76 (1948), pp. 114–155. [MR 10-502]

RINGLEB, P.

1969 ***Untersuchungen über die Kategorie der geordneten Mengen.*** Doctoral Dissertation, Freie Universität Berlin, 1969, 56 pp.

ROBERTS, J. W.

1977 *A compact convex set with no extreme points.* **Polska Akademia Nauk. Instytut Matematyczny. Studia Mathematica**, vol. 60 (1977), pp. 255–266. [MR 57:10595]

ROGERS, H., JR.

1967 **Theory of Recursive Functions and Effective Computability.** McGraw-Hill Book Company, New York-Toronto, 1967, xx+482 pp. [MR 37:61]

RUDIN, M. E.

1981 *Directed sets which converge.* In: **General Topology and Modern Analysis**, University of California, Riverside, 1980, edited by L. F. McAuley and M. M. Rao. Academic Press, 1981, pp. 305–307. [MR 82f:54006]

SCHMIDT, E. T.

1982 **A Survey on Congruence Lattice Representations.** Teubner, Leipzig, 1982, 115 pp. [MR 84:06012]

SCHMIDT, H. -J.

1981 *Hyperspaces of quotient and subspaces I. Hausdorff topological spaces.* **Mathematische Nachrichten**, vol. 104 (1981), pp. 271–280. [MR 84e:54011a]

1981 *Hyperspaces of quotient and subspaces II: Metrizable spaces.* **Mathematische Nachrichten**, vol. 104 (1981), pp. 281–288. [MR 84e:54011b]

SCHMIDT, J.

1953 *Beiträge zur Filtertheorie II.* **Mathematische Nachrichten**, vol. 10 (1953), pp. 197–232. [MR 15-297]

1974 *Each join completion of a partially ordered set is the solution of a universal problem.* **Journal of the Australian Mathematical Society (Series A)**, vol. 17 (1974), pp. 406–413. [MR 50:4415]

SCHNARE, P. S.

1965 *Two definitions of local compactness.* **American Mathematical Monthly**, vol. 72 (1965), pp. 764–765.

SCHREINER, E. A.

1973 *Tight residuated mappings.* In: **Proceedings of the University of Houston Lattice Theory Conference**, Houston, Texas, March 22-24, 1973, edited by S. Fajtlowicz and K. Kaiser. University of Houston, 1973, pp. 519–530. [MR 53:2766]

SCHRÖDER, J.

1977 *Das Wallman-Verfahren und inverse Limites.* **Quaestiones Mathematicae**, vol. 2 (1977), pp. 325–333. [MR 80a:54043]

SCHUBERT, H.

1972 **Categories**. Springer-Verlag, 1972, xi + 385 pp. [MR 50:2286]

SCHWARZ, F.

1981 *"Continuity" properties in lattices of topological structures.* In: **Continuous Lattices**, Proceedings of the Conference on Topological and Categorical Aspects of Continuous Lattices (Workshop IV), University of Bremen, Germany, November 9-11, 1979, edited by B. Banaschewski and R.-E. Hoffmann. **Lecture Notes in Mathematics**, vol. 871, Springer-Verlag, 1981, pp. 335–347.

1982 *Exponential objects in categories of (pre) topological spaces and their natural function spaces.* **La Société Royale du Canada. L'Académie des Sciences. Comptes Rendus Mathématiques**, vol. 4 (1982), pp. 321–326. [MR 84g:18016]

1983 ***Funktionenräume und exponentiale Objekte in punktetrennend initialen Kategorien.*** Doctoral Dissertation, Universität Bremen, 1983, vi+193 pp.

SCHWARZ, F. and WECK, S.

1985 *Scott topology, Isbell topology and continuous convergence.* In: **Continuous Lattices and Their Applications**, edited by R.-E. Hoffmann and K. H. Hofmann. Marcel Dekker, Inc., 1985, pp. 251–272.

SCOTT, D. S.

1970 *Outline of a mathematical theory of computation.* In: **Proceedings of the 4th Annual Princeton Conference on Information Science and Systems**, 1970, pp. 169–176.

1972 *Continuous lattices.* In: **Toposes, Algebraic Geometry and Logic,** Dalhousie University, Halifax, Nova Scotia, January 16-19, 1971, edited by F. W. Lawvere. **Lecture Notes in Mathematics**, vol. 274, Springer-Verlag, 1972, pp. 97–136. [MR 53:7879]

1972 *Lattice theory, data types, and semantics.* In: **Formal Semantics of Programming Languages**, edited by R. Rustin. **Courant Computer Science Symposia**, vol. 2, New York 1970, Prentice-Hall, Inc., 1972, pp. 65–106. [MR 56:7304]

1973 *Models for various type-free calculi.* In: **Logic, Methodology and Philosophy of Science IV**, Bucureşti, 1971, edited by P. Suppes. North-Holland Publishing Company, 1973, pp. 157–187. [MR 57:15987]

1975 *Combinators and classes.* In: **λ-Calculus and Computer Science Theory**, Proceedings of the Symposium Held in Rome, March 25-27, 1975, edited by C. Böhm. **Lecture Notes in Computer Science**, vol. 37, Springer-Verlag, 1975, pp. 1–26. [MR 58:21489]

1975 *Some philosophical issues concerning theories of combinators.* In: **λ-Calculus and Computer Science Theory**, Proceedings of the Symposium Held in Rome, March 25-27, 1975, edited by C. Böhm. **Lecture Notes in Computer Science**, vol. 37, Springer-Verlag, 1975, pp. 346–366. [MR 57:15986]

1976 *Data types as lattices.* **SIAM Journal on Computing**, vol. 5 (1976), pp. 522–587. [MR 55:10262]

1977 *Logic and programming languages.* **Communications of the Association for Computing Machinery**, vol. 20 (1977), pp. 634–641. [MR 56:10114]

1980 *Lambda calculus: Some models, some philosophy.* In: **The Kleene Symposium**, 1978, edited by J. Barwise, H. J. Keisler and K Kunen. North-Holland Publishing Company, 1980, pp. 223–265. [MR 82d:03024]

1980 *Relating theories of the λ-calculus.* In: **To H.B. Curry: Essays on Combinatory Logic, Lambda Calculus and Formalism**, edited by J.P. Seldin and J. R. Hindley. Academic Press, 1980, pp. 403–450. [MR 82d:03025]

1981 ***Lectures on a mathematical theory of computation.*** Technical Report, no. PRG-19, Oxford University Computing Laboratory, 1981, iv+148 pp.

1982 *Domains for denotational semantics.* In: **Automata, Languages and Programming**, Ninth Colloquium Aarhus, Denmark, July 12-16, 1982, edited by M. Nielsen and E. M. Schmidt. **Lecture Notes in Computer Science**, vol. 140, Springer-Verlag, 1982, pp. 577–613. [MR 83m:68029]

1982 *Some ordered sets in computer science.* In: **Ordered Sets**, Proceedings of the NATO Advanced Study Institutute, BANFF, Canada, August 28 to September 12, 1981, edited by I. Rival. **Nato Advanced Study Institute, Series C**, vol. 83, D. Reidel Publishing Company, 1982, pp. 677–718. [MR 83j:06007]

SHIRLEY, E. D. and STRALKA, A. R.

1971 *Homomorphisms on connected topological lattices.* **Duke Mathematical Journal**, vol. 38 (1971), pp. 483–490. [MR 43:5497]

SHMUELY, Z.

1974 *The structure of Galois connections.* **Pacific Journal of Mathematics**, vol. 54 (1974), pp. 209–225. [MR 51:12630]

SIKORSKI, R.

1964 **Boolean Algebras**. Springer-Verlag, Berlin, 1964, ix+237 pp. [MR 31:2178]

SIMMONS, H.

1978 *A framework for topology.* In: **Logic Colloquium '77**, Wrocław, Poland, August 1-12, 1977, edited by A. Macintyre, L. Pacholski and J. Paris. North-Holland Publishing Company, 1978, pp. 239–251. [MR 80b:06008]

1980 *Reticulated rings.* **Journal of Algebra**, vol. 66 (1980), pp. 169–192. [MR 82d:13005]

1982 *Erratum: reticulated rings.* **Journal of Algebra**, vol. 74 (1982), p. 292. [MR 83a:13002]

1982 *A couple of triples.* **Topology and its Applications**, vol. 13 (1982), pp. 201–223. [MR 83f:18006]

SKUCHA, J.

1982 *Topological properties of c-nets.* **Demonstratio Mathematica**, vol. 15 (1982), pp. 405–419. [MR 84f:06007]

SKULA, L.

1969 *On a reflective subcategory of the category of all topological spaces.* **Transactions of the American Mathematical Society**, vol. 142 (1969), pp. 37–41. [MR 40:1969]

SMYTH, M. B.

1977 *Effectively given domains.* **Theoretical Computer Science**, vol. 5 (1977), pp. 257–274. [MR 58:10366]

1978 *Power domains.* **Journal of Computer and System Sciences**, vol. 16 (1978), pp. 23–36. [MR 57:8128]

1983 *The largest cartesian closed category of domains.* **Theoretical Computer Science**, vol. 27 (1983), pp. 109–119.

1983 *Power domains and predicate transformers: a topological view.* In: **ICALP 83**, Proceedings of the 10th International Colloquium on Automata, Languages, and Programming, Barcelona, July 18-22, 1983, edited by J. Díaz. **Lecture Notes in Computer Science**, vol. 154, Springer-Verlag, 1983, pp. 662-675.

SMYTH, M. B. and PLOTKIN, G. D.

1978 ***The category-theoretic solution of recursive domain equations.*** Technical Report, no. 60, Edinburgh, 1978.

1978 *The category-theoretic solution of recursive domain equations.* In: **18th Annual Symposium of Foundations of Computer Science**, Providence, RI, October 31 to November 2, 1977, IEEE Computer Society, 1978, pp. 13–29. [MR 58:8459]

1982 *The category-theoretic solution of recursive domain equations.* **SIAM Journal on Computing**, vol. 11 (1982), pp. 761–783. [MR 84d:68052]

ŠNEPERMAN, L. B.

1968 *On the theory of characters of locally bicompact topological semigroups.* **Mathematics of the USSR Sbornik**, vol. 6 (1968), pp. 471–492. [MR 38:5984]

SOBRAL, M.

1983 *Restricting the comparison functor of an adjunction to projective objects.* **Quaestiones Mathematicae**, vol. 6 (1983), pp. 303–312.

1984 *On adjunctions inducing the same monad.* **Quaestiones Mathematicae**, vol. 7 (1984), pp. 179–201.

1985 *Projectiveness with regard to a right adjoint functor.* In: **Continuous Lattices and Their Applications**, edited by R.-E. Hoffmann and K. H. Hofmann. Marcel Dekker, Inc., 1985, pp. 273–278.

SPEED, T. P.

1972 *Profinite posets.* **Bulletin of the Australian Mathematical Society**, vol. 6 (1972), pp. 177–183. [MR 45:4371]

STEPP, J. W.

1971 *Semilattices which are embeddable in a product of min intervals.* **Proceedings of the American Mathematical Society**, vol. 28 (1971), pp. 81–86. [MR 43:1895]

1971 *Semilattices which are embeddable in a product of min intervals.* **Semigroup Forum**, vol. 2 (1971), pp. 80–82. [MR 44:5260]

1972 *The lattice of ideals of a compact semilattice.* **Semigroup Forum**, vol. 5 (1972), pp. 176–180. [MR 49:4882]

1973 *Topological semilattices which are embeddable in topological lattices.* **The Journal of the London Mathematical Society**, vol. 7 (1973), pp. 76–82. [MR 49:6183]

1975 *Algebraic maximal semilattices.* **Pacific Journal of Mathematics**, vol. 58 (1975), pp. 243–248. [MR 51:12640]

1975 *The free compact lattice generated by a topological semilattice.* **Journal für die Reine und Angewandte Mathematik**, vol. 273 (1975), pp. 77–86. [MR 51:7974]

1980 *Topological semilattices with enough closed subsemilattices to separate points.* **Semigroup Forum**, vol. 20 (1980), pp. 49–54. [MR 81m:06009]

STONE, M. H.

1936 *The theory of representations for Boolean algebras.* **Transactions of the American Mathematical Society**, vol. 40 (1936), pp. 37–111.

1937 *Topological representation of distributive lattices and Brouwerian logics.* **Časopis pro Pěstování Matematiky a Fysiky**, vol. 67 (1937), pp. 1–25.

STOY, J. E.

1977 **Denotational Semantics–The Scott-Strachey Approach to Programming Language Theory.** MIT Press, Cambridge, Massachusetts, 1977, xxxi+414 pp. [MR 58:8460]

STRALKA, A. R.

1970 *Locally convex topological lattices.* **Transactions of the American Mathematical Society**, vol. 151 (1970), pp. 629–640. [MR 41:9216]

1971 *The congruence extension property for compact topological lattices.* **Pacific Journal of Mathematics**, vol. 38 (1971), pp. 795–802. [MR 46:3394]

1972 *The lattice of ideals of a compact semilattice.* **Proceedings of the American Mathematical Society**, vol. 33 (1972), pp. 175–180. [MR 46:7439]

1973 *Distributive topological lattices.* In: **Proceedings of the University of Houston Lattice Theory Conference**, Houston, Texas, March 22-24, 1973, edited by S. Fajtlowicz and K. Kaiser. University of Houston, 1973, pp. 269–276. [MR 55:226]

1974 *Imbedding locally convex lattices into compact lattices.* **Colloquium Mathematicum**, vol. 29 (1974), pp. 147–150. [MR 49:4883]

1977 *Congruence extension and amalgamation in CL.* **Semigroup Forum**, vol. 13 (1977), pp. 355–375. [MR 58:10640]

1979 *Quotients of products of compact chains.* **Bulletin of the London Mathematical Society**, vol. 11 (1979), pp. 1–4. [MR 80h:22004]

1980 *A partially ordered space which is not a Priestley space.* **Semigroup Forum**, vol. 20 (1980), pp. 293–297. [MR 82f: 54051]

1981 *Fundamental congruences on Lawson semilattices.* In: **Continuous Lattices**, Proceedings of the Conference on Topological and Categorical Aspects of Continuous Lattices (Workshop IV), University of Bremen, Germany, November 9-11, 1979, edited by B. Banaschewski and R.-E. Hoffmann. **Lecture Notes in Mathematics**, vol. 871, Springer-Verlag, 1981, pp. 348–359.

STRAUSS, D. P.

1968 *Topological lattices.* **Proceedings of the London Mathematical Society**, vol. 18 (1968), pp. 217–230. [MR 37:3532]

TANG, A.

1979 *Chain properties in Pω.* **Theoretical Computer Science**, vol. 9 (1979), pp. 153–172. [MR 80j:68009]

1981 *Wadge reducibility and Hausdorff difference hierarchy in Pω.* In: **Continuous Lattices**, Proceedings of the Conference on Topological and Categorical Aspects of Continuous Lattices (Workshop IV), University of Bremen, Germany, November 9-11, 1979, edited by B. Banaschewski and R.-E. Hoffmann. **Lecture Notes in Mathematics**, vol. 871, Springer-Verlag, 1981, pp. 360–371.

TARSKI, A.

1930 *Sur les classes d'ensembles closes par rapport à certaines opérations élémentaires.* **Fundamenta Mathematicae**, vol. 16 (1930), pp. 181–304.

1955 *A lattice-theoretical fixpoint theorem and its applications.* **Pacific Journal of Mathematics**, vol. 5 (1955), pp. 285–309. [MR 17:574]

TAYLOR, J. C.

1963 *Filter spaces determined by relations, I.* **Indagationes Mathematicae**, vol. 25 (1963), pp. 7–22. [MR 26:6920]

1963 *Filter spaces determined by relations, II.* **Indagationes Mathematicae**, vol. 25 (1963), pp. 23–40. [MR 26:6921]

THRON, W. J.

1962 *Lattice-equivalence of topological spaces.* **Duke Mathematical Journal**, vol. 29 (1962), pp. 671–679. [MR 26:4307]

TILLER, J.

1980 ***Continuous Lattices and Convexity Theory.*** Doctoral Dissertation, McMaster University, Hamilton, Ontario, 1980, vii+106 pp.

1981 *Augmented compact spaces and continuous lattices.* **Houston Journal of Mathematics**, vol. 7 (1981), pp. 441–453. [MR 82m:06008]

TUNNICLIFFE, W. R.

1974 *A criterion for the existence of an extension of a function on a partially ordered set.* **The Journal of the London Mathematical Society**, vol. 8 (1974), pp. 352–354. [MR 51:5428]

VAN DE VEL, M.

1982 *Finite dimensional convex structures I: general results.* **Topology and its Applications**, vol. 14 (1982), pp. 201–225. [MR 83m:52005]

1982 *Convex structures and dimension.* In: **General Topology and its Relations to Modern Analysis and Algebra V**, (Proceedings, Prague, 1981), Prague, 1981, edited by J. Novák. Heldermann Verlag, 1982, pp. 648–652. [MR 84d:54063]

1983 *Finite-dimensional convex structures II: the invariants.* **Topology and its Applications**, vol. 16 (1983), pp. 81–105. [MR 84m:52015]

1983 *Matching binary convexities.* **Topology and its Applications**, vol. 16 (1983), pp. 207–235.

1983 *Two-dimensional convexities are join-hull commutative.* **Topology and its Applications**, vol. 16 (1983), pp. 181–206. [MR 84m:52002]

1983 *Pseudo-boundaries and pseudo-interiors for topological convexities.* **Dissertationes Mathematicae**, vol. 210 (1983), pp. 1–76. [MR 85c:52002]

1983 *Dimension of convex hyperspaces: nonmetric case.* **Compositio Mathematica**, vol. 50 (1983), pp. 95-108.

1983 *On the rank of a topological convexity.* **Fundamenta Mathematicae**, vol. 119 (1983), pp. 101-132. [MR 85b:52001]

1984 *Binary convexities and distributive lattices.* **Proceedings of the London Mathematical Society**, vol. 48 (1984), pp. 1-33. [MR 85c:52004a]

1984 *Dimension of binary convex structures.* **Proceedings of the London Mathematical Society**, vol. 48 (1984), pp. 34–54. [MR 85c:52004b]

1984 *Dimension of convex hyperspaces.* **Fundamenta Mathematicae**, vol. 122 (1984), pp. 107–127.

1985 *Lattices and semilattices: a convex point of view.* In: **Continuous Lattices and Their Applications**, edited by R.-E. Hoffmann and K. H. Hofmann. Marcel Dekker, Inc., 1985, pp. 279–302.

198? *A selection theorem for topological convex structures.* 198?. (To appear).

VAN MILL, J. and VAN DE VEL, M.

198? *Equality of the Lebesgue and the inductive dimension functions for compact spaces with a uniform convexity.* **Colloquium Mathematicum**, (198?). (To appear).

VIETORIS, L.

1922 *Bereiche zweiter Ordnung.* **Monatshefte für Mathematik und Physik**, vol. 32 (1922), pp. 258–280.

1923 *Kontinua zweiter Ordnung.* **Monatshefte für Mathematik und Physik**, vol. 33 (1923), pp. 49–62.

VUILLEMIN, J.

1974 ***Syntaxe, sémantique et axiomatique d'un langage de programmation simple.*** Thèse D'État, University of Paris, 1974.

WADSWORTH, C.

1976 *The relation between computational and denotational properties for Scott's D_∞-models of the λ-calculus.* **SIAM Journal on Computing**, vol. 5 (1976), pp. 488–521. [MR 58:21493]

WAGNER, F. J.

1957 *Notes on compactification, I.* **Indagationes Mathematicae**, vol. 19 (1957), pp. 171–176. [MR 19:436]

1957 *Notes on compactification, II.* **Indagationes Mathematicae**, vol. 19 (1957), pp. 177–181. [MR 19:436]

WALLACE, A. D.

1945 *A fixed-point theorem.* **Bulletin of the American Mathematical Society**, vol. 51 (1945), pp. 413–416. [MR 6-278]

1954 *Partial order and indecomposability.* **Proceedings of the American Mathematical Society**, vol. 5 (1954), pp. 780–781. [MR 16-275]

1955 *Struct ideals.* **Proceedings of the American Mathematical Society**, vol. 6 (1955), pp. 634–638. [MR 17-179]

1957 *The center of a compact lattice is totally disconnected.* **Pacific Journal of Mathematics**, vol. 7 (1957), pp. 1237–1238. [MR 20:823]

1957 *The peripheral character of central elements of a lattice.* **Proceedings of the American Mathematical Society**, vol. 8 (1957), pp. 596–597. [MR 19-429]

1957 *Two theorems on topological lattices.* **Pacific Journal of Mathematics**, vol. 7 (1957), pp. 1239–1241. [MR 20:824]

1958 *Factoring a lattice.* **Proceedings of the American Mathematical Society**, vol. 9 (1958), pp. 250–252. [MR 20:825]

1961 *Acyclicity of compact connected semigroups.* **Fundamenta Mathematicae**, vol. 50 (1961), pp. 99–105. [MR 24:A2373]

WAND, M.

1979 *Fixed-point constructions in order-enriched categories.* **Theoretical Computer Science**, vol. 8 (1979), pp. 13–30. [MR 80e:18002]

WARD, A. J.

1955 *On relations between certain intrinsic topologies in partially ordered sets.* **Proceedings of the Cambridge Philosophical Society**, vol. 51 (1955), pp. 254–261. [MR 17-67]

1969 *Problem.* In: **Topology and its Applications**, Belgrade, 1969, edited by D. R. Kurepa. Savez Društava Matematičara Fizičara i Astronoma, 1969, pp. 351–352.

1974 *Representations of proximity lattices.* **Annales Universitatis Scientiarum Budapestinensis de Rolando Eötvös Nominatae**, vol. 17 (1974), pp. 41–57. [MR 52:5506]

WARD, L. E., JR.

1954 *Partially ordered topological spaces.* **Proceedings of the American Mathematical Society**, vol. 5 (1954), pp. 144–161. [MR 16-59]

1957 *Completeness in semilattices.* **Canadian Journal of Mathematics**, vol. 9 (1957), pp. 578–582. [MR 19-938]

1965 *Concerning Koch's theorem on the existence of arcs.* **Pacific Journal of Mathematics**, vol. 15 (1965), pp. 347–355. [MR 31:6206]

1965 *On a conjecture of R. J. Koch.* **Pacific Journal of Mathematics**, vol. 15 (1965), pp. 1429–1433. [MR 32:6397]

WECK, S.

1981 *Scott convergence and Scott topology in partially ordered sets I.* In: **Continuous Lattices**, Proceedings of the Conference on Topological and Categorical Aspects of Continuous Lattices (Workshop IV), University of Bremen, Germany, November 9-11, 1979, edited by B. Banaschewski and R.-E. Hoffmann. **Lecture Notes in Mathematics**, vol. 871, Springer-Verlag, 1981, pp. 372–383.

WEIHRAUCH, K. and DEIL, T.

1980 **Berechenbarkeit auf cpo's.** Technical Report, no. 63, RWTH, Aachen, 1980, 101 pp.

WEIHRAUCH, K. and SCHREIBER, U.

1979 *Metric spaces defined by weighted algebraic cpo's.* In: **Fundamentals of Computation Theory**, Proceedings, Berlin/Wendisch-Rietz, 1979, Berlin, 1979, edited by L. Budach. Akademie-Verlag, 1979, pp. 516–522. [MR 81e:68055]

1981 *Embedding metric spaces into cpo's.* **Theoretical Computer Science**, vol. 16 (1981), pp. 5–24. [MR 83i: 06023]

WEINBERG, E. C.

1962 *Completely distributive lattice-ordered groups.* **Pacific Journal of Mathematics**, vol. 12 (1962), pp. 1131–1137. [MR 26:5064]

1962 *Higher degrees of distributivity in lattices of continuous functions.* **Transactions of the American Mathematical Society**, vol. 104 (1962), pp. 334–346. [MR 25:2013]

WIGNER, D.

1979 *Two notes on frames.* **Journal of the Australian Mathematical Society (Series A)**, vol. 28 (1979), pp. 257–268. [MR 81d:06016]

WILKE, G.

1980 *Eine Kennzeichnung topologischer Räume durch Hüllenalgebren.* **Archiv der Mathematik**, vol. 34 (1980), pp. 276–282. [MR 81k:54071]

1981 *Eine Charakterisierung der Dualräume distributiver Supremumshalbverbände.* **Archiv der Mathematik**, vol. 37 (1981), pp. 359–363. [MR 83b:06004]

1983 *Eine Kennzeichnung topologischer Räume durch Vervollständigungen.* **Mathematische Zeitschrift**, vol. 182 (1983), pp. 339–350.

WILKER, P.

1970 *Adjoint product and hom functors in general topology.* **Pacific Journal of Mathematics**, vol. 34 (1970), pp. 269–283. [MR 42:5218]

WILLIAMS, W.

1975 *Semilattices on Peano continua.* **Proceedings of the American Mathematical Society**, vol. 49 (1975), pp. 495–500. [MR 52:4255]

WILSON, R. L.

1977 *Relationships between continuous posets and compact Lawson posets.* **Notices of the American Mathematical Society**, vol. 24 (1977), pp. A-628.

1978 ***Intrinsic topologies on partially ordered sets and results on compact semigroups.*** Doctoral Dissertation, University of Tennessee, 1978, 167 pp.

WINSKEL, G.

1983 *A note on powerdomains and modality.* In: **Foundations of Computation Theory 1983**, Proceedings of the 1983 International FCT-Conference, Borgholm, Sweden, August 21-27, 1983, edited by M. Karpinski. **Lecture Notes in Computer Science**, vol. 158, Springer-Verlag, 1983, pp. 505-514.

WOJDYSLAWSKI, M.

1939 *Rétracts absolus et hyperespaces des continus.* **Fundamenta Mathematicae**, vol. 32 (1939), pp. 184–192.

WOLFF, H.

1981 *Relative injectives and free monads.* **Canadian Mathematical Bulletin**, vol. 24 (1981), pp. 29–35. [MR 82e:18012]

WOLK, E. S.

1958 *Order-compatible topologies on a partially ordered set.* **Proceedings of the American Mathematical Society**, vol. 9 (1958), pp. 524–529. [MR 20:3079]

1961 *On order-convergence.* **Proceedings of the American Mathematical Society**, vol. 12 (1961), pp. 379–384. [MR 25:32]

WRIGHT, J. B., WAGNER, E. G. and THATCHER, J. W.

1978 *A uniform approach to inductive posets and inductive closure.* **Theoretical Computer Science**, vol. 7 (1978), pp. 57–77. [MR 58:404]

WULFSOHN, A.

1972 *A compactification due to Fell.* **Canadian Mathematical Bulletin**, vol. 15 (1972), pp. 145–146. [MR 48:3004]

WYLER, O.

1976 *Algebraic theories of continuous lattices.* Manuscript, Carnegie-Mellon University, 1976, 25 pp.

1977 *Injective spaces and essential extensions in TOP.* **General Topology and its Applications**, vol. 7 (1977), pp. 247–249. [MR 56:16559]

1977 *On continuous lattices as topological algebras.* Manuscript, Carnegie-Mellon University, 1977, 6 pp.

1981 *Algebraic theories of continuous lattices.* In: **Continuous Lattices**, Proceedings of the Conference on Topological and Categorical Aspects of Continuous Lattices (Workshop IV), University of Bremen, Germany, November 9-11, 1979, edited by B. Banaschewski and R.-E. Hoffmann. **Lecture Notes in Mathematics**, vol. 871, Springer-Verlag, 1981, pp. 390–413.

1981 *Dedekind complete posets and Scott topologies.* In: **Continuous Lattices**, Proceedings of the Conference on Topological and Categorical Aspects of Continuous Lattices (Workshop IV), University of Bremen, Germany, November 9-11, 1979, edited by B. Banaschewski and R.-E. Hoffmann. **Lecture Notes in Mathematics**, vol. 871, Springer-Verlag, 1981, pp. 384–389.

YANG, J. -C.

1969 *A theorem on the semigroup of binary relations.* **Proceedings of the American Mathematical Society**, vol. 22 (1969), pp. 134–135. [MR 39:2897]

YEAGER, D. P.

1976 *On the topology of a compact inverse Clifford semigroup.* **Transactions of the American Mathematical Society**, vol. 215 (1976), pp. 253–267. [MR 54:457]

Index